W9-CDP-589

ROBOTICS AND AUTOMATED SYSTEMS

Robert L. Hoekstra, CMfgE

Department Chairman,
Robotics and Automated
Machinery Maintenance
Southern Ohio College

Published by

IE65 **SOUTH-WESTERN PUBLISHING CO.**

CINCINNATI   WEST CHICAGO, IL   DALLAS   PELHAM MANOR, NY   LIVERMORE, CA

NORTH-HOLLAND

Copyright © 1986

by South-Western Publishing Co.

Cincinnati, Ohio

All Rights Reserved

The text of this publication, or any part thereof, may not be reproduced or transmitted in any form or by any means, electronic or mechanical, including photocopying, recording, storage in an information retrieval system, or otherwise, without the prior written permission of the publisher.

ISBN: 0-538-33650-1 (U.S.A. and Canada)
(Reprint) ISBN: 0-444-879021 (outside U.S.A. and Canada)
Library of Congress Catalog Card Number: 84-52266

1 2 3 4 5 6 7 8 9 D 3 2 1 0 9 8 7 6 5

Printed in the United States of America

Distributed outside the U.S.A. and Canada by
North–Holland
(a division of Elsevier Science Publishers BV)
P. O. Box 1991
1000 BZ Amsterdam, The Netherlands

PHOTO CREDITS

The author and publisher wish to express their appreciation to the following organizations and individuals for generously supplying illustrations used in this book:

Chapter 1

Fig. 1.7	Cincinnati Milacron, Inc.
Fig. 1.8	Photo by Ed Dahlin
Fig. 1.9	Cincinnati Milacron, Inc.
Fig. 1.10	Unimation
Fig. 1.11	Vickers Incorporated
Fig. 1.13	Schrader Bellows
Fig. 1.16	Photo by Ed Dahlin
Fig. 1.18	Photo courtesy of Numatics, Inc., Highland, MI
Fig. 1.21	Prab Robots, Inc.
Fig. 1.22	General Motors Corporation
Fig. 1.24	Cincinnati Milacron, Inc.
Fig. 1.25	The DeVilbiss Company
Fig. 1.26	Schrader Bellows
Fig. 1.27	Unimation
Fig. 1.29 (left)	Cincinnati Milacron, Inc.
Fig. 1.29 (right)	Photo by Ed Dahlin
Fig. 1.30	Cincinnati Milacron, Inc.
Fig. 1.31	The DeVilbiss Company
Fig. 1.32	Prab Robots, Inc.
Fig. 1.33	Photo courtesy of Allen-Bradley, Systems Division
Fig. 1.34	Cincinnati Milacron, Inc.
Fig. 1.35	IBM Corporation
Fig. 1.36	Cincinnati Milacron, Inc.
Fig. 1.37	The DeVilbiss Company
Fig. 1.38	Cincinnati Milacron, Inc.

Chapter 3

Fig. 3.1 (right)	Photo courtesy of FESTO CORP.
Fig. 3.21	Courtesy of Barksdale Controls Division, Transamerica Delaval Inc.

Chapter 4

Fig. 4.1	Photo by Ed Dahlin
Fig. 4.2	Electro-Craft Corp.
Fig. 4.23	PMI Motors
Fig. 4.29	Electro-Craft Corp.
Fig. 4.30	Electro-Craft Corp.
Fig. 4.31	Renco Corporation
Fig. 4.47	Warner Electric Brake & Clutch Company
Fig. 4.51	Emhart Machinery Group/ Harmonic Drive Division

Chapter 5

Fig. 5.80	IBM Corporation
Fig. 5.81	National Semiconductor
Fig. 5.82	Photo courtesy of Sperry Corporation
Fig. 5.83	Industrial Training, Inc.
Fig. 5.84	Industrial Training, Inc.
Fig. 5.85	Honeywell, Inc.
Fig. 5.86	Industrial Training, Inc.
Fig. 5.87	Honeywell, Inc.
Fig. 5.88	Industrial Training, Inc.
Fig. 5.89	Courtesy of Bell Laboratories
Fig. 5.90 (top)	Courtesy of Motorola Inc.
Fig. 5.90 (bottom)	National Semiconductor

Chapter 6

Fig. 6.70	National Semiconductor
Fig. 6.72	National Semiconductor

Chapter 8

Fig. 8.20	National Semiconductor

Chapter 9

Fig. 9.54	Vernitron Control Components

Chapter 10

Fig. 10.3	Courtesy of Square D Company
Fig. 10.4	Courtesy of Square D Company
Fig. 10.5	Photo courtesy of Allen-Bradley Company
Fig. 10.6	Eaton Corporation, Cutler-Hammer Products
Fig. 10.8	Courtesy of Square D Company
Fig. 10.11	Photo courtesy of Allen-Bradley Company
Fig. 10.13	Courtesy of Square D Company

PREFACE

Introduction

Robotics and Automated Systems is a text that examines the mechanical, electrical, and electronic components utilized in robots and automated systems. Since the components used in robots and automated systems such as NC/CNC machines are similar, a student who has mastered the systems presented in this text will be prepared to enter industry as an entry-level technician.

The text explains each concept in detail but as concisely as possible. Therefore, the student is not required to sift through long textual passages in order to grasp a concept. The major concepts discussed are reinforced through questions presented at the end of each chapter.

The *Instructor's Manual and Key,* which reinforces the text with additional examples and alternative explanations of difficult concepts, can be utilized by the instructor as an additional resource in preparing lectures. The *Instructor's Manual and Key* also includes the answers to the questions at the end of each chapter as well as additional questions and answers as a resource for class discussion and test preparation.

This text can be utilized as an overview of robotics and automated systems or for a capstone course, or can be integrated with existing courses in hydraulics, pneumatics, industrial electricity/ electronics, and microprocessors. The prerequisites for the student using this text are a knowledge of basic electricity and an understanding of elementary algebra.

Overview

In Chapter 1 the student is introduced to the essential terminology used in robotics and the basic operations of robots.

Following the introduction to robots, the student is instructed in the various power-supply systems which are used to move the robot's manipulator. Chapter 2 introduces the student to hydraulic power-supply systems. The emphasis of this chapter is on the components used in hydraulic systems rather than on theoretical design. The operation of each component is presented, along with the ANSI symbols which represent these components in schematic drawings. Chapter 3 introduces the student to pneumatic power-supply systems. The format of this chapter is similar to that of Chapter 2. Chapter 4 introduces the various types of electric motors used to supply power to robots and/or automated systems.

Following the introduction of the various power-supply systems, the student is instructed in the digital electronic control systems which control the power supply. It is assumed that the student has not previously studied these control systems, and therefore Chapter 5 begins with the basics. The student is introduced to logic gates and their truth tables. In Chapter 6 the student is introduced to flip-flops, the binary system, counting circuits, binary addition and subtraction, decoders, and displays. The chapter closes with the design of a simple robot which incorporates most of the concepts presented in the first six chapters.

Chapter 7 discusses analog circuits, because robot and automated system controls are not limited to digital control. Digital-to-analog (DAC) and analog-to-digital (ADC) conversions are also presented.

With this background, the student is prepared to begin the examination of a robot controller. Since most robot controllers are now microprocessor-based, the basics of microprocessors are presented in Chapter 8. The discussion of microprocessors focuses on systems rather than on assembler programming of the microprocessor chip itself. The technician in industry is responsible for the repair of previously working systems rather than for the design of new systems. Since the focus here is on systems, the microprocessor is described on the basis of a bus structure and the relationships among the various components that make up the working system.

Most robots and automated systems in industry today are designed with a feedback system. A command to move is outputted by the microprocessor to the power supply, and a sensing device senses the actual position of the robot's manipulator and "feeds back" the position to the controller. If the commanded and actual positions do not agree, the system makes the necessary adjustment that will move the manipulator to the commanded position. Such systems, which are referred to as "servo systems," are presented in Chapter 9. Most discussions of servo systems are based on high-level mathematical models. Here the author has avoided high-level mathematical discussion of servo systems and has focused on the systems themselves. In this chapter a servo system, including proportional, integral, and derivative control, is developed utilizing the information presented in the previous chapters.

In Chapter 10, having completed the basic discussion of the components used in robots and automated systems, the author describes interfacing of robots in manufacturing environments. The commonly used sensors that supply the robot with information about its environment are described in detail.

Finally, in Chapter 11, are presented successful applications of robots in industry.

The Author

Robert L. Hoekstra is a Certified Manufacturing Engineer. He is currently the Chairman of the Robotics and Automated Machinery Maintenance Department at Southern Ohio College. Before joining Southern Ohio College in 1983, Mr. Hoekstra was a consultant to industry and developed industrial training programs. His clients included many of the Fortune 500 companies and his work covered a broad range of topics including robotics, process control, geometric tolerancing, statistical quality control, and electrical and electronic machinery maintenance. Mr. Hoekstra has lectured extensively both in the United States and abroad, and has been a guest on more than 50 radio and television stations. He has received more than 30 national and international awards for his work in technical education.

Acknowledgments

The author would like to express his thanks to the many people who have supported the writing of this text through their encouragement and their critical reviews. Unfortunately, space does not allow for a listing of all their names. However, the author would like to express formal thanks to Shirley Hoekstra for encouraging him to write while she assumed many of the responsibilities in running their home that were rightfully his, and to David Favin, Member of the Technical Staff, Bell Laboratories, for his constructive criticism and enthusiastic encouragement. The author would also like to thank the staff and faculty members of Southern Ohio College who offered encouragement and reviewed the manuscript, and to express special thanks to Elmer Smith, President of Southern Ohio College, for his support in the development of the staff, faculty, and facilities of the Robotics and Automated Machinery Maintenance Department, and for his personal support, encouragement, and friendship.

CONTENTS

CHAPTER ONE

Introduction to Industrial Robots

Industrial robots became a reality in the early 1960's when Joseph Engelberger and George Devol teamed up to form a robotics company they called "Unimation."

Engelberger and Devol were not the first to dream of machines that could perform the unskilled, repetitive jobs in manufacturing. The first use of the word "robot" was by the Czechoslovakian philosopher and playwright Karel Capek in his play *R.U.R.* (Rossum's Universal Robot). The word "robot" in Czech means "worker" or "slave." The play was written in 1922.

In Capek's play, Rossum and his son discover the chemical formula for artificial protoplasm. Protoplasm forms the very basis of life. With their compound, Rossum and his son set out to make a robot.

Rossum and his son spend 20 years forming the protoplasm into a robot. After 20 years the Rossums look at what they have created and say, "It's absurd to spend twenty years making a man if we can't make him quicker than nature, you might as well shut up shop."

The young Rossum goes back to work eliminating organs he considers unnecessary for the ideal worker. The young Rossum says, "A man is something that feels happy, plays piano, likes going for a walk, and in fact wants to do a whole lot of things that are unnecessary... but a working machine must not play piano, must not feel happy, must not do a whole lot of other things. Everything that doesn't contribute directly to the progress of work should be eliminated."

A half century later, engineers began building Rossum's robot, not out of artificial protoplasm, but of silicon, hydraulics, pneumatics, and electric motors. Robots that were dreamed of by Capek in 1922, that work but do not feel, that perform unhuman or subhuman jobs in manufacturing plants, are available and are in operation around the world.

The modern robot lacks feeling and emotions just as Rossum's son thought it should. It can only respond to simple "yes/no" questions. The modern robot is normally bolted to the floor. It has one arm and one hand. It is deaf, blind, and dumb. In spite of all of these handicaps, the modern robot performs its assigned task hour after hour without boredom or complaint.

A robot is not simply another automated machine. Automation began during the industrial revolution with machines that performed jobs that formerly had been done by human workers. Such a machine, however, can do only the specific job for which it was designed, whereas a robot can perform a variety of jobs.

A robot must have an arm. The arm must be able to duplicate the movements of a human worker in loading and unloading other automated machines, spraying paint, welding, and performing hundreds of other jobs that cannot be easily done with conventional automated machines.

DEFINITION OF A ROBOT

The Robot Industries Association (RIA) has published a definition for robots in an attempt to clarify which machines are simply automated machines and which machines are truly robots. The RIA definition is as follows:

> "A robot is a reprogrammable multifunctional manipulator designed to move material, parts, tools, or specialized devices through variable programmed motions for the performance of a variety of tasks."

This definition, which is more extensive than the one in the RIA glossary at the end of this book, is an excellent definition of a robot. We will look at this definition, one phrase at a time, so as to understand which machines are in fact robots and which machines are little more than specialized automation.

First, a robot is a "reprogrammable multifunctional manipulator." In this phrase RIA tells us that a robot can be taught ("reprogrammed") to do more than one job by changing the information stored in its memory. A robot can be reprogrammed to load and unload machines, weld, and do many other jobs ("multifunctional"). A robot is a "manipulator." A manipulator is an arm (or hand) that can pick up or move things. At this point we know that a robot is an arm that can be taught to do different jobs.

The definition goes on to say that a robot is "designed to move material, parts, tools, or specialized devices." Material includes wood, steel, plastic, cardboard . . . anything that is used in the manufacture of a product.

A robot can also handle parts that have been manufactured. For example, a robot can load a piece of steel into an automatic lathe and unload a finished part out of the lathe.

In addition to handling material and parts, a robot can be fitted with tools such as grinders, buffers, screwdrivers, and welding torches to perform useful work.

Robots can also be fitted with specialized instruments or devices to do special jobs in a manufacturing plant. Robots can be fitted with television cameras for inspection of parts or products. They can be fitted with lasers to accurately measure the size of parts being manufactured.

The RIA definition closes with the phrase, ". . . through variable programmed motions for the performance of a variety of tasks." This phrase emphasizes the fact that a robot can do many different jobs in a manufacturing plant. The variety of jobs that a robot can do is limited only by the creativity of the application engineer.

JOBS FOR ROBOTS

Jobs performed by robots can be divided into two major categories: hazardous jobs and repetitive jobs.

Hazardous Jobs

Many applications of robots are in jobs that are hazardous to humans. Such jobs may be considered hazardous because of toxic fumes, the weight of the material being handled, the temperature of the material being handled, the danger of working near rotating or press machinery, or environments containing high levels of radiation.

Repetitive Jobs

In addition to taking over hazardous jobs, robots are well suited to doing extremely repetitive jobs that must be done in manufacturing plants. Many jobs in manufacturing plants require a person to act more like a machine than like a human. The job may be to pick a piece up from here and place it there. The same job is done hundreds of times each day. The job requires little or no judgment and little or no skill. This is not said as a criticism of the person who does the job, but is intended simply to point out that many of these jobs exist in industry and must be done to complete the manufacture of products. A robot can be placed at such a work station and can perform the job admirably without complaining or experiencing the fatigue and boredom normally associated with such a job.

Although robots eliminate some jobs in industry, they normally eliminate jobs that humans should never have been asked to do. Machines should perform as machines doing machine jobs, and humans should be placed in jobs that require the use of their ability, creativity, and special skills.

POTENTIAL FOR INCREASED PRODUCTIVITY

In addition to removing people from jobs they should not have been placed in, robots offer companies the opportunity of achieving **increased productivity.** When robots are placed in repetitive jobs they continue to operate at their programmed pace without fatigue. Robots do not take either scheduled or unscheduled breaks from the job. The increase in productivity can result in at least 25% more good parts being produced in an eight-hour shift. This increase in productivity increases the company's profits, which can be reinvested in additional plants and equipment. This increase in productivity results in more jobs in other departments in the plant. With more parts being produced, additional people are needed to deliver the raw materials to the plant, to complete the assembly of the finished products, to sell the finished products, and to deliver the products to their destinations.

ROBOT SPEED

Although robots increase productivity in a manufacturing plant, they are not exceptionally fast. At present, robots normally operate at or near the speed of a human operator. Every major move of a robot normally takes approximately one second. For a robot to pick up a piece of steel from a conveyor and load it into a lathe may require ten different moves taking as much as ten seconds. A human operator can do the same job in the same amount of time. The increase in productivity is a result of the consistency of operation. As the human operator repeats the same job over and over during the workday, he or she begins to slow down. The robot continues to operate at its programmed speed and therefore completes more parts during the workday.

Custom-built automated machines can be built to do the same jobs that robots do. An automated machine can do the same loading operation in less than half the time required by a robot or a human operator. The problem with designing a special machine is that such a machine can perform only the specific job for which it was built. If any change is made in the job, the machine must be completely rebuilt, or the machine must be scrapped and a new machine designed and built. A robot, on the other hand, could be reprogrammed and could start doing the new job the same day.

Custom-built automated machines still have their place in industry. If a company knows that a job will not change for many years, the faster custom-built machine is still a good choice.

Other jobs in factories cannot be done easily with custom-built machinery. For these applications a robot may be a good choice. An example of such an application is spray painting. One company made cabinets for the electronics industry. They made cabinets of many different sizes, all of which needed painting. It was determined that it was not economical for the company to build special spray painting machines for each of the different sizes of enclosures that were being built. Until robots were developed, the company had no choice but to spray the various enclosures by hand.

Spray painting is a hazardous job, because the fumes from many paints are both toxic and explosive. A robot is now doing the job of spraying paint on the enclosures. A robot has been "taught" to spray all the different sizes of enclosures that the company builds. In addition, the robot can operate in the toxic environment of the spray booth without any concern for the long-term effect the fumes might have on a person working in the booth.

FLEXIBLE AUTOMATION

Robots have another advantage: they can be taught to do different jobs in the manufacturing plant. If a robot was originally purchased to load and unload a punch press and the job is no longer needed due to a change in product design, the robot can be moved to another job in the plant. For example, the robot could be moved to the end of the assembly operation and be used to unload the finished enclosures from a conveyor and load them onto a pallet for shipment.

ACCURACY AND REPEATABILITY

One very important characteristic of any robot is the **accuracy** with which it can perform its task. When the robot is programmed to perform a specific task, it is led to specific points and programmed to remember the locations of those points. After programming has been completed, the robot is switched to "run" and the program is executed. Unfortunately, the robot will not go to the exact location of any programmed point. For example, the robot may miss the exact point by 0.025 in. If 0.025 in. is the greatest error by which the robot misses any point during the first execution of the program, the robot is said to have an accuracy of 0.025 in.

In addition to accuracy, we are also concerned with the robot's **repeatability**. The repeatability of a robot is a measure of how closely it returns to its programmed points every time the program is executed. Say, for example, that the robot misses a programmed point by 0.025 in. the first time the program is executed and that, during the next execution of the program, the robot misses the point it reached during the previous cycle by 0.010 in. Although the robot is a total of 0.035 in. from the original programmed point, its accuracy is 0.025 in. and its repeatability is 0.010 in.

THE MAJOR PARTS OF A ROBOT

The major parts of a robot are the **manipulator**, the **power supply,** and the **controller**.

The manipulator is used to pick up material, parts, or special tools used in manufacturing. The power supply supplies the power to move the manipulator. The controller controls the power supply so that the manipulator can be taught to perform its task.

AXES OF ROBOT MOVEMENT

The various movements that the manipulator of a robot can make are defined by its **degrees of freedom** or **axes**. If a robot's manipulator can rotate, the robot is said to be a single-axis robot. If the manipulator can move up and down as well as rotate, the robot is called a two-axis robot. If, in addition to the rotational movement and the up-and-down movement, the manipulator can also extend its arm, or "reach," the robot is said to be a three-axis robot (Figure 1.1). Most industrial robots have all three **major axes** (rotational, up and down, and reach) as well as some minor axes of movement.

The **minor axes** of a robot are found in the robot's **wrist**. The wrist of a robot is attached to the end of the robot's arm. There are three possible movements or axes of a robot wrist: **pitch**, **roll**, and **yaw** (Figure 1.2). The pitch movement bends the wrist up and down. The roll movement is the twisting of the wrist. The yaw movement is the side-to-side movement of the wrist.

The combination of the major axes and the minor axes gives the robot six possible movements (six axes or six degrees of freedom). Many industrial robots are equipped with all six axes. Some robots, however, have the three major axes but only one or two of the minor axes.

WORK ENVELOPE

The total area that the end of the robot's arm can reach is called the **work envelope.** Figure 1.3 shows the work envelope of a robot. The view from the top of the robot is called the **plan view.** The view

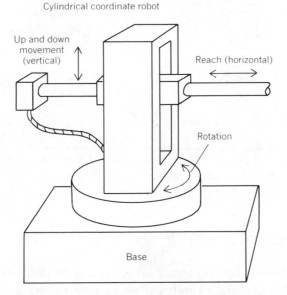

Figure 1.1 Three-axis robot with cylindrical (post-type) manipulator, illustrating two linear axes, and one rotational axis, of movement

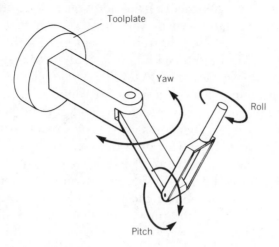

Figure 1.2   The three possible movements or axes of a robot wrist

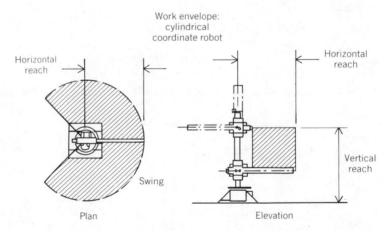

Figure 1.3   Work envelope of a cylindrical coordinate robot

from the side of the robot is called the **elevation view.** Combining the plan view and the elevation view would show the total volume that the end of the robot's arm can reach.

CLASSIFICATION OF ROBOTS BY WORK ENVELOPE

Robots can be classified according to their work envelopes into four types: the **cylindrical coordinate robot,** the **rectangular coordinate robot,** the **spherical coordinate robot,** and the **jointed arm coordinate robot.**

Cylindrical Coordinate Robot

If the end of the robot's arm forms a cylinder or a portion of a cylinder as the arm moves through the work envelope, the robot is called a **cylindrical coordinate robot.** The work envelope of a cylindrical coordinate robot is shown in Figure 1.3. The elevation view (right) shows that the arm can rise and fall. This is called the vertical axis. The arm can also go in and out. This is called the horizontal axis. The combination of the horizontal axis and the vertical axis forms a rectangle. The shaded area shows the combined vertical and horizontal reach of the robot when viewed from the side (elevation view).

Looking at the same robot from the top (plan view; left) we see that the robot pivots or turns at the center of its base. This pivoting forms a circle or a portion of a circle. From the plan view we can also see the horizontal axis. The total area that the end of the robot's arm can reach is shown by the shaded area.

Combining the two views of the robot (plan view and elevation view) would show that the work envelope of the cylindrical coordinate robot forms a cylinder or portion of a cylinder.

Rectangular Coordinate Robot

The **rectangular coordinate robot** and the cylindrical coordinate robot appear to be identical when viewed from the side (elevation view). The rectangular coordinate robot has a vertical axis. The robot's arm rises and falls on the vertical column. The rectangular coordinate robot also has a horizontal axis. The horizontal axis moves the robot's arm in and out (Figure 1.4).

The difference between the rectangular coordinate robot and the cylindrical coordinate robot can be seen in the plan view (Figure 1.4). The rectangular coordinate robot does not pivot on its base; rather, it slides back and forth on its base. The sliding back and forth in combination with the horizontal axis, which can also be seen from the plan view, forms a rectangle. By combining the plan view and the elevation view we can see that the work envelope for a rectangular coordinate robot forms a box.

Spherical Coordinate Robot

The **spherical coordinate robot** (Figure 1.5) resembles the cylindrical coordinate robot when viewed from the top (plan view). The spherical coordinate robot pivots on its base, forming an arc. The horizontal axis of the spherical coordinate robot is also seen in

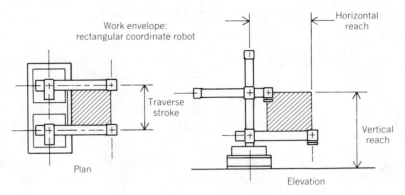

Figure 1.4   Work envelope of a rectangular coordinate robot

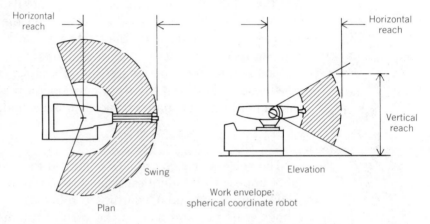

Figure 1.5   Work envelope of a spherical coordinate robot

the plan view. The shaded area in the plan view shows the total area, as viewed from the top, that the end of the robot's arm can reach.

When the spherical coordinate robot is viewed from the side (elevation view), the difference between the cylindrical coordinate robot and the spherical coordinate robot is apparent. The spherical coordinate robot does not rise and fall on a vertical axis as the cylindrical coordinate robot does; instead, it pivots, forming an arc.

By combining the plan view and the elevation view we can see that the work envelope of the spherical coordinate robot forms part of a "sphere" or a ball.

The spherical coordinate robot is sometimes referred to as a "polar coordinate robot." We will use the term "spherical coordinate robot" throughout this text.

The spherical coordinate robot was one of the first coordinate systems built by robot manufacturers. Although some of the other coordinate systems are easier to apply in some situations, the spherical coordinate robot is still popular today.

Jointed Arm Coordinate Robot

The **jointed arm coordinate robot** has the most complex work envelope (see elevation view in Figure 1.6). The joints of the jointed arm robot are often referred to as the **waist**, the **shoulder**, and the **elbow**.

The jointed arm coordinate robot is sometimes called an **anthropomorphic robot.** The word "anthropomorphic" means "taking on a human form." In this text, however, we will not use the term "anthropomorphic" but rather will use the term "jointed arm."

In the plan view of the jointed arm robot (Figure 1.6), the work envelope appears to be the same as those of the cylindrical and spherical coordinate robots.

The advantage of the jointed arm robot is its ability to reach up, down, and back on itself. This combination of movements is not possible using any of the other coordinate systems.

The versatility of the jointed arm robot makes it easier to apply in many situations. For example, loading of a turning center requires the ability to reach down into the machine (Figure 1.7). Although this operation can be done by most other robots, it is easier with the jointed arm. If we were to attempt this operation with a

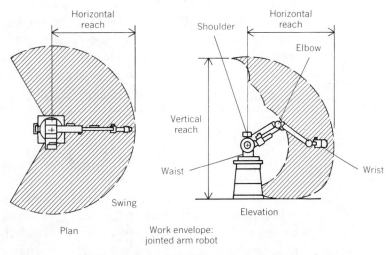

Figure 1.6   Work envelope of a jointed arm coordinate robot

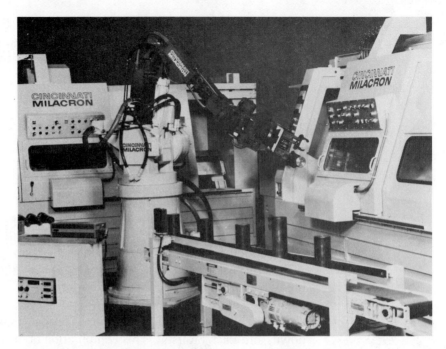

Figure 1.7   Loading of a turning center using a jointed arm robot

spherical coordinate robot we would have to build a special platform that would allow the robot to reach down into the turning center.

The major disadvantage of the jointed arm robot is its cost. The versatility of the jointed arm does not come free. In general, the jointed arm robot is the most expensive of all the coordinate systems. It is more difficult to build the manipulator of a jointed arm robot, and therefore it costs more. The controller for a jointed arm robot is also more complex, and this also adds to the cost.

There are many applications that do not require the versatility of the jointed arm robot. In these applications the designer should consider using some other form of robot.

The LERT System

Identifying robots as cylindrical, rectangular, spherical, and jointed arm is adequate for a general description of the work envelope; however, with the number of different robots on the market, this simple classification is no longer adequate for describing the work envelopes and degrees of versatility of the various robots. A new system which is being used by some engineers is called **LERT**.

Every movement of the robot can be identified as **linear, extensional, rotational**, or **twisting**. Figure 1.4 shows a rectangular coordinate robot. In the elevation view it can be seen that the robot arm rises and falls on a vertical column. This sliding is called a "linear" movement. Any movement along a straight line, with the exception of an "extensional" ("reach") movement, is called a linear movement. Another type of linear movement is side-to-side movement, which can be seen in the plan view in Figure 1.4.

The plan view and the elevation view of the rectangular coordinate robot also show an example of an "extensional" movement. The extensional movement is the "reach" of the robot. This is the horizontal axis of the robot (Figure 1.4). The cylindrical coordinate robot and the spherical coordinate robot also have the ability to reach. This is shown in Figures 1.3 and 1.5.

Another movement that is possible for many robots is the "rotational" movement. Figure 1.6 shows the plan view and the elevation view of a jointed arm robot. The movement of the shoulder axis illustrates the rotational movement. The robot's arm can go up and down by pivoting at the shoulder. This is a rotational movement. Another example of a rotational movement is at the base of the jointed arm robot. As the robot turns at its base the arm swings through an arc.

The final possible movement is the "twisting" movement. Twisting is similar to the rotational movement. Twisting, however, does not create an arc. The twisting movement is a turn about the center point. The twisting movement is best illustrated by the wrist of a robot (Figure 1.2). The pitch and yaw movements are rotational movements, but the roll movement is an example of a twisting movement. When the wrist moves through a pitch or a yaw, an arc is formed. On the other hand, when the wrist rolls it turns on its own center, and an arc is not formed.

One of the easiest ways to conceptualize the difference between the twisting and rotational movements is to examine the work envelope. If the movement increases the work envelope and causes the end of the arm to form an arc it is a rotational movement, but if the movement does not create a larger work envelope it is a twist.

Look again at Figure 1.4, which illustrates a rectangular coordinate robot. Starting at the base there is a linear movement which moves the robot from side to side. There is also a linear movement which raises and lowers the arm on the vertical axis. Finally, there is an extensional movement: the horizontal axis or reach movement.

The movements starting at the base of the rectangular coordinate robot are linear, linear, and extensional, which can be abbreviated as L,L,E. To simplify the notation of a rectangular coordinate robot even further it can be written as L^2E. The superscript 2 indicates that there are two linear movements.

The LERT notation for a spherical coordinate robot is R^2E. This notation tells us that there are two rotational movements followed by an extensional movement. The spherical coordinate robot rotates at the base (R), rotates at the shoulder (R), and has an extensional (E) movement for reach.

If we were to add a wrist to a spherical coordinate robot (R^2E) the LERT notation would be R^2ER^2T. Beginning at the base there is a rotational movement, followed by a rotational movement at the shoulder and an extensional movement. In addition, there are two rotational movements—the pitch and yaw—and finally a twisting movement.

When using the LERT system to define the movements of a robot, be certain to label the movements in proper order. The movement at the base is always first, the movement closest to the base is second, and the movement next closest to the base is third.

POWER SUPPLIES

The manipulator (arm) of a robot must have power supplied to it. The power moves the manipulator through its programmed moves. There are three different types of power supplies in common use: **hydraulic**, **pneumatic**, and **electric**.

Hydraulic Power Supplies

Hydraulic power supplies provide manipulator movement by pumping oil or some other fluid through pipes or hoses to hydraulic cylinders or hydraulic motors (Figure 1.8).

As fluid is pumped into a cylinder, the rod of the cylinder extends. If the cylinder is mounted as shown in Figure 1.9, the robot's arm will rise. If fluid is allowed to leave the cylinder, the arm will go down.

A hydraulic cylinder can also be used for an extensional movement. Figure 1.10 shows a spherical coordinate robot (R^2E). As hydraulic fluid is pumped into the cylinder the manipulator extends, and as fluid is allowed to flow out of the cylinder it retracts.

Hydraulic motors can be used for rotational and twisting movements. As fluid is pumped into the hydraulic motor, the shaft of the

Figure 1.8 Hydraulic power supply

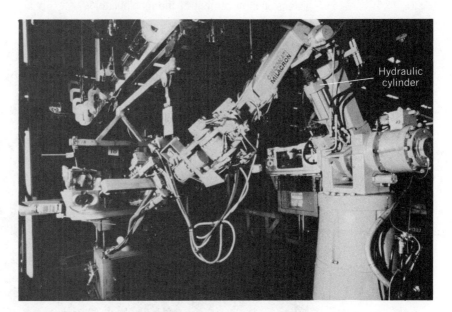

Figure 1.9 Cincinnati Milacron T3 robot

motor turns (Figure 1.11). If the motor is connected to the base of a manipulator, the manipulator rotates.

One major advantage of using hydraulics for supplying power to the manipulator of a robot is that a hydraulically powered manipula-

Figure 1.10   Spherical coordinate robot in which a hydraulic cylinder is used for extensional movement of the manipulator

Figure 1.11   Hydraulic motor

tor does not arc and spark as it operates. (Electrically powered manipulators normally generate sparks.)

The hydraulically powered robot can be placed in a volatile (explosive) environment such as a spray paint booth. Since the robot does not generate sparks while it is operating, the possibility of an explosion is significantly reduced.

Another advantage of the hydraulically powered robot is its load-carrying capability. Hydraulic robots can be constructed that can lift far heavier loads than can electric robots. Hydraulic robots have been constructed that can lift more than 30,000 lb.

One disadvantage of a hydraulic robot is the additional service that is required to keep it in good operating condition. To maintain a hydraulic robot, hydraulic fluid levels must be checked on a regular basis, and the hydraulic filter must be changed periodically. The filter removes dirt and other particles that contaminate the hydraulic fluid. Many hydraulic robot manufacturers recommend daily checks of the hydraulic fluid level and the hydraulic filter. Many also recommend visual inspection of all hydraulic lines, cylinders, and seals. (See the manufacturer's maintenance manual for specific recommendations.)

Another possible disadvantage of the hydraulic robot is the warm-up period required. Some hydraulic robots require from 15 to 30 min of warm-up time before they can be put into service each morning. If the robot is not warmed up according to the manufacturer's recommendations, the robot will not move smoothly and accurately. (Hydraulic systems are described in detail in Chapter 2.)

Pneumatic Power Supplies

Pneumatic power supplies provide the power to move the robot's manipulator through the use of compressed air. Pneumatic power supplies operate much the same as hydraulic power supplies. The difference between hydraulic and pneumatic power supplies lies in the medium used to transmit the power to the manipulator. Hydraulic power supplies utilize the flow of oil to generate movement whereas pneumatic power supplies utilize the flow of compressed air to generate movement.

A pneumatic power supply directs air through pipes or hoses to cylinders. The compressed air enters the end of a cylinder. As the air pressure in the cylinder goes up, the piston in the cylinder moves, extending the rod (Figure 1.12).

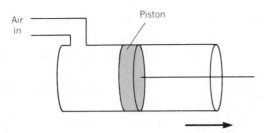

Figure 1.12   Cylinder in a pneumatic power supply

The construction of a pneumatically powered robot is much the same as that of a hydraulically powered robot (Figure 1.13). To raise the arm of the robot, air is pumped into the cylinder. At first the air pressure in the cylinder is not high enough to overcome the weight of the arm and the load. The process of pumping air into the cylinder would take far too long if we relied solely on the compressor to supply air to the cylinder. To increase the speed of the robot, an air compressor is used to pump air into a storage tank (Figure 1.14). When the valve between the storage tank and the manipulator is opened, air rushes into the cylinder and the arm moves quickly. (Pneumatic systems and their schematic symbols are covered in detail in Chapter 3.)

The major advantage of the pneumatic power supply is its simplicity. The robot is normally operated from "shop air." Most manufacturing plants have a central air compressor and storage tank. The air is piped to various machines in the plant that require compressed air. It is a simple matter to tap into the shop air lines to supply the compressed air to the robot.

Figure 1.13   Pneumatically powered robot

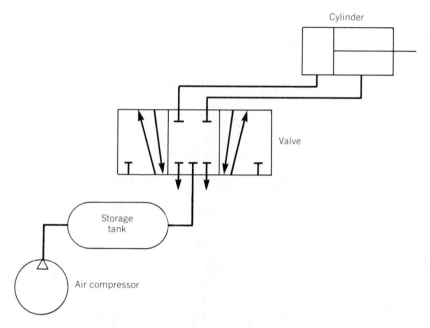

Figure 1.14 Schematic illustration of a pneumatic power supply

Another advantage of pneumatic power over hydraulic power is that if a leak should develop in the system air is pumped out through the leak rather than the oil used in a hydraulic system.

The major disadvantage of the pneumatic robot lies in its "sponginess." If the weight at the end of the robot's arm increases, the arm has a tendency to sag. The sag reduces the robot's ability to return to the same point every time. This problem can be overcome by adding special brakes and control systems; however, such added systems reduce the simplicity of the pneumatic robot and thus diminish its major advantage.

Electric Power Supplies

Electrically powered robots have become very popular. They offer simplicity in design and maintenance. Their major disadvantage in the past was their low load-carrying capability, but with the development of more powerful motors the load-carrying capability of the electric robot has been increased.

Electrically powered robot manipulators do not require the plumbing of shop air or the installation of a hydraulic power supply. They do not require the warm-up time of the hydraulic robot, and they do not suffer from the sag problem associated with the pneumatic robot.

In Chapter 4 we will study in detail the electric servo motors used to supply power to robot manipulators.

CLASSIFICATION OF ROBOTS BY POWER-SUPPLY CONTROL

There are two methods of controlling the power supplies of robots: **nonservo control** and **servo control.**

Nonservo Controlled Robots

The nonservo control is the simpler of the two controls. The nonservo control sends a signal to the power supply that turns it fully on or fully off. In response, the manipulator moves until it runs into a stop. Figure 1.15 shows a plan view of a nonservo spherical coordinate robot. The arm rotates. The arm can move from point A to point B. There are no other points at which the arm can be stopped with any degree of certainty. The arm literally "bangs" into the stops at points A and B, and hence this type of robot is called a **bang-bang robot.** The stops are normally blocks of steel bolted to the base of the robot.

Nonservo Pneumatic Robots.   The manipulator of a nonservo robot is powered by either hydraulics or pneumatics. For our first example we will consider a pneumatic robot (Figure 1.16).

To move the manipulator from point A to point B, air is fed into the cylinder through air line 1 (Figure 1.17). As air is fed to the cylinder through air line 1, line 2 is disconnected from the system and air is allowed to escape to the atmosphere. The robot's manipulator begins to move from point A to point B. The robot's arm continues to move until it "bangs" into the mechanical stop at point B.

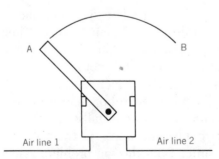

Figure 1.15   Plan view of a nonservo spherical coordinate robot

Figure 1.16 Nonservo pneumatic robot

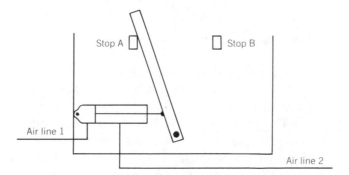

Figure 1.17 Plan view of the cylinder and manipulator of a nonservo pneumatic robot with no intermediate stop

To move the arm from point B to point A, air is fed to the cylinder through air line 2 and air line 1 is disconnected from the system. As air is fed to the system the manipulator begins to move. It continues to move until it "bangs" into the mechanical stop at point A.

To control a robot by hooking up one air line and disconnecting the other line is very inefficient. An easier method of controlling a robot is the installation of **solenoid valves** (Figure 1.18). Solenoid valves control the flow of air to and from the cylinder automatically. When a voltage is applied to the valve, the valve turns on and air is allowed to flow from the air compressor's storage tank through the

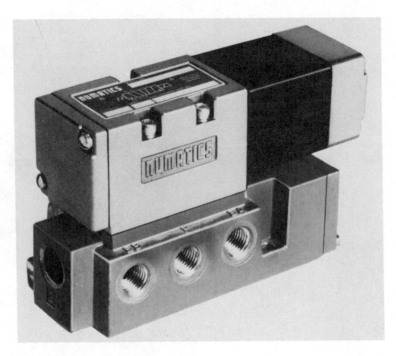

Figure 1.18   Solenoid valve used for control of a nonservo pneumatic robot

valve and to the cylinder on the manipulator. Solenoid valves are discussed in detail in Chapter 2 under the heading "Directional Control Valves."

The pneumatic bang-bang robots discussed up to this point have been limited to two end stops. They can only move from point A to point B or vice versa. They cannot be stopped at any intermediate point. There are some pneumatic robots that can be stopped at more than the two end points (Figure 1.19). Adding an intermediate stop to the robot is done by adding another air cylinder to the system. If

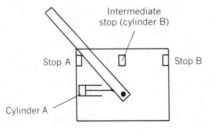

Figure 1.19   Plan view of the cylinders and manipulator of a nonservo pneumatic robot with one intermediate stop

air is fed to the new cylinder (cylinder B in Figure 1.19), an intermediate stop is pushed into the path of the robot's manipulator. When air is fed to cylinder A, the robot's arm moves from point A and continues to move until it bumps into the intermediate stop. When the intermediate stop is pulled out of the path of the manipulator, the manipulator will continue its move until it "bangs" into the mechanical stop at point B.

Nonservo Hydraulic Robots. A hydraulic power system can also be used for a nonservo robot. If we choose to use a hydraulic power system rather than a pneumatic system, additional control must be added. Figure 1.20 shows the plan view of a nonservo hydraulic robot. This system includes cylinders that are similar to those used in a pneumatic robot. The major difference between a hydraulically powered robot and a pneumatically powered robot is that the hydraulic system pumps oil rather than air. Hydraulic fluid (oil) is not compressible. When the manipulator is moved from point A to point B, oil is pumped into the cylinder. When the manipulator "bangs" into the end stop at point B it hits the stop very hard. The force of the manipulator hitting the end stop may be great enough to jar the mechanical stop out of place. This is obviously not desirable. This problem can be eliminated if a **limit switch** is added to the system (Figure 1.21). When the arm bumps the limit switch, an electrical signal is sent to the hydraulic system and the flow of the hydraulic fluid is reduced. The robot's manipulator then coasts to a stop against the mechanical end stop.

Another method of slowing the robot's arm is through the use of a shock absorber. When the arm approaches the end stop, the shock absorber slows the arm before it bangs into the end stop.

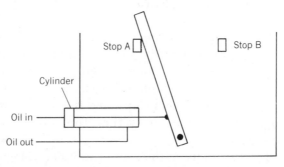

Figure 1.20 Plan view of a nonservo hydraulic robot

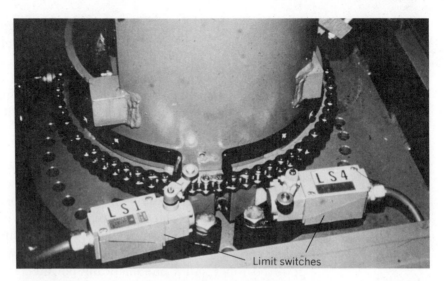

Figure 1.21   Limit switches used in a nonservo hydraulic robot to reduce the force with which the manipulator hits the end stops

Nonservo Electric Robots.   Nonservo electric robots can be built using electric motors, although this is not common in industry. There are many problems in the use of electric motors in nonservo robots. The major problem is that electric motors do not survive the shock of the manipulator banging into the end stop. This problem could be overcome if special clutches were used, but addition of the special clutches would complicate the design, construction, and maintenance of the robot. Hydraulic and pneumatic robots are simple and perform well, and most designers of industrial robots have chosen to ignore electric motors as an option for powering nonservo robots.

Servo Controlled Robots

Servo controlled robots are more versatile than their nonservo counterparts. Servo robots can be programmed to stop at any point within their work envelope. Figure 1.22 shows a servo controlled robot. The robot's manipulator can be programmed to stop at any point between the end stops, and the manipulator will return to the programmed point every time the robot is directed to execute the program.

The versatility of the servo controlled robot does not come without cost. The control necessary to operate a servo robot is far more

Figure 1.22 Servo controlled robot

complex than that of the nonservo robot. The basic difference between the servo robot and the nonservo robot is a **feedback system.** Figure 1.23 demonstrates the operation of the feedback system. The controller says, "The manipulator should be at point B." An electronic device mounted on the manipulator checks the position of the manipulator and sends a signal to the comparator that says, "I am at position A." An "error" exists. The controller has said that the manipulator should be at point B and the electronic device on the manipulator tells the comparator that it is at point A. The compara-

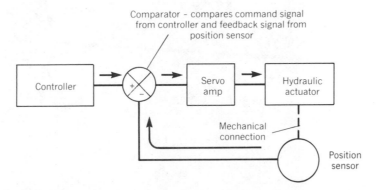

Figure 1.23 Schematic illustration of the feedback system of a servo controlled robot

tor recognizes that an error exists and sends a signal to a motor or
cylinder to move the manipulator to correct the error. The comparator continues to monitor the progress of the manipulator as the
error gets smaller. When the error has finally been reduced to zero
the manipulator stops moving.

Any point within the total work envelope can be identified, and
taught to the robot. The controller will move the manipulator from
its current position to the next taught position by examining the
error and moving the manipulator until the error is eliminated.

In Chapter 9 we will examine in detail the specific devices used
to monitor the manipulator's position and the circuitry of the controller used to correct the "errors."

Power Supplies for Servo Controlled Robots   Servo controlled
robots can be powered electrically, hydraulically, or pneumatically. The most popular power system for small- and medium-load-
carrying servo robots is electric power. The robot's joints are moved
by special motors called servo motors. Most servo motors used in
robots are powered by direct current (Figure 1.24), although alternating current motors are used in some robots.

Figure 1.24   Servo controlled robot powered by a direct-current electric motor

Hydraulically powered servo robots are used to carry heavy loads. Many can carry loads of over 100 lb, and some can carry loads in excess of 2000 lb. Hydraulic servo robots are also used for spraying paint (Figure 1.25) and for applications in other volatile (explosive) environments. Hydraulic robots do not arc or spark during operation, which makes them safer when the potential for explosion exists.

Finally, servo controlled robots may be powered by pneumatics. Pneumatic power supplies are not commonly used for servo controlled robots because they tend to "sag" under the load. It is also very difficult to move the arm smoothly and stop it precisely at the point that eliminates the "error." These problems have been overcome by one company and are being worked on by others.

CLASSIFICATION OF ROBOTS BY MOTION CONTROL

There are three major classifications of motion for robots: **pick-and-place, point-to-point,** and **continuous path.** These terms describe the movements the manipulator can make within its work envelope.

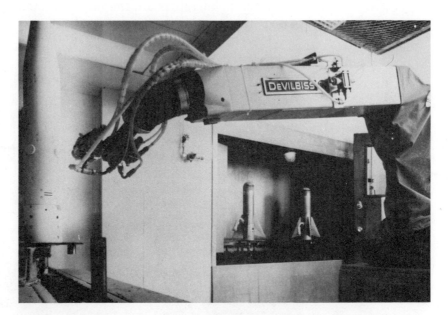

Figure 1.25 Servo controlled hydraulic robot used for spraying paint

Pick-and-place Robots

Pick-and-place robots are nonservo robots. They are normally limited to two end points; however, some pick-and-place robots also include one or two **intermediate stops.**

Pick-and-place robots were named by the job they normally perform in industry. They pick up parts or materials at one location and place them at another location. A pick-and-place robot can be used to unload a conveyor or transfer line. It also can be used for simple press loading and unloading applications (Figure 1.26).

Pick-and-place robots are the simplest of all robots. This is not to say that because they are simple they lack value. Pick-and-place robots are an excellent choice for simple jobs. They are the least expensive of the three types. There is no reason to buy more capacity than is necessary to do a job.

Pick-and-place robots also have the advantage of being the easiest to maintain. They can normally be serviced by the plant electrical or machine repair personnel. In general, a pick-and-place robot will operate longer between failures, and when it does fail the repairs will normally be simple and fast.

Point-to-point Robots

Point-to-point robots are the most common of the three classifications. The point-to-point robot is a servo controlled robot. It can

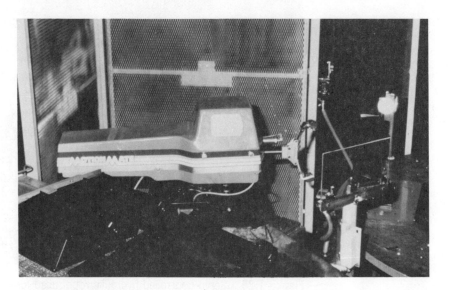

Figure 1.26 Pick-and-place robot

be programmed (taught) to move from any point within its work envelope to any other point within the work envelope. This versatility greatly expands its potential applications.

Point-to-point robots can be used in simple machine loading and unloading applications (Figure 1.27) as well as the more complex applications such as spot welding (resistance welding), assembly, grinding, inspection, palletizing, and depalletizing.

The point-to-point robot can move more than one of its axes at a time. For example, the spherical coordinate robot can rotate about its base at the same time it is reaching. In a more complex application it is possible to have the robot moving all of its major axes and all of its minor axes at the same time.

Although the point-to-point robot can move to any point within its work envelope, it does not necessarily move in a straight line between two points. Figure 1.28 shows the path that a point-to-point robot might take between points A and B. The vertical movement is much shorter than the horizontal movement required to move from point A to point B. When directed by the control, the robot's manipulator will begin to move from point A toward point B. The manipulator will begin to rise on the vertical axis and to reach on the horizontal axis. Since the distance the manipulator must travel along the vertical axis is much shorter than the distance it must travel

Figure 1.27 - Point-to-point robot

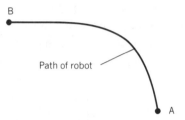

B

Path of robot

A

Figure 1.28 Illustration of the path of the manipulator of a point-to-point robot as it moves from one point to another (combined horizontal and vertical movement)

along the horizontal axis, the manipulator will complete the required vertical movement long before it completes the horizontal movement. The resulting path will be some form of an arc, but the exact shape of the arc is not predictable during the programming of the robot.

To program a point-to-point robot the programmer pushes buttons on a **teach pendant** (Figure 1.29). The teach pendant is much like the control used to move an overhead crane in a plant. To rotate the robot about its base, a button is pushed and the robot turns. To extend the manipulator, another button is pushed. When the robot has been led to the desired point, the programmer pushes a button to record that point in the robot's memory. The programmer then moves the manipulator to the next desired point by pushing the appropriate buttons. The path the programmer takes to get the manipulator to the next point is not remembered by the robot. When

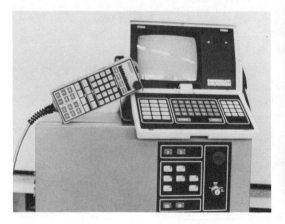

Figure 1.29 Teach pendant used for programming a point-to-point robot

the manipulator is finally brought to the desired point, the button to record the point is again pushed, and the second point is recorded into the robot's memory. When the program is played back, the robot will move from the first point in its memory to the second point, then to the third point, and so forth, until it has moved to all the points it has been taught. After the last point has been reached by the robot, the controller moves the manipulator back to the first point in the memory and the entire program is repeated.

Controlled Path Robots

The simple control method of the point-to-point robot makes it difficult if not impossible to predict the exact path of the manipulator between two taught points. A specialized control method that is part of the general category of a point-to-point robot is the **controlled path robot.** The controlled path robot ensures that the robot will always describe a straight line between the taught points. Figure 1.30 shows a multiple-exposure photograph of a robot that is under the control of a controlled path controller. As can be seen in this illustration, the robot moves in a straight line between the two taught points.

Figure 1.30 Controlled path robot. Dotted line shows path of robot manipulator.

Controlled path robot control has value when the robot's manipulator must move in a perfectly straight line. Although assembly operations can be accomplished with the simpler point-to-point control, programming of a controlled path robot to perform such operations is much easier. For example, if the robot is to put a shaft into a bearing, any deviation from a straight line could cause the shaft to score the bearing or could bend the shaft. Use of a controlled path robot ensures that the robot will slip the shaft into the bearing in a straight line. On the other hand, it is possible to use the point-to-point robot. In order to ensure that the robot will move in a straight line it is necessary to program many points along the path. The more points that are programmed, the straighter the path.

Other applications that are simplified through the use of a controlled path robot are arc welding, drilling, polishing, and assembly.

The method for programming the controlled path robot is identical to that for programming the point-to-point robot.

Continuous Path Robots

The **continuous path robot** can be programmed to follow any path desired. Its path can be an arc, a circle, or a straight line.

The continuous path robot is programmed differently than the point-to-point robot and the controlled path robot. Rather than leading the robot to the point desired by pushing buttons on a teach pendant, the manipulator of a continuous path robot is programmed by grabbing hold of the robot's arm and actually leading the arm through the path that we wish the robot to remember. The robot remembers not only the exact path through which the programmer leads the manipulator but also the speed at which the programmer moves the manipulator. If the programmer should move the arm too slowly, the speed can be adjusted at the control console. Changing the speed does not affect the path of the robot's arm.

The continuous path robot is really a form of the point-to-point robot. As the robot is taught the desired program, the control examines the location of the manipulator hundreds of times per second and stores each point in memory for playback at a later time. It is like programming thousands of individual points into the memory of a standard point-to-point robot. The continuous path robot is often used for spray painting (Figure 1.31), arc welding, or any other operation that requires constant control of the robot's path.

The major difference between the continuous path control and the standard point-to-point control is the control's ability to remem-

Figure 1.31   Continuous path robot

ber thousands of programmed points. Many point-to-point controls are limited to several hundred points of memory.

ROBOT CONTROLS

There are five basic types of robot controls in use today. These controls are the **drum controller,** the **air logic controller,** the **programmable controller,** the **microprocessor control,** and the **minicomputer.** All of these controls have the same purpose: to direct the motion of the robot's manipulator. Each of these controls has its own advantages and disadvantages. We will cover them one at a time.

The Drum Controller

The drum controller (Figure 1.32) is the oldest type of controller used in robotics. It is simple and reliable. The disadvantage of this controller is its limited versatility. The drum controller is used to program the motions of the pick-and-place robot. It cannot be used for point-to-point, controlled path, or continuous path robots.

The drum controller is similar to a music box. A music box has a drum with small pegs sticking out of it. As the drum rotates, the pegs bend spring steel. As the peg clears the spring steel, the spring

 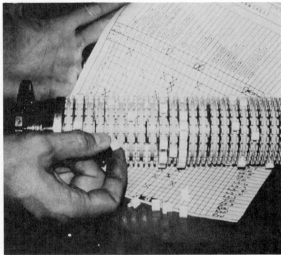

Figure 1.32   Drum controller

steel is released and it vibrates, sounding a note. The drum controller is also a drum. It has hundreds of holes in it. Small pegs are inserted into the holes. As the drum rotates, the small pegs close switches. The switches are wired to hydraulic or pneumatic valves. When a switch closes, the valve is opened and the robot's manipulator moves.

After one movement is completed, the drum is advanced one step. Another peg closes another switch and the manipulator moves again.

The Air Logic Controller

The air logic controller is also used for simple pick-and-place robots. An air logic controller is made up of a group of small pneumatic valves and timers. The valves and timers are connected together with small pieces of tubing. The sequence of hoses hooking the valves and timers together controls the opening and closing of the robot's main valves. When a main valve opens, the robot's manipulator moves. Air logic controllers are used only for pneumatic pick-and-place robots.

The Programmable Controller

The programmable controller is an electronic version of the drum controller and the air logic controller. The memory of the programmable controller is stored electronically rather than in the pegs of a drum controller or the air lines of an air logic controller. The programmable controller is also used with simple pick-and-place robots.

Many programmable controllers are programmed through a keyboard that looks similar to a typewriter. The order in which switches are to be closed is entered into the controller through the keyboard (Figure 1.33).

Many programmable controllers also have a **CRT (cathode ray tube)** to display the program that has been entered into its memory.

Programmable controllers are used for many other functions in the automated factory. Programmable controllers are used to control the conveyor systems, machinery, and storage systems. They are occasionally wired to other robot controls to allow more information to be fed to the robot from the outside world and to allow the robot to control other machinery around it.

Figure 1.33 Programmable controller

The Microprocessor Control

The microprocessor control (Figure 1.34) is the most common of all robot controls. It can be used with pick-and-place robots, point-to-point robots, controlled path robots, and continuous path robots.

The microprocessor control is a specialized computer designed specifically to control a robot. The microprocessor control contains memory to hold the robot's program, special circuits to interpret the instructions contained in the robot's memory, and electronic circuits for powering the valves or motors that move the robot's manipulator.

The Minicomputer

In recent years some robot manufacturers have been purchasing minicomputers (or microcomputers) and adapting them to control their robots. The output of the small computer (Figure 1.35) is fed into a specialized electronic control to control the robot's manipulator. For example, many minicomputers can be hooked up to a printer to print out information. Rather than being hooked up to a printer, the output can be processed by additional electronics, amplified, and used to open and close valves on the robot's manipulator.

ROBOT PROGRAMMING

Robots can be programmed or taught to do a job by placing pegs in the holes of a drum, connecting small air hoses between valves,

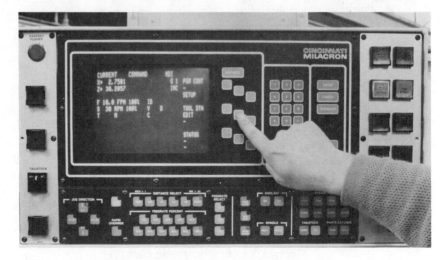

Figure 1.34   Microprocessor control

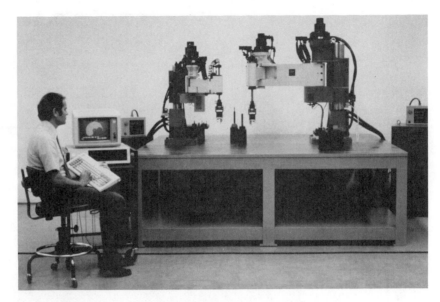

Figure 1.35 Minicomputer for robot control

or entering instructions through the keyboard of a programmable controller. The more common methods of teaching a robot are the **teach pendant, lead through programming,** and **computer terminal programming.**

The Teach Pendant

The teach pendant (Figure 1.36) is the most popular method of programming a robot. The program is taught to the robot by leading the robot's manipulator to the desired position by pushing buttons on the teach pendant. If we want to extend the robot's arm, a button is pushed to extend the arm. If we want the base of the robot to rotate clockwise, we push a different button. When the arm is finally at the desired position the "record" or "learn" button is pushed. This position of the robot's arm is entered into the controller's memory. The robot's manipulator is led to the next desired point by pressing buttons. When this next position is reached, the record button is pressed again, and the position is entered into the memory of the robot as the next point in the program. None of the moves made by the programmer between the first point and the second point is remembered by the robot. The robot remembers only the final points. When the program is played back, the robot's manipulator moves smoothly from the first point stored in its memory to the next point stored in its memory.

Figure 1.36   Teach pendant used for robot programming

Most teach pendants also allow the operator to enter speed commands into the robot's memory. The speed from point to point can be accurately controlled. It is also possible to program the acceleration and deceleration for each move.

After the robot has been taught all the moves it needs to complete its job, the operator normally "steps through the program." The robot's manipulator moves slowly from the first point to the second point. The programmer steps the robot through the program, checking each point. If one of the points is not correct, the programmer can "edit" the program, making the required small corrections.

After the program has been checked and edited, the robot is ready to run. The robot will move smoothly from point to point at the taught speed and will repeat the exact motion hour after hour.

Lead Through Programming

Lead through programming is normally used to program continuous path robots. To program a robot using the lead through method, the robot is switched to the "program" or "learn" mode. The programmer grasps the manipulator of the robot and pushes the manipulator through the desired path (Figure 1.37). For example, continuous path robots are often taught to spray paint. A skilled painter switches the robot to the programming mode, grasps the arm of the robot, and sprays a sample piece. The robot control remembers every move the operator makes. When the program is played back the robot will exactly duplicate every move it was taught by the painter. If it is discovered that the painter made a small error while teaching the robot, the program can be edited.

Some continuous path robots have a "teach arm." The teach arm is a small version of the robot's manipulator. The programmer moves the teach arm, and the robot's manipulator follows.

Most continuous path robots also allow for speed control. The programmer can teach the robot at a slow speed, being certain to make each move precisely. The programmer can then enter a speed command to increase the speed at which the robot operates.

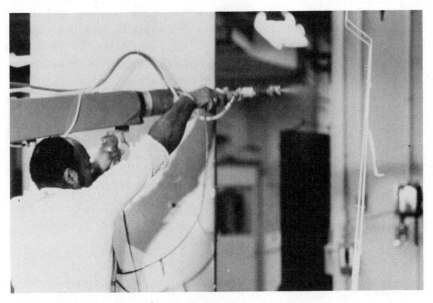

Figure 1.37 Lead through programming

Figure 1.38   Computer terminal programming

Computer Terminal Programming

The computer terminal programming method (Figure 1.38) is not as popular as the teach pendant and the lead through method. Programming a robot using the computer terminal is much like programming a computer to do bookkeeping. The "language" used is designed with special commands to move the robot manipulator. This method of programming allows the programmer to make the robot move more precisely than any of the other methods. For some robots, a teach pendant can be used first and then the program can be modified using the computer terminal to make the robot perform precisely.

In the future, this method of programming will become more popular as the jobs that robots are asked to do become more difficult.

QUESTIONS FOR CHAPTER 1

1. For what two types of jobs are robots normally used?

2. How do robots increase productivity?

3. How fast are robots?

4. What is meant by "degrees of freedom"?

5. Define "work envelope."

6. Draw plan and elevation views of a cylindrical coordinate robot, a rectangular coordinate robot, a spherical coordinate robot, and a jointed arm coordinate robot.

7. Using the LERT system, write the description of a cylindrical coordinate robot.

8. What are the three types of power supplies for robots and what are the advantages and disadvantages of each?

9. Define the terms "nonservo" and "servo."

10. Describe the difference between the paths that a point-to-point robot and a controlled path robot will take between the same two programmed points.

11. How are point-to-point and controlled path robots normally programmed?

12. How is a continuous path robot normally programmed?

13. For what application is the continuous path robot normally used?

14. What is the most common type of control used for industrial robots?

CHAPTER TWO

Hydraulic Systems

Hydraulic systems have been used in industry for many years. Hydraulic systems can lift and move large loads with seemingly little effort. Hydraulic systems use fluid such as oil to transmit energy rather than the more complicated mechanical linkages that use gears, cams, levers, and rods to transmit energy.

It was these advantages that led robot designers to use hydraulics to power robots. Hydraulic systems are used in both servo and nonservo robots.

HYDROSTATICS AND HYDRODYNAMICS

The field of hydraulics is divided into two major categories: **hydrostatics** and **hydrodynamics**.

Hydrostatics

Hydrostatics is based on the principle that a contained fluid, under pressure, transmits pressure equally in all directions. Figure 2.1 shows a container filled with water. If a pressure is applied to the water in the container by placing a piston in the top and placing a weight on top of the piston (Figure 2.2), the pressure in the container will be equal throughout. There will be the same pres-

Figure 2.1   Container of water

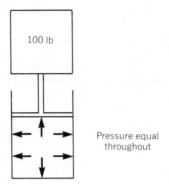

Figure 2.2   Container of water with piston and weight

sure pushing against the sides, bottom, and top. This is known as **Pascal's Law.**

It is important to remember that fluids are not compressible. The volume of the water in the container does not change as the pressure in the system is increased.

If we connect the first container to another container of the same size with a hose and place a piston in the second container, the 100-lb weight on the top of the first container will cause the piston in the second container to push upward with a force of 100 lb. The pressure in the first container is transmitted to the second container. The pressure throughout the entire system is equal. The piston in the first container is being pushed downward with a force of 100 lb, causing the piston in the second container to be pushed upward with a force of 100 lb.

The principle of pressure being equally distributed in a container is used in automobile jacks, industrial cranes, industrial robots, and many of the machine tools that are common in industry.

Hydrodynamics

Hydrodynamic systems use the principle that fluid in motion transmits force. Hydroelectric power plants use this principle to generate the power to turn the electric generators. The water falling over a dam turns the water wheel. The water wheel is connected to the generator with a shaft, and the generator is turned.

Although hydrodynamic systems can be used to produce usable energy they are not common in industry. One application that has become commonplace is the torque converter.

FLOW AND PRESSURE

Two terms that we must be familiar with in the study of hydraulic systems are **flow** and **pressure**. These terms are easy to understand but are often confused by troubleshooters servicing hydraulic systems, causing incorrect diagnoses.

Flow

Flow is simply the amount of fluid which is moving in a system. Flow is measured in gallons per minute (gal/min) or in liters per minute (L/min). Figure 2.3 shows a pipe with water flowing out of it. If the water coming out of the pipe can fill a 5-gal bucket in 1 min, there is a flow of 5 gal/min. If we double the flow in the system we will be able to fill the 5-gal bucket in 30 sec, and there will be a flow of 5 gal/30 sec. Because we normally express flow in terms of minutes, we would say that we have a flow of 10 gal/min. If we had a similar system but wished to express the flow using the metric system rather than the English system, we would express the flow rate in liters per minute (L/min). If a 5-L bucket is filled in 1 min, the flow rate is 5 L/min.

Pressure

Pressure is defined as force divided by area. Pressure is normally expressed in pounds per square inch (lb/in.^{2}, or psi) in the English system or in kilograms per square centimeter (kg/cm^2) in the metric system.

Figure 2.4 shows a block that has a surface area of 1 in.^{2}. If the block weighs 1 lb it will exert a pressure of 1 psi.

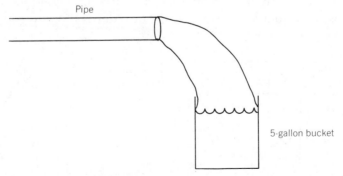

Pipe

5-gallon bucket

Figure 2.3 Illustration of fluid flow

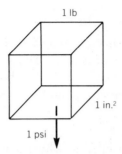

Figure 2.4   Illustration of pressure exerted by a 1-lb block with a surface area of 1 in.^{2}

Figure 2.5 shows a block that also has an area of 1 in.^{2} but weighs 5 lb. This block will exert a force of 5 psi.

Figure 2.6 shows a block that has a surface area of 5 in.^{2}. If the block weighs 10 lb it will exert a force of 10 lb/5 in.^{2}. This is not the normal way of expressing pressure. To convert to psi, divide the 10-lb weight by the 5 in.^{2} of surface area: 10 lb/5 in.^{2} = 2 lb/in.^{2} or 2 psi.

Earlier we said that pressure in a hydraulic system is distributed equally throughout the system. Figure 2.7 shows a container that is filled with fluid. At the top of the container is a piston. If the area of the piston is 10 in.^{2} and a weight of 100 lb is placed on top of the piston, the pressure in the system can be found by dividing the weight by the area of the piston: 100 lb/10 in.^{2} = 10 psi.

MECHANICAL ADVANTAGE

Figure 2.8 shows two hydraulic cylinders connected together with a hose. A hydraulic cylinder is a round container with a piston

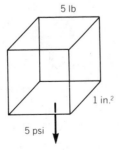

Figure 2.5   Illustration of pressure exerted by a 5-lb block with a surface area of 1 in.^{2}

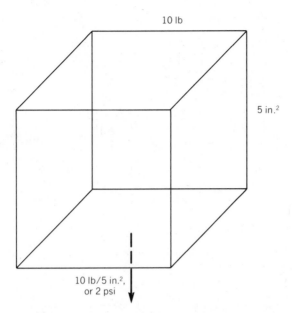

Figure 2.6 Illustration of pressure exerted by a 10-lb block with a surface area of 5 in.²

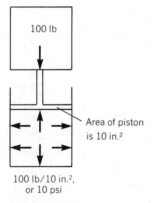

Figure 2.7 Container of fluid with piston and weight

inside. Both cylinders in Figure 2.8 are the same size. If a 10-lb force is exerted on piston A the piston moves downward. The fluid in cylinder A is forced through the hose connecting cylinders A and B and the piston in cylinder B moves upward. Piston B moves upward the same distance that piston A is pushed downward.

If we place a 100-lb weight on piston A and a 100-lb weight on piston B the pistons do not move (Figure 2.9). The fluid in the sys-

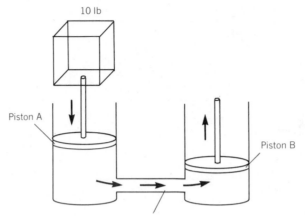

Figure 2.8 Two-cylinder hydraulic system. Fluid flows from cylinder A to cylinder B.

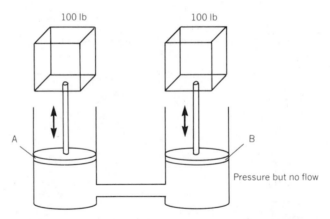

Figure 2.9 Two-cylinder hydraulic system. Pressure, but no flow.

tem is not moving from one cylinder to the other. There is "no flow"; however, there is pressure. The pressure is caused by the downward force of the weights. The formula for computing the pressure in the system is introduced later in this chapter.

If an additional 10-lb weight is added to piston A there is both flow and pressure (Figure 2.10). The extra weight on piston A unbalances the system and piston A moves downward. Fluid flows from cylinder A to cylinder B. As piston A moves downward, piston B moves upward. In this system there is both pressure and flow.

In Figure 2.11 the piston in cylinder B has twice the area of the piston in cylinder A. If a 100-lb weight is placed on piston A, then a

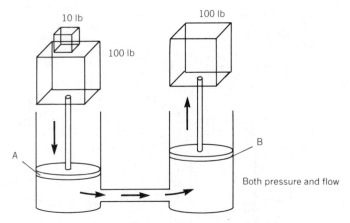

Figure 2.10 Two-cylinder hydraulic system. Pressure and flow.

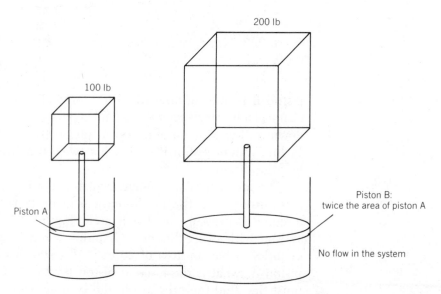

Figure 2.11 Two-cylinder hydraulic system. Pressure, but no flow.

200-lb weight must be placed on piston B to balance the system. This is an example of **mechanical advantage.** A 100-lb weight is balancing a 200-lb weight. In this system, as in the system described in Figure 2.9, there is no flow but there is pressure.

If a 10-lb weight is added to the 100-lb weight, fluid begins to flow in the system (Figure 2.12). Piston A moves downward and piston B moves upward. Recall that our system has a mechanical advantage. Piston A is moving downward with a force of 110 lb, and

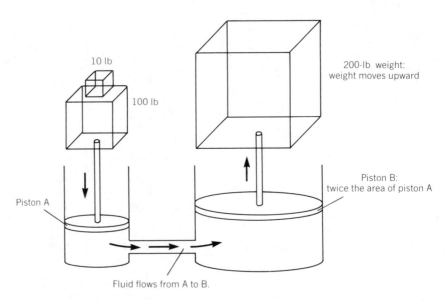

Figure 2.12   Two-cylinder hydraulic system. Fluid flows from cylinder A to cylinder B.

piston B is moving upward with a force of 220 lb. To gain this mechanical advantage something had to be sacrificed. For every 2 in. of downward movement of piston A, piston B moves upward only 1 in. In this system piston B is exerting twice the force but moving only half the distance.

The mechanical advantage available in a hydraulic system can be computed using the mathematical formula $F = PA$, where F is force, P is pressure, and A is area. Figure 2.13 shows a hydraulic system with two cylinders. Cylinder A has an area of 10 in.^{2} and cylinder B has an area of 20 in.^{2}. If a 200-lb weight is placed on piston A, what is the force at piston B? To compute the answer we must first find the pressure in the system. The formula $F = PA$ can be rearranged so that $P = F/A$. Substituting into the formula we have $P = 200$ lb/10 in.^{2}, and therefore $P = 20$ psi. We know that we have a pressure of 20 psi at every point in the system. Using the formula $F = PA$ we can compute the upward force of piston B: $F = PA = 20 \times 20 = 400$ lb. We have a force of 400 lb at piston B.

Now re-examine Figure 2.9. What is the pressure in this system if each of the two pistons has an area of 2 in.^{2}? First compute the pressure resulting from the 100-lb weight on piston A. The formula is $P_1 = F/A$, where P_1 is the pressure in the system, F is the force, and A is the area of the piston:

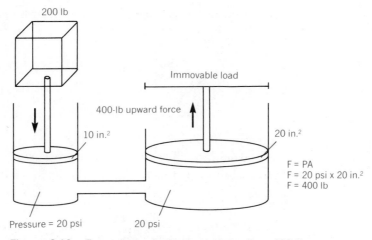

200 lb

Immovable load

400-lb upward force

10 in.²

20 in.²

F = PA
F = 20 psi x 20 in.²
F = 400 lb

Pressure = 20 psi   20 psi

Figure 2.13   Two-cylinder hydraulic system. F = 400 lb.

$$P_1 = \frac{100}{2}$$

$$P_1 = 50 \text{ psi}$$

Now compute the pressure resulting from the weight on piston B:

$$P_2 = \frac{100}{2}$$

$$P_2 = 50 \text{ psi}$$

Note that pressure P_1 is equal to pressure P_2. The system is in balance, and the pistons are not moving. The total pressure in the system is 50 psi, not 100 psi. As an aid to understanding why the pressure is only 50 psi, recall that pressure in a system is the result of a resistance. If there were no weight on top of piston B, the weight on piston A would push piston A downward and piston B would move upward. There would be flow but there would be no resistance to that flow, and the pressure would be zero. As weight is added to piston B, the pressure in the system will rise and the rate of flow will decrease. When the weight on piston B is 100 lb, flow is at zero and pressure is at its maximum — that is, a total of 50 psi as a result of the resistance.

Look again at Figure 2.10. A 10-lb weight has been added to piston A. (Each of the two pistons has an area of 2 in.².) The total pressure that potentially can be created can be calculated using the formula P = F/A:

$$P_1 = \frac{110}{2}$$

$$P_1 = 55 \text{ psi}$$

The pressure is 55 psi.

Now calculate the "resistance" (pressure) caused by placing a weight on top of piston B. Again, the formula P = F/A is applied:

$$P_2 = \frac{100}{2}$$

$$P_2 = 50 \text{ psi}$$

The resistance to the flow of fluid is 50 psi. Since piston A is creating a pressure of 55 psi and piston B is resisting the flow of fluid by creating a "back pressure" of 50 psi, there is a **pressure differential** of 5 psi (55 psi − 50 psi = 5 psi). This pressure differential causes flow. Piston A moves downward while piston B moves upward.

It appears that two different pressures exist in this system: what is the actual pressure? It is important to remember that pressure is caused only by resistance to the flow of the fluid. The resistance to the flow of fluid from cylinder A to cylinder B is caused by the weight on piston B as well as by the friction resulting from the fluid being forced through the pipe that connects the two cylinders. If a pressure gauge is placed in cylinder A, the pressure will read 55 psi. If a pressure gauge is placed in cylinder B, the pressure will read 50 psi. The pressure in cylinder B is only 50 psi since the resistance to the flow of fluid is caused only by the weight on cylinder B. There are in fact two different pressures in this system. It is this difference in pressure between cylinder A and cylinder B that causes the fluid to flow.

Now look again at Figure 2.12. Earlier we said that the 110-lb load on piston A would cause the 200-lb load on piston B to move upward. We can apply the same formulas and reasoning and prove that this is in fact the case. For this example we will assume that piston A has an area of 2 in.^{2} and that piston B has an area of 4 in.^{2}. First we will compute the pressure caused by the 110-lb weight on piston A:

$$P_1 = \frac{110}{2}$$

$$P_1 = 55 \text{ psi}$$

Now we will compute the pressure caused by the 200-lb weight on piston B:

$$P_2 = \frac{200}{4}$$

$$P_2 = 50 \text{ psi}$$

Since the two pressures are not equal, a pressure differential exists and there is flow in the system. Because fluid is flowing from cylinder A to cylinder B, we can further conclude that the piston in cylinder A must be moving downward while the piston in cylinder B is moving upward.

Since we know that fluid is flowing in the system we can now determine how far piston B will move upward for every inch that piston A moves downward. Recall that the fluid in a hydraulic system is noncompressible. Because the fluid is noncompressible, one cubic inch of fluid flows into cylinder B for every cubic inch of fluid that flows out of cylinder A. This can be expressed as:

Fluid out of cylinder A = Fluid into cylinder B

The amount (volume) of fluid that flows out of cylinder A can be found by multiplying the area of piston A by the distance the piston travels. This can be expressed mathematically as:

$$\text{Vol} = A \times D$$

where Vol is the volume of fluid, A is the area of the piston, and D is the distance traveled by the piston.

Since the volume of fluid flowing out of cylinder A is the same as the volume of fluid flowing into cylinder B, we can set the volume flowing out of cylinder A equal to the volume entering cylinder B. This can be expressed mathematically as:

$$A_1 \times D_1 = A_2 \times D_2$$

where A_1 is the area of the piston in cylinder A, D_1 is the distance traveled by piston A, A_2 is the area of the piston in cylinder B, and D_2 is the distance traveled by piston B.

Recall that piston A has an area of 2 in.^{2} and piston B has an area of 4 in.^{2}. We will assume for this problem that piston A has moved 1 in. How far has piston B moved? By substitution:

$$A_1 \times D_1 = A_2 \times D_2$$

$$2 \times 1 = 4 \times D_2$$

$$4D_2 = 2$$

$$D_2 = \frac{2}{4}$$

$$D_2 = \frac{1}{2}$$

Thus, for every 1 in. of downward travel of piston A, piston B moves upward 1/2 in., and the upward force of piston B is twice the downward force of piston A. It is this increase of force and reduction of movement that is referred to as "mechanical advantage."

HYDRAULIC TANKS

Hydraulic tanks perform the obvious function of holding the hydraulic oil for the system. In addition to being a reservoir, the tank also helps dissipate the heat that builds up in the system as it is operating.

When a hydraulic system is operating, oil is being pumped through the system. The oil is being forced through hoses, valves, and cylinders, causing friction. The friction heats up the oil. If the oil should become too hot the system could be destroyed. The hydraulic tank acts like a radiator, dissipating the heat.

The tank is not simply a box or a drum. Figure 2.14 shows a cutaway view of a hydraulic tank. The tank contains **baffles**. The baffles perform two functions. First, the baffles prevent the oil that enters the tank from going directly to the outlet of the tank. The incoming hot oil mixes with the cooler oil in the tank, ensuring that the oil will not overheat. Second, the baffles reduce turbulence in the tank. If there were no baffles, the incoming oil would cause waves in the tank. This churning would mix air into the oil. If air becomes mixed with the hydraulic oil the system becomes "spongy." If air

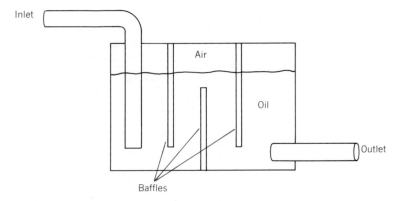

Figure 2.14 Cutaway view of a hydraulic tank

should become mixed in with the hydraulic oil of a robot, the manipulator will sag when it is picking up a heavy load and it will jerk as it moves the load.

Air mixed in with hydraulic oil can also cause damage to the hydraulic components. This is referred to as **cavitation**. The air bubbles are compressed and then expand in the system while the robot is operating, causing pitting of valves, pumps, and motors. When pitting occurs, the life of the hydraulic components is significantly reduced.

Vented Tanks

Hydraulic tanks can be either vented or pressurized. A vented tank allows air from the atmosphere to enter and leave the tank. As the hydraulic oil in the tank heats up it expands. Air in the tank also expands and contracts. The heated oil warms the air and the air expands (Figure 2.15). As the oil and air expand, air is forced out of the tank through the tank vent. When the oil cools, the air and the oil both contract, and air is drawn back into the tank through the vent.

To keep dirt from entering the tank through the vent, the vent normally has filter material in the cap. The fill pipe for the tank also has a strainer to keep dirt from entering the tank while it is being refilled with oil.

Pressurized Tanks

A pressurized tank does not have a vent, but rather has a cap much like the cap on an automobile radiator. As the hydraulic fluid

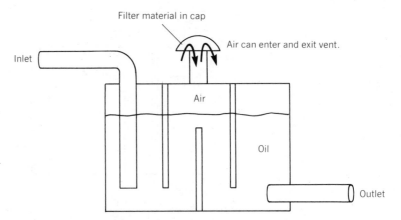

Figure 2.15   Cutaway view of a vented tank

heats during operation, the air pressure in the tank is not allowed to escape to the atmosphere. This raises the pressure in the tank a few pounds. The added pressure forces oil out of the tank to the hydraulic pump, ensuring that there is always an adequate supply of oil at the pump.

The pressure cap of a pressurized hydraulic tank is shown in Figure 2.16. The cap contains a spring. Should the pressure in the tank rise above a safe level, the spring is compressed, opening a vent which allows some of the pressure to escape to the atmosphere.

There is also another valve in the pressure cap. It is called an atmosphere valve. When the oil in the hydraulic system cools, it contracts, causing a vacuum in the tank. The atmosphere valve opens and allows air to flow back into the tank.

The symbol used for hydraulic tanks in blueprints (schematics) is shown in Figure 2.17.

FILTERS

Hydraulic systems cannot tolerate dirt. Dirt causes the close-fitting parts of the hydraulic pumps, motors, and valves to wear quickly. To prevent damage due to dirt, hydraulic systems have filters. The filter may be mounted inside or outside the tank. In some cases the manufacturer of the system may mount filters both inside and outside the tank.

The filter not only filters out dirt that may have entered the system through the atmosphere but also filters out steel or brass parti-

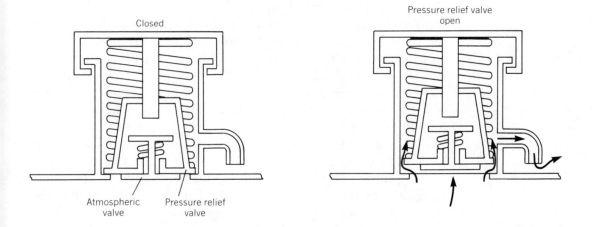

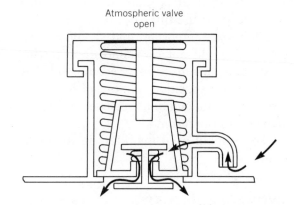

Figure 2.16 Cutaway views of a pressurized tank cap

Figure 2.17 Schematic symbol for a hydraulic tank

cles that wear off the pumps, motors, and valves. If these particles are not filtered out of the system they will accelerate the natural wear of the system and will cause the system to fail prematurely.

Filters are rated according to the amount of pressure they can handle and the size of the smallest particle they will trap. When replacing a filter, be certain to replace the old filter with the one recommended by the manufacturer. If you install a filter that allows passage of excessively large particles, the system will be damaged. On the other hand, if you replace a filter with one that filters out particles that are smaller than necessary, the system pressure may rise and the filter will fill up quickly, requiring more frequent replacement. Always consult your owner's manual to be certain you are replacing the worn-out filter with a filter of the proper type and rating.

Many hydraulic robot systems now incorporate filter-failure alarms. In some systems, a small window in the top of the filter housing turns red when the filter needs to be replaced. In others, the filter contains an electronic sensor that is connected to the robot controller. When the filter needs replacement, a warning light on the control console turns on. Many systems are designed so that if the warning light is ignored the system turns off and cannot be restarted until the filter has been replaced.

Filters are important and should not be ignored, and electronic systems should not be relied upon for indication of filter failure. Most robot manufacturers recommend that filters be replaced routinely after a specified number of hours of operation. The manufacturer's recommendations regarding filters should be understood and followed to the letter.

The symbol used for filters in hydraulic schematics is shown in Figure 2.18.

HYDRAULIC PUMPS

In order to discuss the various types of pumps in common use there are several terms that must be understood. The first term is **displacement**. Displacement is the amount of fluid a pump will discharge each time the pump shaft is turned one full revolution. The

Figure 2.18   Schematic symbol for a filter

displacement of a pump is measured in cubic inches. If a pump discharges 10 cubic inches (10 in.^{3}) of fluid for each revolution of the pump's shaft, its displacement is 10 in.^{3}.

There are two basic classifications of hydraulic pumps: **nonpositive displacement pumps** and **positive displacement pumps.**

Nonpositive Displacement Pumps

Nonpositive displacement pumps are not sealed between the inlet and the outlet of the pump. Figure 2.19 shows an example of a nonpositive displacement pump. As the propeller turns, fluid is drawn into the pipe and is discharged at the other end. As fluid is drawn into the pump and discharged, pressure is created. The pressure generated by the pump is a function of the speed at which the propeller is turning. As the speed of the propeller is increased, the output pressure increases. If the outlet of the pump is restricted, the flow is reduced. The propeller in the pump slips in the fluid. The back pressure created by restriction of the outlet allows fluid to leak backward past the propeller. At this point the propeller is "slipping." The greater the restriction, the lower the flow. If the outlet end of the pipe is blocked, the propeller continues to turn. The pressure remains at maximum and the flow is zero. If the outlet of the pump is opened, the pump will again be at maximum flow.

An example of this type of pump is a sump pump. When the outlet of the sump pump is blocked, the propeller continues to turn. Flow is stopped and the pressure is at maximum. Nonpositive

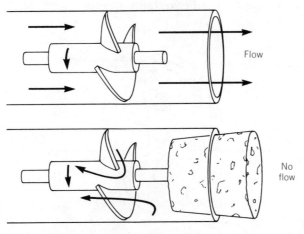

Figure 2.19   Nonpositive displacement pump

displacement pumps are not used in hydraulic systems because hydraulic systems require both high pressure and high flow.

Positive Displacement Pumps

The positive displacement pump differs from the nonpositive displacement pump. The nonpositive displacement pump creates pressure as a result of the propeller being turned. The positive displacement pump creates flow rather than pressure. In a positive displacement pump, pressure is the result of a restriction at the outlet. When the outlet of a positive displacement pump is restricted, the flow remains constant while the pressure begins to rise. If there is no restriction at the outlet of the pump, there is no pressure — but there is flow.

Positive displacement pumps have a positive seal between the inlet and the outlet. When fluid is drawn into the pump the fluid is moved from one sealed chamber to the next. Figure 2.20 shows an example of a positive displacement pump. This type of pump is called a **gear pump.** (We will be covering the operation of gear pumps in detail a little later in this chapter.) As fluid is drawn into the inlet of the pump it is picked up by the gears and moved toward the outlet. If the outlet of the pump is blocked, the gears will continue to turn, the pump will continue to pump fluid, and the pressure in the pump will rise very rapidly. If the blockage at the outlet of the pump is not released, pressure will continue to build until the pump housing cracks.

Positive displacement pumps are used in hydraulic systems because hydraulic systems require flow. The "flow" of fluid into a

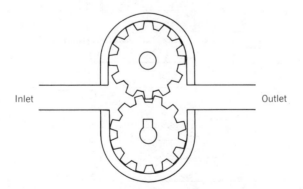

Figure 2.20   Positive displacement pump (gear pump)

cylinder causes the cylinder to extend. The pressure in the system is simply a result of the load that the cylinder is attempting to move.

Fixed and Variable Positive Displacement Pumps

Within the category of positive displacement pumps there are two subcategories: **fixed positive displacement pumps** and **variable positive displacement pumps.**

The fixed positive displacement pump pumps a specific volume of fluid for each revolution of the pump shaft. If the pump is designed as a 20-in.^{3} pump it will pump 20 in.^{3} with each revolution.

The variable positive displacement pump can be adjusted to different outputs for each revolution. A variable displacement pump may have a maximum volume of 20 in.^{3}, but its output can be reduced. The displacement of the pump can be adjusted from a displacement of 0 in.^{3} to a maximum displacement of 20 in.^{3}.

Pump Volume

The term **pump volume** should not be confused with the term **displacement**. The displacement of a pump has been defined as the amount of fluid that can be moved with one revolution of the pump. Pump volume is the amount of fluid that can be moved in one minute. The volume of a pump is directly proportional to the speed at which the shaft is turned and the displacement of the pump. For example, a pump with a displacement of 10 in.^{3} turning at 231 revolutions per minute (rpm) will have a volume of 10 gal/min. A pump with the same displacement (10 in.^{3}) turning at 462 rpm will have a volume of 20 gal/min. As the speed of the pump increases, the volume of the pump increases.

Pump volume is affected not only by the rotational speed (rpm) of the pump's shaft but also by the pressure. If we could build an ideal positive displacement pump, pressure would not have any effect, but since we cannot build such a pump we must take pressure into consideration. Most positive displacement pumps will continue to pump until the pressure exceeds several thousand pounds. At pressures above 2000 psi, some fluid begins to leak backward through the pump, and the pump begins to act a little like a nonpositive displacement pump. Figure 2.21 shows a graph of pump volume versus pressure. In this graph the volume of the pump remains constant until the pressure exceeds 2000 psi. Above 2000 psi, the volume of the pump begins to fall.

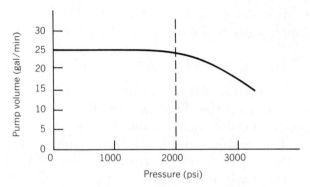

Figure 2.21   Graph of pump volume vs. pressure

It is important when specifying pump volume that you state both the rotational speed (rpm) and the working pressure.

Gear Pumps

A gear pump is a positive displacement pump (Figure 2.22). The gear pump is also a fixed displacement pump. As the gears in the pump are turned, fluid is drawn into the inlet and discharged from the outlet. Each revolution of the gears produces the same amount of flow. The volume of a gear pump is controlled by the size of the pump and the speed at which the pump's shaft is turned. As the speed of the shaft is increased, the volume increases.

The gears in Figure 2.22 are turning in the direction of the arrows. The inlet of the pump is on the left and the outlet of the pump is on the right. As the gears turn, the size of the chamber at the inlet becomes larger, causing a vacuum. Atmospheric pressure in the tank forces fluid into the chamber, and the fluid is carried

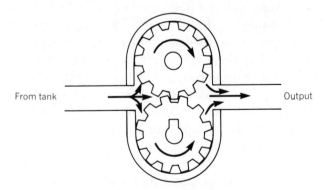

Figure 2.22   Cutaway view of a gear pump

around the outside of the gears to the outlet of the pump (it does not pass between the gears).

It is important that the gears fit tightly in the pump case. If there is clearance between the gears and the pump case, fluid will leak between the gears and the case, reducing flow. Some gear pumps are equipped with **wear plates** to help seal the gears. A small amount of fluid is taken from the outlet of the pump and directed to the outside of the wear plate. As the flow of the pump is restricted due to a load, the pressure in the system rises, forcing the wear plates more tightly against the gears. When the pump is no longer pushing a load, the pressure goes down and the wear plates relax, reducing wear (Figure 2.23).

The hydraulic fluid being pumped is a lubricant. This reduces the wear of the gears and extends the life of the pump.

Vane Pumps

Vane pumps, which are also very common in industry, are positive displacement pumps and are available in both fixed and variable displacement types. We will look at the fixed displacement type first.

A vane pump consists of three main elements: the **pump housing,** the **rotor,** and the **vanes** (Figure 2.24). The vanes fit into slots in the rotor, and the entire rotor assembly fits into the pump housing. The

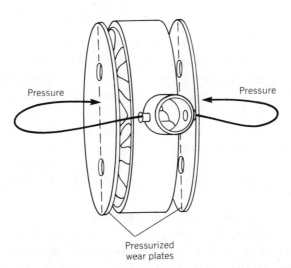

Figure 2.23 View of a gear pump showing pressurized wear plates

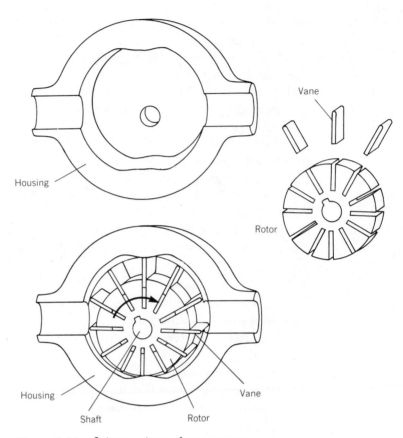

Figure 2.24   Cutaway views of a vane pump

rotor is mounted in the pump housing off center. As the rotor is turned, the vanes slide in and out of the slots.

Since the rotor is not centered in the pump housing, a vacuum is created as the rotor turns (Figure 2.25). The atmospheric pressure in the tank forces fluid into the pump's intake chamber. As the rotor continues to turn, the size of the chamber is reduced, forcing the fluid out of the pump.

The vane pump is self-compensating for wear. As the vanes wear they simply slide out of their slots a little farther due to centrifugal force.

Some vane pumps have passages in the rotor that allow oil to be directed to the bottom of the vanes. As the rotor turns, a small amount of oil is directed to the bottom of the vanes through oil passages in the rotor (Figure 2.26). There is no pressure at the inlet, and the vanes slide along the outside of the pump housing without

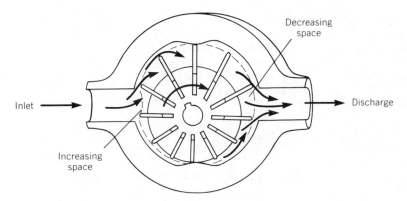

Figure 2.25 Cutaway view illustrating the operation of a vane pump

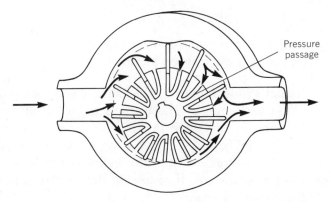

Figure 2.26 Cutaway view of a vane pump with pressure passages

exerting very much pressure against the housing. As the oil is carried toward the outlet, pressure may begin to build if the pump is attempting to move a load. The pressure in the pump is directed to the bottom of the vanes, pushing them against the pump housing and improving the seal between the vanes and the housing.

A variable displacement vane pump can be built by centering the rotor in the pump housing and placing a moveable ring inside the pump housing between the case and the rotor (Figure 2.27). As the ring is pushed off center the chamber begins to resemble the housing of a fixed displacement pump. As the ring is moved farther off center the displacement of the pump increases. A variable displacement pump can be built to automatically adjust itself to the demand of the system. If the system requires only a small amount of oil to move its load, the pump's output is low. As the demand for oil

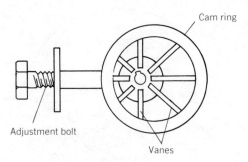

Figure 2.27   Variable displacement vane pump

increases, the ring in the pump moves farther off center and the output of the pump increases.

Piston Pumps

There are two different types of piston pumps: **radial piston pumps** and **axial piston pumps.**

Radial Piston Pumps.   The radial piston pump operates somewhat like a vane pump. The rotor is mounted off center in the pump housing (Figure 2.28). As the rotor turns, the pistons move in and out, pumping oil. As a piston moves out, oil from the inlet is drawn into the piston. As the rotor continues to turn, the housing forces the piston inward, thus forcing the oil out of the piston and out of the pump's discharge.

Axial Piston Pumps.   There are two types of axial piston pumps: the **bent axis axial piston pump** and the **in-line axial piston pump.** The

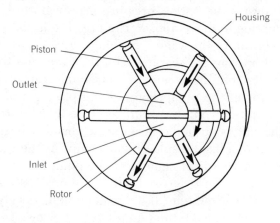

Figure 2.28   Radial piston pump

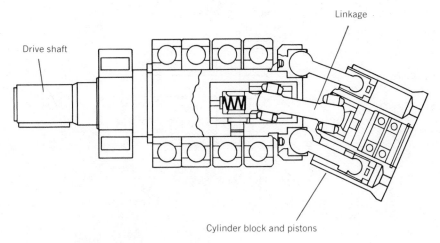

Figure 2.29 Bent axis axial piston pump

bent axis axial piston pump is shown in Figure 2.29, and the in-line axial piston pump is shown in Figure 2.30.

The bent axis axial piston pump is driven by turning the input shaft. The input shaft has a universal joint that is connected to the cylinder block. When the input shaft is turned the cylinder block rotates. Against the outside surface of the cylinder block is a **valve plate.** The valve plate has an opening for fluid intake and another opening for fluid discharge (Figure 2.31).

Figure 2.30 In-line axial piston pump

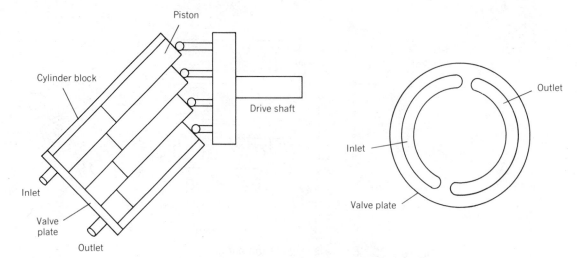

Figure 2.31   View of bent axis axial piston pump and valve plate

The pistons are connected to a collar by means of connecting rods. The collar of the pump remains stationary as the input shaft is rotated. The ends of the connecting rods slide in a groove cut in the collar. When the drive shaft is turned, the cylinder block, the pistons, and the connecting rods all rotate as a unit. The pump collar and the valve plate are the only parts of the pump that do not rotate. When the shaft is turned the pistons are pulled out of and pushed into the cylinder block. As the pistons are pulled out of the cylinder block they draw oil in through the inlet, and as they are pushed into the cylinder block they push oil out through the outlet port.

Figure 2.32 shows an in-line axial piston pump. The in-line axial piston pump is similar to the bent axis axial piston pump. The major difference is that the bent axis axial piston pump is a fixed displacement pump and the in-line axial piston pump is a variable displacement pump.

The shaft of the in-line axial piston pump is "in line" with the cylinder block. The pistons are connected to a **swash plate.** When the swash plate is parallel to the cylinder block and the shaft is rotated, the pistons do not move into and out of the cylinder block. The pump displacement is zero and no fluid is pumped.

If the swash plate is tipped while the input shaft is turned, the pump will pump fluid (Figure 2.33). As the shaft is turned the cylinder block rotates. The pistons are forced into and out of the cylinder block as in the bent axis axial piston pump.

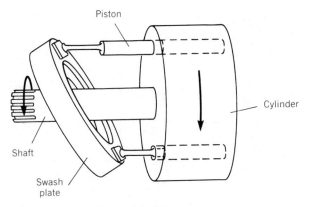

Figure 2.32   In-line axial piston pump

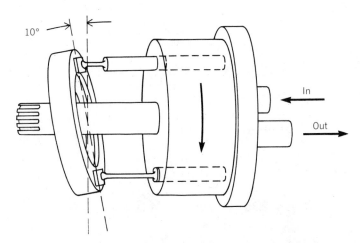

Figure 2.33   In-line axial piston pump with swash plate tilted at 10°

If the swash plate is tilted farther the displacement of the pump is increased (Figure 2.34). The pistons are forced deeper into the cylinder block and withdrawn farther.

The in-line axial piston pump has a valve plate identical to the valve plate of the bent axis axial piston pump (Figure 2.35). Notice that the openings of the valve plate have not been labeled as inlet or outlet ports. The reason for this is that the ports can be either inlet or outlet ports. When the swash plate is tilted in one direction, one port becomes the inlet and the other port becomes the outlet. If the swash plate is tilted in the other direction the inlet and outlet ports

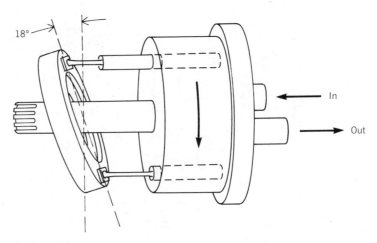

Figure 2.34 In-line axial piston pump with swash plate tilted at 18°

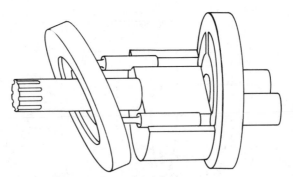

Figure 2.35 View of in-line axial piston pump showing valve plate

reverse: the inlet port becomes the outlet port and the outlet port becomes the inlet port.

The schematic symbol for a variable displacement pump is shown in Figure 2.36.

One advantage of a variable displacement pump such as the in-line axial piston pump is the ability to vary the displacement of the pump automatically. When the hydraulic system has a high demand for fluid, the pump displacement is at maximum. When the system is in neutral (does not require fluid flow), the swash plate is parallel to the cylinder block, and no fluid is pumped. The pump is said to be "destroked."

Figure 2.37 shows a variable displacement axial piston pump with a small cylinder connected to the swash plate. The cylinder has

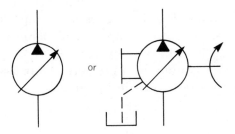

Figure 2.36 Schematic symbol for a variable displacement pump

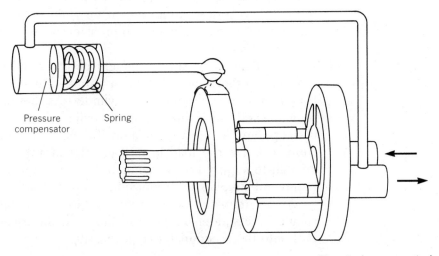

Figure 2.37 Variable displacement axial piston pump with cylinder connected to swash plate

a spring that attempts to push the piston to one end of the cylinder. If the spring is allowed to push the piston, the swash plate will tilt. The spring will attempt to push the piston to the extreme. With the piston at the extreme, the swash plate is tilted as far as possible and the displacement of the pump is at its maximum. Each revolution of the cylinder block will pump the maximum amount of fluid.

There is a small tube connected from the outlet of the pump to the other end of the cylinder. As pressure rises in the system, the pressure in the cylinder rises, pushing the piston toward the spring end of the cylinder. As the piston moves against the spring pressure, the swash plate is also being pushed toward the neutral position. This reduces the displacement of the pump, and less fluid is pumped.

This type of pump is called a **pressure compensated axial piston pump.** It is called "pressure compensated" because the output pressure of the pump is automatically adjusted.

The schematic symbol for a pressure compensated variable displacement pump is presented in Figure 2.38.

ACCUMULATORS

High volumes of fluid are sometimes required in hydraulic systems. If this sudden requirement for fluid is more than the pump can supply, it can be made up on a temporary basis if the system includes an **accumulator**.

An accumulator can be thought of as a small pressurized tank. Figure 2.39 shows an accumulator plumbed to a pump. The outlet of the pump is also plumbed to a cylinder. As the pump pumps fluid, some of the fluid is directed to the cylinder while the excess output of the pump fills the accumulator. If there is a sudden need for more fluid to move the load faster, and the pump volume is not high enough to supply the demand, the accumulator takes over for the pump and supplies the required amount of fluid.

An accumulator can also be used as a shock absorber. If there should be a sudden increase in the load, the excess fluid will flow back into the accumulator (Figure 2.40).

Types of Accumulators

There are three types of accumulators: spring loaded, weighted, and gas pressurized accumulators (Figure 2.41). All of these accumulators perform the same function. The only difference among the three types of accumulators is the method of supplying pressure.

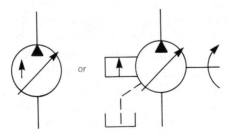

Figure 2.38   Schematic symbol for a pressure compensated variable displacement pump

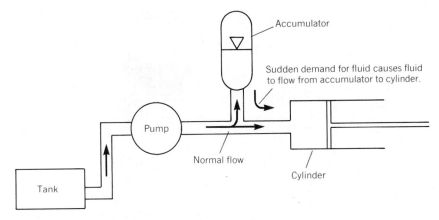

Figure 2.39 Hydraulic system with accumulator

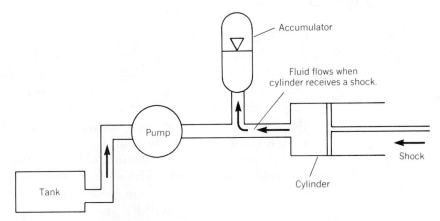

Figure 2.40 Illustration of how an accumulator can act as a shock absorber

The **spring loaded accumulator** can be mounted in any direction. It functions as well upside down as it does when mounted upright. The disadvantage of the spring loaded accumulator is its uneven pressure. When fluid is first pumped into the accumulator, the spring is easily compressed. As more fluid is pumped into the accumulator, the spring exerts more and more pressure. When the accumulator is supplying fluid to the system, the pressure varies. At first the accumulator supplies fluid under high pressure; then, as the spring extends, the output pressure of the accumulator drops.

The **weighted accumulator** delivers an even pressure throughout its entire stroke. The pressure required to fill the accumulator is the same when it begins to fill and remains the same as it continues to

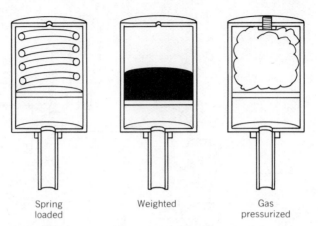

Spring   Weighted   Gas
loaded   pressurized

Figure 2.41 Cutaway views of spring loaded, weighted, and gas pressurized accumulators

fill. When the weighted accumulator is delivering fluid to the system, its fluid pressure is always the same. The weighted accumulator has two major disadvantages. First, if it is necessary to have an accumulator that can deliver a high pressure, the weight must be very large. The second disadvantage of the weighted accumulator is the requirement that it must always be mounted upright since the weight is supplying the pressure.

The third type of accumulator is the **gas pressurized accumulator.** The gas pressurized accumulator is the most popular type of accumulator. Since gas is used to exert the pressure, the output pressure will be almost even throughout the entire stroke. The gas pressurized accumulator can be mounted in any direction. It works as well when mounted upside down as when mounted upright.

At times it is necessary to replace the gas charge in a gas pressurized accumulator. Never charge the gas pressurized accumulator with any gas except dry nitrogen. If acetylene, oxygen, or any other

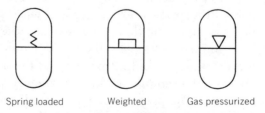

Spring loaded   Weighted   Gas pressurized

Figure 2.42 Schematic symbols for spring loaded, weighted, and gas pressurized accumulators

gas that is not **inert** is used, you may be creating a bomb. An accumulator charged with acetylene or oxygen can explode with a tremendous force. Never risk your life or the lives of your fellow workers by making this mistake.

The schematic symbols for spring loaded, weighted, and gas pressurized accumulators are shown in Figure 2.42.

ACTUATORS

For a hydraulic system to do work, the fluid is directed to an **actuator**. There are two basic types of actuators: **linear actuators** and **rotary actuators.** Linear actuators produce straight line motion. Rotary actuators produce rotary motion.

Linear Actuators

A hydraulic cylinder is a linear actuator. When hydraulic fluid is pumped into a cylinder, the fluid forces the piston to move. The piston is connected to a piston rod. When the piston is moved by hydraulic fluid, the piston rod moves. The cylinder is made to do work by connecting the piston rod to a "load" (Figure 2.43).

The total distance that the piston can move is called its **stroke**. Cylinders are available with many different stroke lengths. The total stroke may be as little as a fraction of an inch or more than 20 ft.

Cylinders also come in a variety of diameters. The diameter of the cylinder is a primary factor in determining the amount of force that the cylinder rod can exert. The formula for determining the force that a cylinder rod can exert is $F = PA$, where F is the force, P is the system pressure, and A is the area of the piston. If a hydraulic system has a piston with an area of 10 in.^{2} and the pressure in the system is 200 psi, the total force exerted by the rod is:

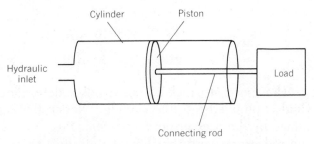

Figure 2.43 Hydraulic cylinder (linear actuator)

$$F = PA$$

$$F = 200 \times 10$$

$$F = 2000 \text{ lb}$$

If the area of the piston is doubled (20 in.^{2}), the force will also double:

$$A = 20 \text{ in.}^2$$

$$P = 200 \text{ psi}$$

$$F = PA$$

$$F = 200 \times 20$$

$$F = 4000 \text{ lb}$$

The force can also be altered in a hydraulic system by increasing the pressure. If we take our original example where the area of the piston was 10 in.^{2} and the pressure was 200 psi, and increase the pressure to 400 psi, the force will be:

$$F = PA$$

$$F = 400 \times 10$$

$$F = 4000 \text{ lb}$$

It is important to remember that altering either the pressure or the area of the piston affects the force of the cylinder.

Cylinder volume is found by multiplying the stroke of the piston by the area of the piston. The formula is V = SA, where V is volume in cubic inches (in.^{3}), S is stroke in inches (in.), and A is area in square inches (in.^{2}). If a piston has a stroke of 10 in. and an area of 5 in.^{2}, then:

$$V = SA$$

$$V = 10 \times 5$$

$$V = 50 \text{ in.}^3$$

The speed at which a rod extends is determined by how quickly fluid is pumped into the cylinder. Notice that speed is not determined by the pressure in the system. Pressure affects the force that the rod exerts, and the flow of fluid determines the speed of the system. It is a common mistake to think that increasing the pressure in a system will increase the speed. The only way to increase the speed is to increase the volume of fluid pumped into the cylinder.

Cylinder Design

A hydraulic cylinder contains a piston that is inside the cylinder body (Figure 2.44). The cylinder also has caps that close off the cylinder body. One of the caps has an opening for the piston rod: this is called the **head end** or the **rod end.** The cap on the other end of the cylinder is called the **cap end** or the **blind end.**

Hydraulic cylinders should not leak hydraulic fluid. To prevent hydraulic fluid from leaking out of the cylinder, the rod end contains a seal. The seal is called a **rod gland.** The rod gland not only prevents hydraulic fluid from leaking out of the cylinder but also supports the piston rod (Figure 2.45).

It is also important to prevent fluid from leaking past the piston. When fluid bypasses the piston, velocity is lost. One type of seal that is used to prevent such leakage is **piston rings,** which are similar to those found on automotive engine pistons. The piston rings help seal the piston but are not perfect. They normally leak from 1 to 3 in.^{3} of fluid per minute. Although piston ring seals leak some fluid past the piston, they are rugged and seldom need replacement.

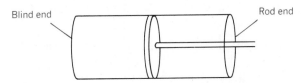

Figure 2.44 Hydraulic cylinder, showing rod and blind ends

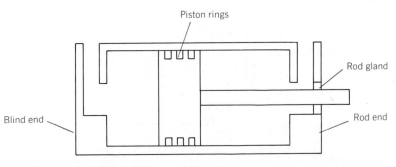

Figure 2.45 Hydraulic cylinder with rod gland

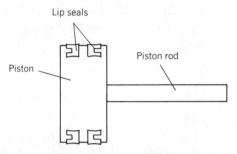

Figure 2.46 Piston of a hydraulic cylinder, showing lip seals

Another type of seal that is used to seal the piston is called a **lip seal** (Figure 2.46). The lip seal is made of rubber and does not leak when new. However, a lip seal is not as durable as a piston ring and requires more frequent replacement.

In addition to piston seals, some cylinders have cushions built in to help absorb the shock of the piston striking the cap. The cushion closes off the outlet or inlet port before the piston strikes the cap (Figure 2.47). As the piston is pushed toward the blind end of the

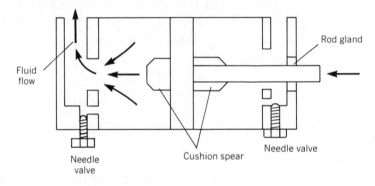

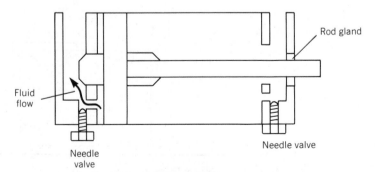

Figure 2.47 Illustration of the action of a cushion spear in a hydraulic cylinder

cylinder, a **cushion spear** closes off the main outlet port. Fluid is forced through a small opening. Since the opening is small, the hydraulic fluid cannot leave the cylinder as quickly, thus slowing the movement of the piston. The same design is used at the rod end of the cylinder.

The schematic symbol for an adjustable cushion cylinder is shown in Figure 2.48.

Common Cylinder Types

There are many common types of cylinders. Each has its own type of application. We will briefly cover the most common types.

Single Acting Cylinder.   In the single acting cylinder (Figure 2.49), fluid is pumped into the blind end of the cylinder only. The piston retracts when fluid is allowed to escape from the blind end and the load forces the piston back toward the blind end.

Single Acting Spring Return Cylinder.   The single acting spring return cylinder operates the same as a single acting cylinder but has a spring between the piston and the rod end of the cylinder to return the piston to the blind end when fluid is allowed to escape. This type of cylinder is rarely used in hydraulic systems.

Double Acting Cylinder.   The double acting cylinder (Figure 2.50) allows fluid to be pumped into both the blind end and the rod end of the cylinder. When fluid is pumped into the blind end of the cylinder the rod extends. When fluid is pumped into the rod end of the cylinder the rod retracts.

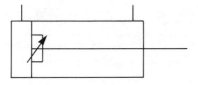

Figure 2.48   Schematic symbol for an adjustable cushion cylinder

Figure 2.49   Schematic symbol for a single acting cylinder

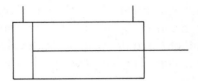

Figure 2.50   Schematic symbol for a double acting cylinder

The retraction force of a double acting cylinder is not as high as the extension force. The piston rod reduces the area of the piston that is exposed to fluid pressure. Recall that force equals pressure times area. Since the area of the piston is less than that of the blind end, the retraction force is reduced. Some double acting cylinders have rods that reduce the piston area at the rod end by 50%, so the retraction force is also reduced by 50% (Figure 2.51).

Although the piston rod reduces the retraction force of the cylinder, the speed at which it retracts increases as long as the volume of fluid entering the cylinder remains constant. Recall that the speed of a piston is determined by the flow of fluid into the cylinder. If the piston rod reduces the area of the piston by 50%, the volume of the cylinder is also reduced by 50%, and therefore the retraction force is reduced by 50% but the speed is doubled.

Double Acting Double Rod Cylinder.   The double acting double rod cylinder (Figure 2.52) has equal force and equal speed in both directions. The piston rod extends from both ends of the cylinder. The rod reduces the piston area by the same amount on both sides of the piston, and thus the speed and force are equal in both directions.

Hydraulic Motors (Rotary Actuators)

Hydraulic motors create rotary motion rather than linear motion. Most hydraulic motors are similar to hydraulic pumps. The common

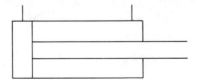

Figure 2.51   Schematic symbol for a double acting cylinder with reduced piston area at rod end

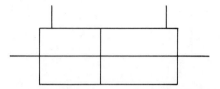

Figure 2.52   Schematic symbol for a double acting double rod cylinder

types of hydraulic motors are the gear motor, the vane motor, and the piston motor.

The **gear motor** is similar to the gear pump. As fluid is pumped into the inlet of the gear motor, the gears are turned by the flow of the hydraulic fluid. Fluid is carried around the outside of the case, flows out the outlet port, and returns to the tank (Figure 2.53). The identity of the inlet and outlet ports of most gear motors is arbitrary. If fluid is pumped into the outlet port, the motor will turn in the opposite direction.

One of the gears in the gear motor is connected to a shaft, and the shaft drives the load. The schematic symbol for a gear motor is shown in Figure 2.54. This symbol is the same as that for a gear pump except for the arrowhead (the arrowhead points outward for a pump and inward for a motor).

Another common hydraulic motor is the **vane motor.** The vane motor is very similar to the vane pump. Fluid is pumped into the motor. The exposed area of the vanes is greater at the top of the motor. Since the area exposed to the flow of fluid is greater near the top of the pump, the force is also greater. The force against the vanes drives the rotor. The rotor is connected to a shaft, and the shaft is connected to the load (Figure 2.55).

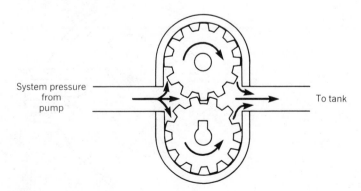

Figure 2.53   Gear motor

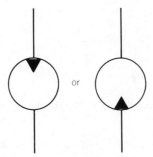

Figure 2.54   Schematic symbol for a gear motor

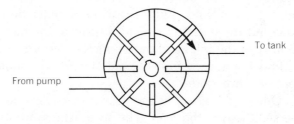

Figure 2.55   Vane motor

It is important that the vanes seat tightly against the vane motor case. To ensure that the vanes are pushed against the motor case, the vanes are pushed out of the rotor with springs, or a small passage is cut into the rotor to direct fluid to the bottom of the vanes, driving them out (Figure 2.56).

The schematic symbol for a vane motor is the same as that for a gear motor (Figure 2.54).

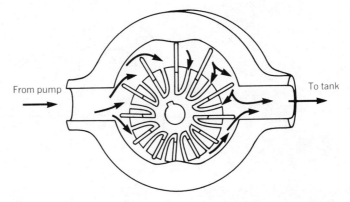

Figure 2.56   Cutaway view of a vane motor with pressure passages

If the rotor is mounted off center in the vane motor case, as shown in Figure 2.55, the pressure at the top of the motor puts an unequal pressure on the shaft. This **side load** is very hard on bearings and seals. A **balanced vane motor** has been developed to maintain equal loading on the shaft.

In a balanced vane motor, the rotor is centered in the vane motor case. The vane motor case is cut in the shape of an ellipse rather than a circle. Fluid flow is directed so that the pressure is equal around the rotor, eliminating the side thrust on the shaft of the vane motor.

Another common type of hydraulic motor is the **piston motor.** It is similar to the variable displacement piston pump (Figure 2.57). Some piston motors have the capability to be reversed without reversing the inlet and the outlet. To reverse the direction of a piston motor, the swash plate can be tipped past center. When the swash plate is tipped in one direction the motor rotates in one direction, and when it is tipped in the other direction the shaft direction reverses. The schematic symbol for a piston motor is shown in Figure 2.58.

Rotary motion can also be created using cylinders (Figure 2.59). As the cylinder extends, the crank is pushed. The crank is connected to a shaft.

Another very common type of rotary actuator is the **rack-and-pinion actuator** (Figure 2.60). The rack-and-pinion actuator gives uniform torque in both directions. It also has the advantage of delivering high torque. The rack-and-pinion actuator can be operated at a low speed without the need for additional reduction gearing. These advantages make it an excellent choice for powering the base rotation of a robot.

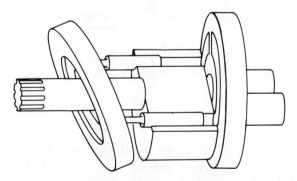

Figure 2.57 In-line axial piston motor

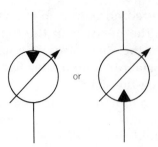

Figure 2.58 Schematic symbol for a variable displacement motor

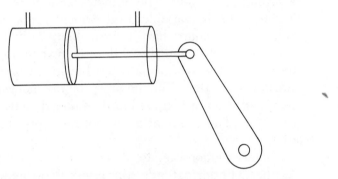

Figure 2.59 Creation of rotary motion using a cylinder

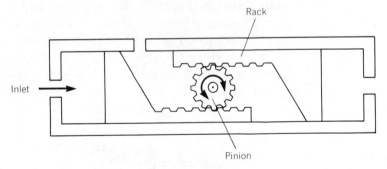

Figure 2.60 Rack-and-pinion actuator

RELIEF VALVES

At this point we have taken a brief look at hydraulic tanks, filters, pumps, accumulators, and actuators. These are the basic elements that supply fluid and create mechanical motion in a hydraulic system. These elements could be plumbed together to form a simple

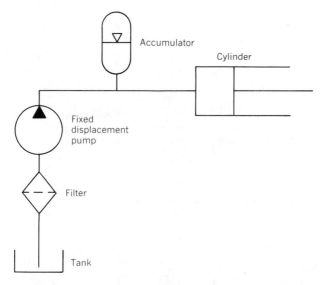

Figure 2.61 Schematic illustration of a simple hydraulic system

hydraulic system. Such a system is shown in Figure 2.61. The problem with this system is that we cannot control it. If the positive displacement pump is being turned by an electric motor, fluid will be pumped and the cylinder will move. After the piston in the cylinder has moved through its total stroke and the accumulator has been filled, the hydraulic fluid will continue to be pumped. Pressure in the system will quickly rise, and if the electric drive motor is not turned off the pressure will cause something in the system to explode.

If a valve is added to the system to allow fluid to be returned to the tank when the system pressure has reached a maximum, the system will be protected. The **relief valve** was designed for just such a purpose.

Figure 2.62 shows a simple relief valve. The valve consists of a **valve body,** a **spool**, a **spring**, and a **bolt** for adjusting the spring tension.

The inlet to the relief valve is plumbed to the outlet of the pump. If the pressure in the system is normal, fluid passes through the valve and the valve has no effect on the operation of the system. If the pressure in the system becomes excessively high, the spool will be pushed up, opening a passage and allowing hydraulic fluid to flow back to the tank (Figure 2.63). This reduces the pressure in the system. When the pressure drops low enough, the spring in the

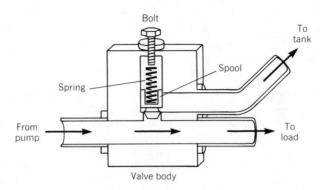

Figure 2.62   Simple relief valve under normal pressure

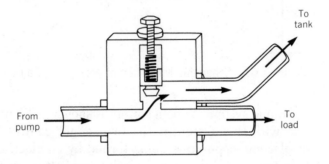

Figure 2.63   Simple relief valve under high pressure

relief valve will push the spool down, closing off the passage to the tank, and the pressure will begin to rise again.

The amount of pressure needed to open the relief valve can be adjusted. If the bolt is turned into the valve body it will compress the valve spring. The farther the bolt is turned in, the higher the system pressure must be before the pressure overcomes the spring tension and opens the passage to the tank. The schematic symbol for a simple relief valve is shown in Figure 2.64.

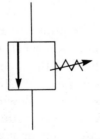

Figure 2.64   Schematic symbol for a simple relief valve

Before looking at other types of relief valves it is important to understand how pressure and flow relate to each other in a hydraulic system.

FLUID FLOW IN A HYDRAULIC SYSTEM

The flow of fluid in a hydraulic system does not cause pressure; rather, **pressure is caused by restriction of flow.** Figure 2.65 shows a tank, a pump, and a pressure gauge. The pump is a positive displacement type. The pressure gauge in the system reads 0 psi. All of the pump's volume is passing through the system without any restriction.

If the outlet of the pump is restricted and a second pressure gauge is added to the system after the restrictor, the first gauge will show a pressure while the second gauge will read 0 (Figure 2.66). The difference in pressure between the first gauge and the second gauge is referred to as **pressure drop.**

Recall that we said that the system has a positive displacement pump. The pump puts out the same amount of fluid with or without the restrictor in the system. With the restrictor in the system, pressure increases in front of the restrictor but the volume of fluid

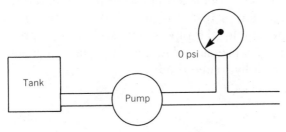

Figure 2.65   Tank, pump, and pressure gauge

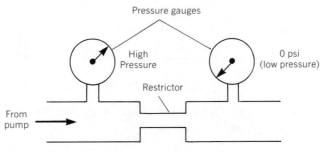

Figure 2.66   Pressure drop across a restrictor

being pumped is not changed. Since the same amount of fluid is being pumped, the velocity (speed) of the fluid is higher after the restrictor than before the restrictor.

Now consider what will happen if the output of the system becomes completely blocked (Figure 2.67). At first, the pressure in front of the restrictor will be higher than that after the restrictor. This situation will not last very long. With the output of the system blocked, the pressure at the output of the restrictor will rise quickly. The pressure in front of the restrictor and after the restrictor will become equal. As the pressure rises, the flow of fluid decreases. When the pressure in front of the restrictor and the pressure after the restrictor become equal, flow is at 0. The **pressure differential** determines the amount of fluid that passes through the restrictor.

If we add a relief valve to the system, the restrictor will reduce the flow. Suppose that the relief valve is set at 100 psi. If the flow out of the pump is high enough and the restrictor is small enough, the relief valve will open. The pressure will not go above 100 psi. The excess flow is diverted to the tank (Figure 2.68).

When the restrictor is closed further the pressure does not go up, because the relief valve is holding the system at 100 psi. Less fluid is being pushed through the restrictor, and more fluid is being diverted to the tank. The output of the restrictor is less (Figure 2.69).

In a hydraulic system we can control the flow of fluid by using a restrictor. A restrictor in a hydraulic system can be as simple as a pipe nipple (the pipe nipple will cause a pressure drop) or as complex as a pressure flow compensator which automatically adjusts the flow of hydraulic fluid.

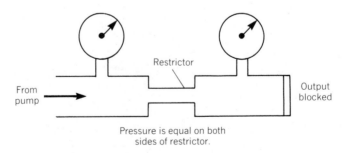

Figure 2.67 With output of system blocked, there is no pressure drop across restrictor, and flow is zero.

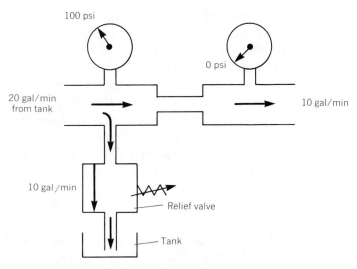

Figure 2.68 Flow through restrictor and flow returned to tank are each 10 gal/min.

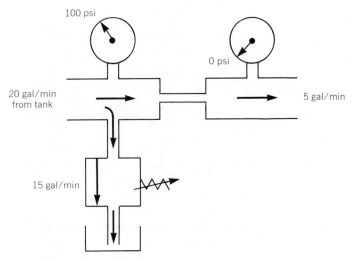

Figure 2.69 With smaller restrictor, more fluid is returned to tank.

FLOW CONTROL VALVES

There are three factors that affect the flow of hydraulic fluid. They are the **size of the restrictor,** the **pressure differential across the restrictor,** and the **temperature of the fluid.** In our discussion of flow control valves we will ignore temperature as a factor; however, you should be aware that more fluid will pass through a restrictor when the fluid is hot than when it is cold.

The hole in a restrictor that allows fluid to pass is called an **orifice**. There are **fixed orifice restrictors** (in which the size of the orifice cannot be changed) and **variable orifice restrictors** (in which the size of the orifice can be changed). A simple valve used for turning a garden hose on and off is a type of variable orifice restrictor. Normally this type of valve is not used in a hydraulic system. The more common type of variable orifice restrictor is the **needle valve.**

The needle valve is shown in Figure 2.70, along with its schematic symbol. To adjust the flow through the restrictor, the knob can be turned in one direction to reduce the size of the orifice and in the other direction to increase the size of the orifice.

Figure 2.71 shows a simple hydraulic system. The system includes a tank, a positive displacement pump, a relief valve, a vari-

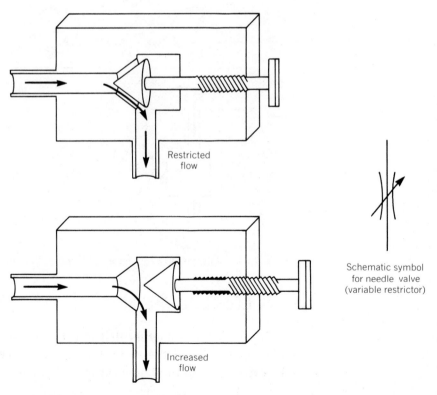

Restricted
flow

Increased
flow

Schematic symbol
for needle valve
(variable restrictor)

Figure 2.70   Needle valve

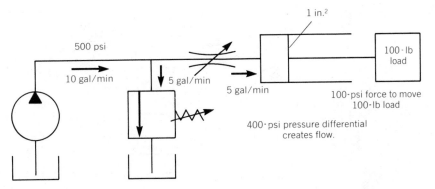

Figure 2.71   Simple hydraulic system acting on 100-lb load, with relief valve set at 500 psi

able orifice restrictor, and a cylinder. The cylinder is attempting to push a 100-lb load. The relief valve is set at 500 psi. The output of the pump is 10 gal/min. When the system is turned on, pressure rises to 500 psi and 5 gal/min passes through the relief valve back to the tank. Since the system is attempting to push a 100-lb load, there is a 400-lb pressure differential. This pressure differential causes flow, and the load is pushed. When the piston strikes the cylinder cap, no more fluid can enter the cylinder. The pressure on the cylinder side of the restrictor quickly rises to 500 psi, and all the fluid passes through the relief valve and back to the tank.

If the system is the same as in the previous example except that the needle in the flow control is turned in, making the orifice smaller, the load will move more slowly. Less fluid can pass through the orifice, and more fluid is returned to the tank through the relief valve.

Now we will return the needle valve to the original setting. Everything is the same as in the first example except that the load is increased from 100 to 300 lb (Figure 2.72). When the pump is started, the pressure rises to 500 psi on the pump side of the restrictor. The relief valve opens and some fluid is pumped to the tank. Since the load is 300 lb there is only a 200-psi pressure differential. Less fluid moves through the restrictor, and the load moves more slowly than in the first example. The pressure differential determines the amount of fluid passing through the restrictor.

Now we will adjust the relief valve pressure (Figure 2.73). The relief valve is adjusted to 700 psi. The load is still 300 lb, and the needle valve is not changed. When the pump is started the pressure rises to 700 psi. The load is 300 lb, leaving a pressure differential

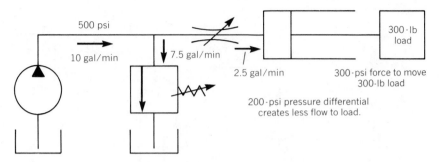

Figure 2.72   Same system as that shown in Figure 2.71, but with load increased to 300 lb

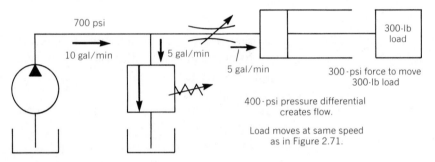

Figure 2.73   Same system as that shown in Figure 2.72, but with relief valve set at 700 psi

of 400 psi. This is the same pressure differential as in our first example. The 300-lb load moves at the same speed as the 100-lb load moved in the first example.

You can see that the size of the orifice in a restrictor and the pressure differential across the restrictor determine the flow in a hydraulic system.

PILOT OPERATED RELIEF VALVES

There are several problems with the simple relief valve described earlier. As you know, when pump volume increases and the system has a restrictor, the pressure differential across the restrictor increases. The simple relief valve acts like a restrictor. As system pressure increases, the relief valve spool opens. This opening acts like an orifice in a restrictor. The more fluid that tries to pass through this orifice, the higher the system pressure.

Another problem with the simple relief valve is vibration of the spool. As the system pressure rises, the spool opens, allowing fluid to return to the tank. So much fluid is returned to the tank that the system pressure falls below the pressure needed to keep the valve open. The spool slams shut and the system pressure again quickly rises, opening the passage to the tank. Not only is this type of relief valve noisy, but the pulsing of the system pressure as the valve opens and closes can damage other hydraulic components.

The **pilot operated relief valve** overcomes these problems. The pilot operated relief valve, which is shown in Figure 2.74 along with its schematic symbol, is similar to the standard relief valve but has several extra parts. The pilot operated relief valve has a main spool. There is a small hole bored through the center of the spool. This forms an orifice. There is also a smaller spool called a pilot spool. There is a spring that holds the main spool closed. The spring tension on this spool cannot be adjusted. There is also a spring that holds the pilot spool closed. The tension on this spring can be adjusted by a screw in the top of the valve.

When the pump is turned on, fluid begins to flow in the system and the load (the weight that the system is attempting to move) causes pressure to rise in the system. The pressure begins to work against the load. Fluid also passes through the hole bored in the main spool. This small opening forms an orifice. The pressure that is created in the chamber between the main spool and the pilot spool

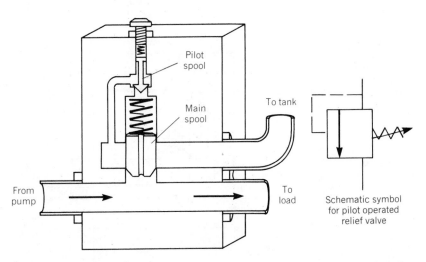

Figure 2.74 Pilot operated relief valve with pilot closed (normal pressure)

forces the main spool to seat. The main spool spring also helps to keep the main spool seated. System pressure continues to rise, and the pressure in the chamber between the main spool and the pilot spool also increases. Finally, the pressure in the chamber becomes so high that the pilot opens (Figure 2.75). When the pilot opens, fluid flows through a small passage to the tank. The pressure in the chamber goes down. The system pressure pushing on the bottom of the main spool is higher than the pressure in the chamber, and the main spool opens. A large passage is opened to the tank to relieve the system pressure (Figure 2.76). The main spool will move only enough to balance the chamber pressure and system pressure. If flow is increased by turning the pump faster, the pilot operated relief valve will open farther, always maintaining the system pressure. The main spool is said to **modulate**. It moves only enough to hold the system pressure at the preset level. The simple relief valve, on the other hand, changes from fully closed to fully open.

PRESSURE COMPENSATED FLOW CONTROL VALVES

Earlier in this chapter we learned that pressure differential affects flow. If the load on a system is increased, the pressure differential decreases and the load moves more slowly. This problem can be overcome by using a **pressure compensated flow control valve.**

Figure 2.77 shows a pressure compensated flow control valve. The valve is made up of a valve body, a spool, a spring, and a needle

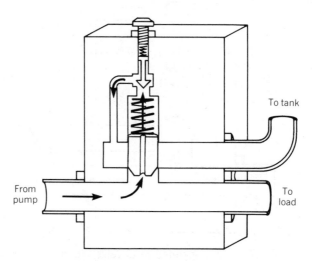

Figure 2.75   Pilot operated relief valve with pilot open (high pressure)

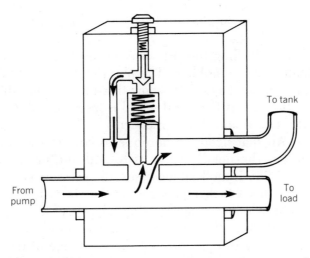

Figure 2.76   Pilot operated relief valve with main spool open

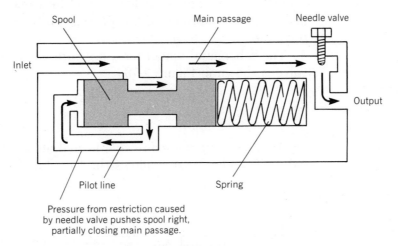

Figure 2.77   Pressure compensated flow control valve

valve. The spring pushes the spool to the left. This opens a passage for fluid through the valve body to the needle valve and finally to the load.

As pressure increases in front of the needle valve, pressure in the pilot line increases. This increases the pressure on the spool end. If this pressure becomes high enough, the spool will slide to the right, compressing the spring and partially closing off the main passage to the needle valve. This reduces the flow of fluid to the needle valve, and the pressure drops. The spool modulates (slides back and forth), keeping the pressure in front of the needle valve constant.

Controlling the pressure on the front side of an orifice solves only half the problem. The purpose of this valve is to maintain flow at a constant rate. We know that the pressure differential across an orifice affects flow. If the load is increased, the pressure differential is reduced and flow is reduced.

All that needs to be done to control the pressure differential across the needle valve is to add one more pilot passage (Figure 2.78). If the load increases, the pressure caused by the increased load puts a higher pressure on the right side of the spool, through the new pilot passage, pushing the spool to the left. This opens the passage from the pump to the needle valve, allowing more flow and increasing the pressure on the front side of the needle valve. The pressure increase on the front side of the needle valve is exactly the same as the pressure increase on the output of the needle valve caused by the increased load, and the pressure differential is maintained. Maintaining the pressure differential maintains the flow rate.

The schematic symbol for a pressure compensated flow control valve is shown in Figure 2.79.

CHECK VALVES

Another very common valve in hydraulic systems is the **check valve.** The check valve allows fluid to flow in only one direction

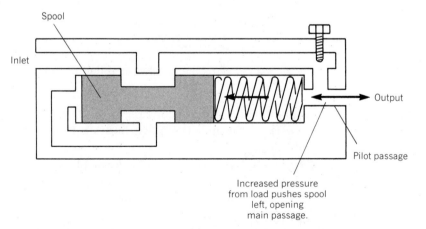

Figure 2.78   Pressure compensated flow control valve with additional pilot passage

Figure 2.79 Schematic symbol for a pressure compensated flow control valve

(Figure 2.80). Fluid coming in from the left side exerts pressure against the spool. When the pressure becomes high enough, the spool compresses the spring that is holding it closed and fluid flows through the opening.

If fluid attempts to flow in the reverse direction (from right to left), the fluid exerts pressure on the spool, forcing the spool tighter into the seat, and no fluid can flow.

The schematic symbol for a check valve is shown in Figure 2.81.

An example of the use of a check valve is shown in Figure 2.82. A pump draws fluid from a tank and through a filter. The outlet of the pump is plumbed to a relief valve, a check valve, an accumula-

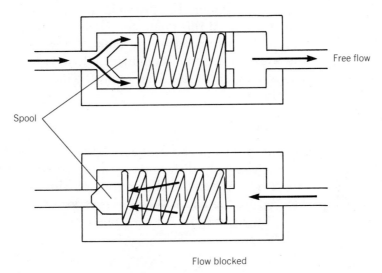

Figure 2.80 Check valve

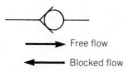

Figure 2.81 Schematic symbol for a check valve

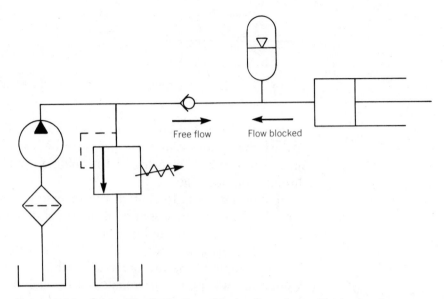

Free flow   Flow blocked

Figure 2.82 Schematic illustration of hydraulic system with check valve

tor, and a cylinder. The piston in the cylinder moves until it hits the cylinder cap. At this point the relief valve opens and the fluid being pumped is returned to the tank. When the pump is turned off, the pressure in the accumulator attempts to flow back to the tank but is stopped by the check valve. The check valve maintains the system pressure.

Pilot Operated Check Valve

The **pilot operated check valve** is similar to a normal check valve. The only difference is a pilot line and pilot piston (Figure 2.83). When fluid attempts to flow from left to right it is blocked by the check valve. If pressure is applied to the pilot passage, the main spool is forced open and fluid can flow back through the valve.

Counterbalance Valve

The **counterbalance valve** is a special form of the pilot operated check valve (Figure 2.84). A counterbalance valve can be placed anywhere in the system but normally is mounted directly on an actuator.

Figure 2.85 shows a counterbalance valve mounted on the base of a cylinder. When fluid is pumped through line A, the pressure in

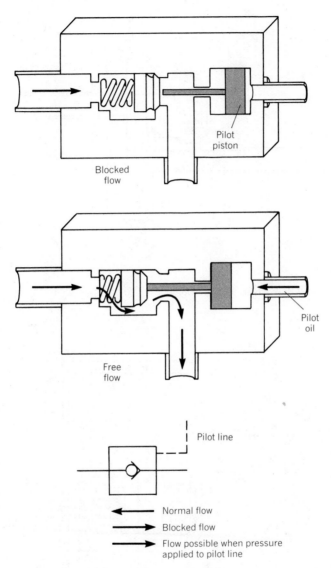

Figure 2.83 Pilot operated check valve

the line opens the check valve and fluid flows into the blind end of the cylinder. When the pump is turned off, the fluid is blocked from flowing out of the cylinder by the check valve. Even if the cylinder is supporting a 1000-lb load and line A breaks, the load will not fall because the check valve blocks flow from the blind end of the cylinder (Figure 2.86).

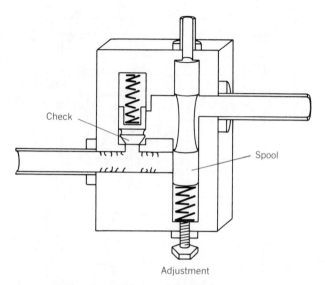

Figure 2.84   Counterbalance valve

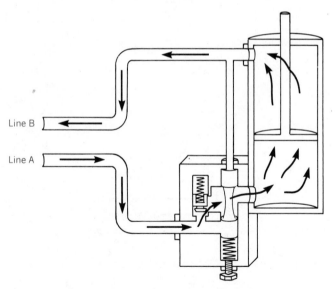

Figure 2.85   Counterbalance valve mounted on a cylinder

To get the cylinder to move downward, fluid is pumped into line B. Notice that there is a pilot line connecting line B to a spool in the counterbalance valve. When fluid is pumped into line B, pressure builds at the rod end of the cylinder and in the pilot line. The pressure in the pilot line forces the spool downward. This

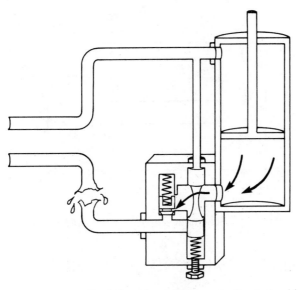

Figure 2.86   Fluid flow from blind end of cylinder blocked by counterbalance valve

opens a passage for fluid to flow from the blind end of the cylinder through line A to the tank (Figure 2.87).

The schematic symbol for a counterbalance valve is shown in Figure 2.88.

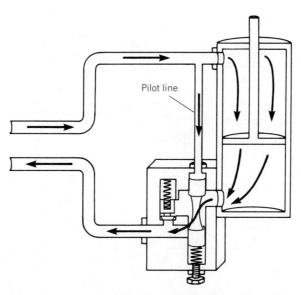

Pilot line

Figure 2.87   Pilot pressure opens counterbalance valve and piston moves down.

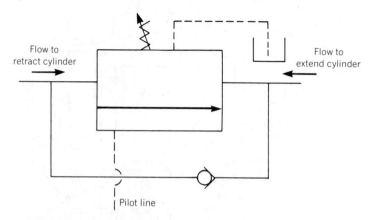

Figure 2.88   Schematic symbol for a counterbalance valve

DIRECTIONAL CONTROL VALVES

In the discussion of counterbalance valves above, we said that fluid was pumped into line A and returned to the tank through line B. We also showed what happens when fluid is pumped into line B and returns to the tank through line A. The switching of fluid flow is done with a **directional control valve.**

A directional control valve has two major parts: the valve body and the spool (Figure 2.89).

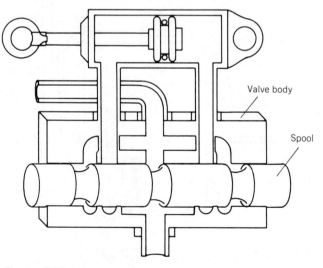

Figure 2.89   Directional control valve

Many directional control valves have two inlets and two outlets. One inlet is plumbed to the pump and is labeled "P." The other inlet is plumbed to the tank and is labeled "T." The outlets are plumbed to the cylinder and are labeled "A" and "B" (Figure 2.90).

The inlets to a directional control valve are called the **center**. The outlets of a directional control valve are called the **ports**. (Various manufacturers use different terms for the valve center and port.)

To control fluid using a directional control valve, the spool is moved in the valve body. Fluid entering the center is directed to a port. As shown in Figure 2.91, fluid from the pump is directed to the blind end of the cylinder through the directional control valve, and the fluid in the rod end of the cylinder passes through the valve and back to the tank.

If the spool in the directional control valve is shifted in the other direction, fluid from the pump is directed to the rod end of the cylinder while fluid in the blind end of the cylinder passes through the directional control valve back to the tank (Figure 2.92).

If the spool is centered in the valve body, the valve is in neutral. When the valve is in neutral, fluid either is blocked from entering

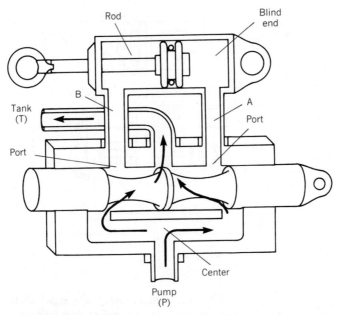

Figure 2.90 Directional control valve with two inlets and two outlets (pump inlet is labeled P, tank inlet is labeled T, and valve outlets are labeled A and B)

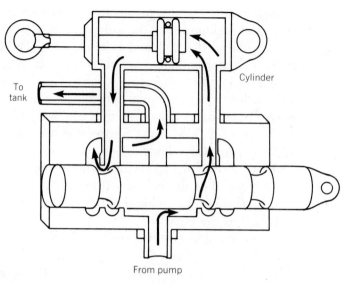

Figure 2.91   Directional control valve with spool shifted to the right. Fluid flows from pump, through valve, to blind end of cylinder, and from rod end of cylinder to tank.

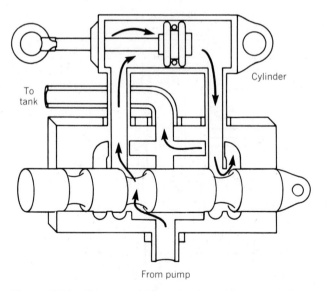

Figure 2.92   Directional control valve with spool shifted to the left. Fluid flows from pump, through valve, to rod end of cylinder, and from blind end of cylinder to tank.

the valve or is allowed to enter the valve and return to the tank. If the fluid is blocked by the spool, the valve is said to have a **closed center** (Figure 2.93). If fluid can enter the valve and return to the tank, the valve is said to have an **open center** (Figure 2.94).

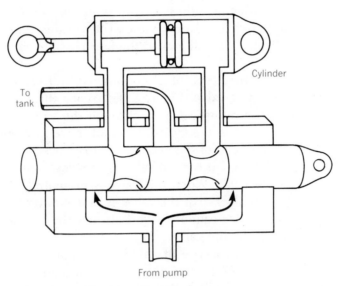

Figure 2.93 Closed center valve

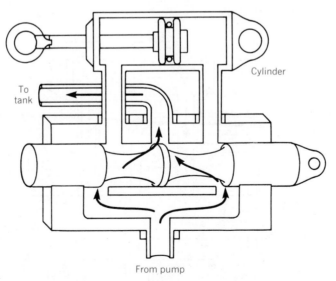

Figure 2.94 Open center valve

When the valve is in neutral, fluid may be blocked at the port or it may be able to flow through the valve and back out the other port line. If fluid is blocked by the port, the valve is called a **closed port** valve. If fluid is allowed to pass through the valve it is said to be an **open port** valve.

Combining either an open center or a closed center with an open port or a closed port yields the various types of valves that are available. We will look at the four possible combinations to see how they operate.

The first type of directional control valve is the **open center, closed port valve** (Figure 2.95). When this valve is in neutral, fluid from the pump can pass through the valve and back to the tank. Also, fluid in the cylinder cannot move from either the rod end or the blind end of the cylinder—it is blocked by the valve.

The second type of valve is the **closed center, closed port valve** (Figure 2.96). When this valve is in neutral, fluid coming from the pump is blocked at the valve (it cannot enter the valve). The fluid in the cylinder cannot move between the rod and blind ends—it is blocked by the valve.

The third type of valve is the **open center, open port valve** (Figure 2.97). When this valve is in neutral, fluid from the pump can move through the valve and back to the tank. Also, fluid can move between the rod and blind ends of the cylinder through the valve.

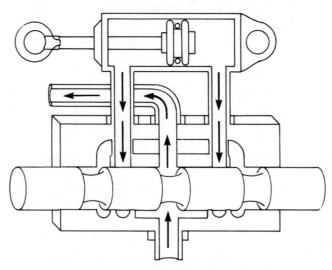

Figure 2.95   Open center, closed port directional control valve

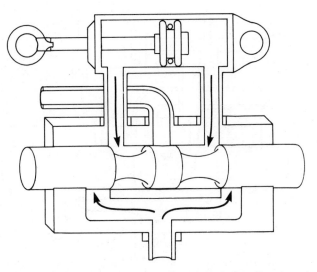

Figure 2.96   Closed center, closed port directional control valve

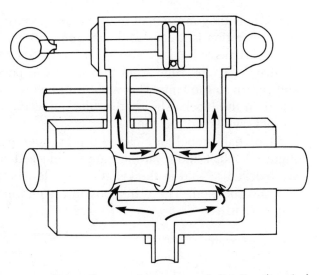

Figure 2.97   Open center, open port directional control valve

The fourth type of valve is the **closed center, open port valve** (Figure 2.98). When this type of valve is in neutral, fluid from the pump is blocked by the valve. Fluid can flow between the rod and blind ends of the cylinder through the valve.

So far we have discussed the operation of directional control valves when they are in neutral. Now we will examine their opera-

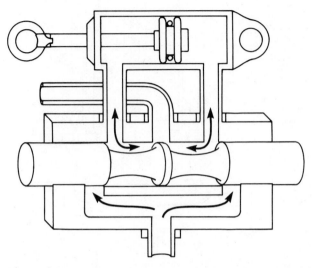

Figure 2.98 Closed center, open port directional control valve

tion when they are not in neutral. The easiest way to understand their operation is to look at their schematic symbols.

A partial symbol for an open center, closed port valve is shown in Figure 2.99. The fluid from the pump can pass through the valve and return to the tank. This is shown as a connection from line P to line T in the block. The fluid is blocked at the port. Lines A and B are not connected together.

A partial symbol for a closed center, open port valve is shown in Figure 2.100. Fluid from the pump is blocked at the valve (P and T are blocked). Fluid in the cylinder can flow between the blind end and the rod end. The other possible combinations are shown in Figure 2.101.

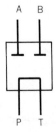

Figure 2.99 Partial schematic symbol for an open center, closed port directional control valve

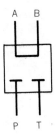

Figure 2.100 Partial schematic symbol for a closed center, open port directional control valve

Closed center, closed port   Open center, open port

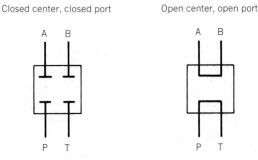

Figure 2.101 Partial schematic symbols for closed center, closed port and open center, open port directional control valves

To show how fluid flows through the valve when the spool is moved, additional boxes are added to the symbol. Figure 2.102 shows the complete symbol for an open center, closed port valve. For illustration of the operation of the valve, the symbol will be shifted. In hydraulic schematics the boxes are never shifted. One must imagine what happens when the spool is shifted.

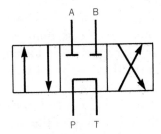

Figure 2.102 Complete schematic symbol for an open center, closed port directional control valve

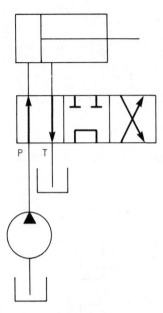

Figure 2.103   Open center, closed port directional control valve with spool shifted to the right

In Figure 2.103 the spool has been shifted to the right. (For illustration, the symbol has been shifted to the right.) Fluid from the pump is directed to the blind end of the cylinder. Fluid from the rod end of the cylinder is returned to the tank.

In Figure 2.104 the spool has been shifted to the left. Fluid from the pump is directed to the rod end of the cylinder, and fluid from the base end of the cylinder is directed to the tank.

The spool in a directional control valve can be moved by pushing a lever. The symbol in Figure 2.105 shows a spool controlled by a hand lever. The spool in the directional control valve shown in Figure 2.106 is controlled by a foot pedal.

Directional control valves can also be controlled by an **electric solenoid** (Figure 2.107). A solenoid is a coil of wire. When electricity is applied to a coil of wire, a magnetic field is created. The magnetic field pulls the spool. Notice that the spool is held in the center by springs until the solenoid is activated.

Solenoid activated directional control valves are used in non-servo robots to control the movement of the actuator. To control the movement of a servo controlled robot the directional control valve must have the capability of being partially opened. This is not possible with a solenoid controlled valve. A servo control valve has the capability of being partially opened.

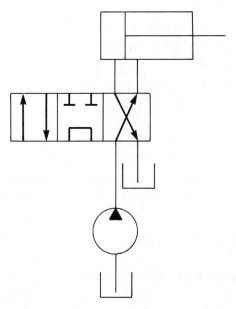

Figure 2.104 Open center, closed port directional control valve with spool shifted to the left

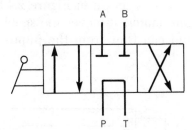

Figure 2.105 Schematic symbol for an open center, closed port directional control valve with spool controlled by a hand lever

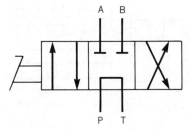

Figure 2.106 Schematic symbol for an open center, closed port directional control valve with spool controlled by a foot pedal

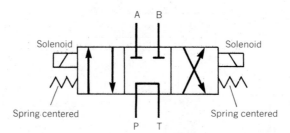

Solenoid controlled, spring centered
Open center, closed port

Figure 2.107   Schematic symbol for an open center, closed port directional control valve with spool controlled by an electric solenoid

SERVO CONTROL VALVES

A **servo control valve** is an electrically controlled valve. The servo valve can be opened a small amount by applying a small voltage. As the voltage is increased the valve opens farther. This differs from a solenoid valve in that a solenoid valve is either fully open or fully closed.

A simple servo valve is shown in Figure 2.108. When voltage is applied to the torque motor, it moves the spool, opening a passage through which fluid can flow from the pump to the cylinder. If

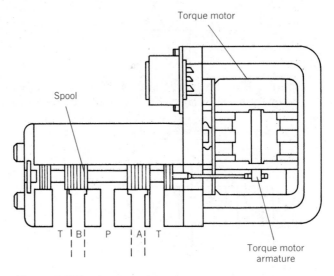

Figure 2.108   Single-stage spool-type servo valve

the voltage is reversed, the spool moves in the other direction. The direction in which the spool moves is a function of the direction in which the voltage is applied to the motor, and the amount the spool moves is a function of the amount of voltage that is applied to the motor.

The size of simple servo valves is limited by the amount of torque that can be developed by a torque motor. As the size of the valve is increased to allow for greater flow, more power is needed to move the spool. A solution to this problem is the **two-stage servo valve.**

Four types of two-stage servo valves are in common usage: the **spool, single flapper, double flapper,** and **jet pipe** valves.

A two-stage spool-type servo valve is shown in Figure 2.109. The valve has a main spool and a pilot spool. The small pilot spool is moved by a torque motor. By moving the small pilot spool, fluid is directed to the main spool. When the pilot spool is moved to the right, fluid is directed to the right side of the main spool while fluid on the left side of the spool is allowed to drain to the tank. The increased pressure on the right side of the main spool shifts it to the left. Fluid from the pump is directed to port A.

If the voltage applied to the torque motor is reversed, the pilot spool is pushed to the left by the motor. Fluid is directed to the left

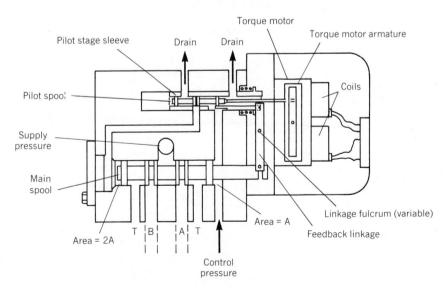

Figure 2.109 Two-stage spool-type servo valve

side of the main spool, and the main spool shifts to the right. Fluid from the pump is directed to port B.

In addition to the pilot spool and the main spool, the valve also contains a feedback linkage. The feedback linkage prevents the main spool from shifting too far. If there were no feedback linkage, a small movement of the pilot spool would cause the main spool to move too far due to the uneven hydraulic pressures on the main spool. Once the main spool had shifted even a small amount, the uneven pressures would keep it moving until the valve was fully open. The feedback linkage senses the excessive movement of the main spool and shifts the pilot spool to equalize the pressures on the main spool, thereby stopping the movement of the main spool. All two-stage servo valves have feedback linkage between the pilot stage and the main spool.

The single flapper servo valve is similar to the spool-type servo valve. The difference lies in the first stage (pilot stage). A single flapper has one fixed orifice (Figure 2.110). When the flapper is away from the port there is a large pressure drop and no pilot pressure is directed to the main spool. As the flapper moves closer to the port, pressure is directed to the main spool, and the spool shifts.

The double flapper servo valve has two fixed orifices (Figure 2.111). The flapper is centered between the ports. When the flapper is pushed toward the port on the right side, pressure on the right side of the main spool increases and pressure on the left side of

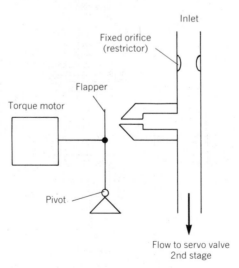

Figure 2.110 Single flapper servo valve

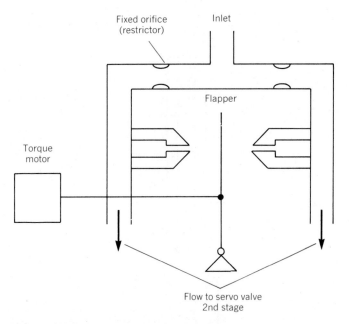

Figure 2.111   Double flapper servo valve

the main spool is reduced. The spool shifts to the left. When the flapper is moved to the left, pressure on the left side of the main spool increases and pressure on the right side of the spool drops. The main spool shifts to the right.

The final type of servo valve is the jet pipe. The jet pipe first stage is shown in Figure 2.112. Pilot fluid is fed through a nozzle. When the nozzle is pulled to the left by a torque motor, pilot fluid pressure is directed to the left side of the main spool. When the nozzle is pushed to the right, fluid is directed to the right side of the main spool.

The servo valve is used to control fluid flow in a servo robot. A hydraulic schematic of a servo controlled hydraulic valve is shown in Figure 2.113.

HYDRAULIC SCHEMATIC OF A "BANG BANG" ROBOT

Figure 2.114 is a schematic diagram of a simple hydraulic "bang bang" robot. Many of the components discussed in this chapter are included in this schematic. This illustration (along with the electronic control) also appears in Chapter 6 (Figure 6.80) and will be discussed in detail there.

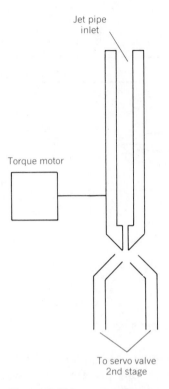

Jet pipe
inlet

Torque motor

To servo valve
2nd stage

Figure 2.112   Jet pipe servo valve

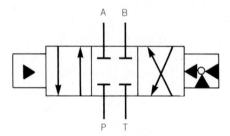

A   B

P   T

Figure 2.113   Schematic symbol for a servo controlled valve

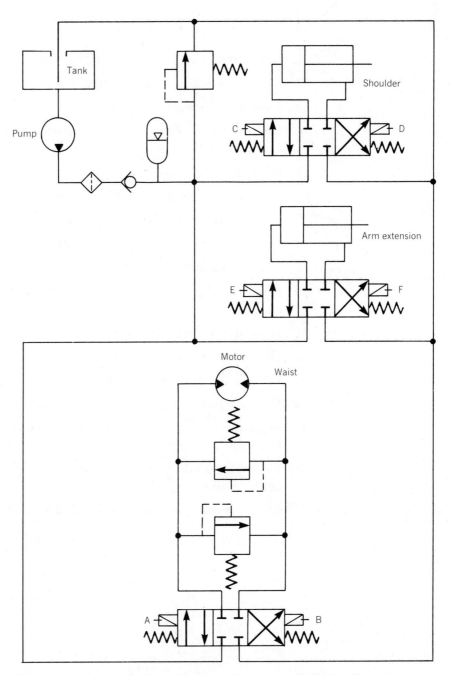

Figure 2.114 Hydraulic schematic of a simple "bang bang" robot

QUESTIONS FOR CHAPTER 2

1. Define hydrostatics and hydrodynamics.

2. To increase speed in a hydraulic system, _____ is increased.

3. Pressure is measured in _____ .

4. The pressure in a cylinder is 450 psi and the area of the piston is 4 in.^{2}. What is the force?

5. Draw the symbol for a filter.

6. Define the term "positive displacement."

7. What is the major advantage of the in-line axial piston pump?

8. Draw the symbol for a hydraulic pump.

9. What is the purpose of an accumulator?

10. What are the four types of accumulators? Draw the symbol for each.

11. Draw the symbol for a double acting cylinder.

12. Draw the symbol for a gear motor.

13. What is the purpose of a relief valve?

14. Does pressure differential affect flow rate? Explain.

15. Draw the symbol for a pressure compensated flow control valve and explain its operation.

16. Draw the symbol for an open center, closed port directional control valve.

17. Describe the operation of a double flapper servo valve.

CHAPTER THREE

Pneumatic Systems

In the previous chapter we discussed hydraulic systems, in which liquids are used for supplying power to robots. Gas can also be used as a power supply. When we are using the power of a compressed gas to do work, we call the power system **pneumatic**, from the Greek word *pneuma,* meaning "breath."

The difference between a hydraulic system and a pneumatic system is the fluid that is used to transmit energy. A hydraulic system uses oil or other liquid to transmit force whereas a pneumatic system uses a gas. A gas is also a "fluid." It will flow from a high-pressure area to a low-pressure area. Because of the similarity between hydraulic systems and pneumatic systems, many of the components in a pneumatic system are similar to those found in a hydraulic system. Figure 3.1 shows a hydraulic directional control valve and a pneumatic directional control valve. They are almost identical. The only difference is the addition of rubber seals around the spool of the pneumatic valve to keep it from leaking. Seals are not necessary in the hydraulic valve because the molecules that make up the oil are too large to leak between the spool and the valve body. The much smaller gas molecules in a pneumatic system will leak between the spool and the valve body if **O rings** are not included in the design.

Another difference between hydraulic and pneumatic systems is the greater compressibility of the air in a pneumatic system. Hydraulic fluid compresses very little. The compressibility of gas can be both an advantage and a disadvantage. The advantage of the compressibility of gas is that it acts as a natural shock absorber. If the load should suddenly increase, the air in the system will compress like a spring, absorbing the load. A disadvantage is that the air can compress when it is not desirable. The arm of a pneumatic robot will "sag" as the load is increased. The sagging of the arm impairs the repeatability of the robot.

Figure 3.1 Hydraulic (left) and pneumatic (right) directional control valves

Another difference between hydraulic and pneumatic systems is the way in which the return fluid is handled. In a hydraulic system the return fluid (oil) is directed back to the tank (Figure 3.2), whereas in a pneumatic system the return fluid (air) is simply exhausted to the atmosphere (Figure 3.3).

The formula F = PA presented in the hydraulics chapter also holds true for pneumatic systems. The speed of a pneumatic actuator is not a function of pressure. Pressure in a pneumatic system causes force. If the pressure in a pneumatic or hydraulic system is 125 psi and the area is 2 in.^{2}, then the force is:

$$F = PA$$
$$F = 125 \times 2$$
$$F = 250 \text{ lb}$$

KINETIC THEORY OF GASES

The kinetic theory of gases states that molecules of a gas move rapidly and in a straight line until they collide with something which causes changes in the direction in which and speed at which they are traveling. Figure 3.4 shows a container filled with gas. (The gas that we will use is air.) The circles represent the molecules of the

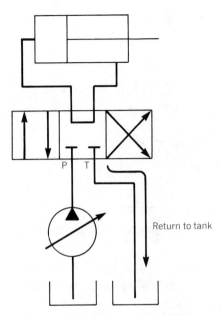

Figure 3.2   Fluid return in a hydraulic system

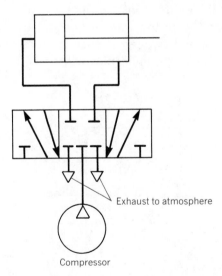

Figure 3.3   Fluid return (exhaust) in a pneumatic system

air. The molecules travel in a straight line until they strike the wall of the container and then bounce off, heading in another direction.

The impact of these molecules hitting the wall causes pressure against the inside wall. Remember that there is air outside the con-

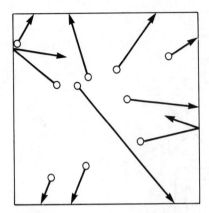

Figure 3.4   Movement of contained gas (air) molecules

tainer too. Molecules are also striking the outside of the container. Pressure on the outside wall is equal to that on the inside wall of the container (Figure 3.5).

Figure 3.6 shows two views of a piston inside a cylinder. If the piston is pushed downward, the space inside the container is reduced. The cylinder still contains the same number of air molecules as it did before the piston was pushed downward, but the volume inside the cylinder has been reduced. Since the space inside the cylinder is reduced, the molecules will strike the walls of the cylin-

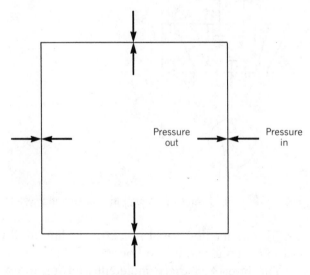

Figure 3.5   Equalization of pressure inside and outside a container of air

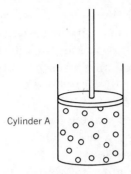

Cylinder A

Same number of molecules in cylinders A and B.
B has higher pressure.

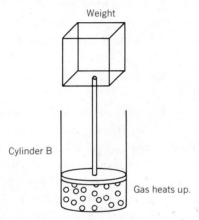

Weight

Cylinder B

Gas heats up.

Figure 3.6 Effects of compression on pressure and temperature of gas molecules

der more often than they did before the piston was pushed downward. This causes the pressure inside the cylinder to be greater than that outside the cylinder. The molecules of air outside the cylinder are still striking the outside walls of the cylinder the same number of times each second as they did before the piston was pushed downward, but the molecules inside the cylinder are striking the inside surface of the container more often, and the pressure inside the cylinder is now higher than the pressure outside the cylinder. The air inside the cylinder is said to be **compressed**.

Not only has the pressure inside the cylinder gone up as a result of the piston being pushed downward, but the temperature has also gone up. When the molecules strike the wall of the container, some of their energy is converted to heat. The more often the sides of the cylinder are struck by the bouncing molecules the higher the temperature.

Recall from our earlier discussion of hydraulics that a contained fluid under pressure transmits its force equally in all directions. Since air is also a fluid, the pressure inside the cylinder is equal throughout. There is the same pressure pushing against every square inch of the cylinder. If we remove the weight that is pushing the piston downward (Figure 3.7), the piston will return to its original position. The molecules striking the piston push the piston back up until it is being struck by molecules the same number of times on the outside as on the inside.

When the piston in Figure 3.7 returns to its original position, the air inside the cylinder cools rapidly because the inside surface area of the cylinder is no longer being struck as often by the molecules. This temperature change can be dramatic. In some situations, the temperature can drop below freezing, and ice can form on the cylinder.

Change of State

Pressure in a pneumatic system can be increased by compressing the gas. The pressure can also be increased by raising the temperature of the confined gas. Figure 3.8 shows a container over a flame. As the gas inside the container is heated by the flame, the gas molecules begin to move faster. As the velocity of the molecules increases, the pressure in the system increases because the molecules are striking the container more often.

If the container were placed on a block of ice, the molecules would vibrate more slowly. The colder the container the more slowly the molecules would move. If we could cool the container enough, the air in the container would become a liquid. This is called **change of state.**

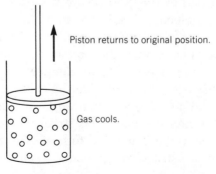

Piston returns to original position.

Gas cools.

Figure 3.7 Return of piston to original position upon removal of weight

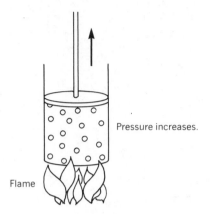

Pressure increases.

Flame

Figure 3.8 Container over a flame (heat causes pressure to increase)

Air in the atmosphere always contains water vapor in the gaseous state. When the air cools, the water vapor turns into its liquid state. In our previous example we said that if the air in the container were cooled enough the air would turn into a liquid. Long before the air inside the container turns into liquid the water vapor in the air inside the container will "condense," turning into a liquid. The change of state from water vapor to liquid requires a drop in temperature of only a few degrees.

Warm air can hold more moisture than cool air. When we compress air it becomes warmer and takes on more moisture. When it is allowed to expand it cools, and the water vapor turns back into a liquid. For this reason, water often forms in a pneumatic system, causing rust and corrosion. Later in this chapter we will describe some methods of removing the water to protect the pneumatic system.

Flow and Pressure

The rate of flow in a pneumatic system is measured in cubic feet per minute (ft³/min). As in a hydraulic system, pressure is measured in pounds per square inch (psi).

A pressure differential causes flow. In the connecting containers shown in Figure 3.9, the gas at higher pressure will flow to the lower-pressure container. When the pressure in the two containers becomes equal, the air will stop flowing.

It is important not to confuse the effects of flow and pressure in a hydraulic or pneumatic system. The rate of flow determines the speed. Figure 3.10 is a schematic diagram of a cylinder plumbed to a hydraulic pump. The greater the flow rate the faster the pis-

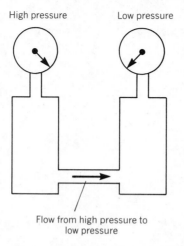

Flow from high pressure to
low pressure

Figure 3.9 Connecting containers, showing flow from high-pressure container to low-pressure container

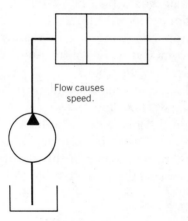

Flow causes
speed.

Figure 3.10   Hydraulic system (cylinder and hydraulic pump)

ton moves. If we substitute a compressor for the hydraulic pump (Figure 3.11), the flow rate of the compressor will determine the speed of the piston.

Flow causes speed, and pressure causes force. The higher the pressure the higher the force. As the pressure is increased, the cylinder can do more work. The increase in pressure does not affect the speed.

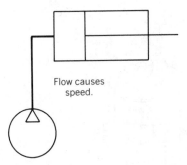

Flow causes
speed.

Figure 3.11   Pneumatic system (cylinder and compressor)

COMPRESSORS

There are two basic types of compressors: **displacement compressors** and **dynamic compressors.** Displacement compressors are similar to the positive displacement pumps studied in the previous chapter. Dynamic compressors are similar to the nonpositive displacement pumps described in the previous chapter.

An example of a dynamic compressor is a fan. The air is accelerated by the spinning blades. The fast-moving air is converted to pressure. In a displacement compressor, on the other hand, pressure is created by moving a piston in a cylinder to reduce the volume of air in the cylinder.

Displacement Compressors

Reciprocating Piston Compressor.   The reciprocating piston compressor is the most common type of compressor. Figure 3.12 shows the operation of the reciprocating compressor. Starting with the piston at the top of the cylinder, the crankshaft pulls the piston downward, creating a vacuum in the expanding area above the piston. As the piston is pulled downward, the intake valve is opened. Air from the atmosphere quickly rushes in to fill the vacuum.

When the piston reaches the bottom of the cylinder, the intake valve closes and the crankshaft pushes the piston upward. The area above the piston is reduced, and the air in the cylinder is compressed.

When the piston nears the top of the cylinder, the outlet valve opens and the compressed air rushes out of the cylinder. When the piston reaches the top of the cylinder, the outlet valve closes and

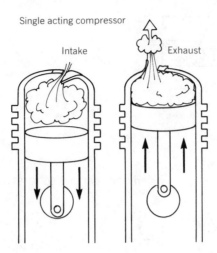

Figure 3.12   Operation of a reciprocating piston compressor

the inlet valve opens. The piston begins its downward travel and the process is repeated.

Diaphragm Compressor.   A diaphragm compressor operates in much the same way as a reciprocating piston compressor (Figure 3.13). The diaphragm is pulled downward and pushed upward by a crankshaft and connecting rod, but instead of sliding up and down inside the cylinder as the piston does, the diaphragm is simply distorted by the movement of the connecting rod and the crankshaft. When the crankshaft and the connecting rod push upward on the diaphragm, the diaphragm forms a **concave** surface, which reduces the volume inside the compressor. When the connecting rod and the crankshaft pull downward on the diaphragm, the diaphragm becomes **convex**, which increases the volume inside the compressor.

When the diaphragm is pulled downward, air rushes into the compressor. When the diaphragm is pushed upward, the air is compressed and exits through the outlet valve of the compressor.

The diaphragm compressor is often smaller than the reciprocating piston compressor and cannot compress as large a volume of air as most reciprocating piston compressors can.

Multistage Compressor.   Earlier we said that the temperature of a gas rises as the gas is compressed. As the gas is heated, it expands, which makes the process of compressing the gas even more difficult. For this reason, industrial systems that require compressed air at more than 100 psi normally incorporate multistage compressors.

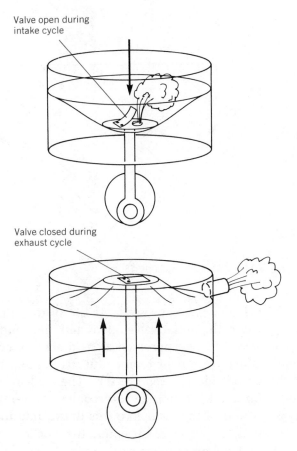

Valve open during
intake cycle

Valve closed during
exhaust cycle

Figure 3.13 Operation of a diaphragm compressor

The multistage compressor overcomes some of the problems of the simpler single-stage compressor. To illustrate the operation of multistage compressors we will use a two-stage compressor. It is possible to add more stages to increase the compression of the gas even further.

Figure 3.14 shows a two-stage compressor. The compressor has two cylinders, one larger than the other. The pistons in both cylinders are connected to a single crankshaft. While the piston in the larger cylinder is being pulled downward, the piston in the smaller cylinder is being pushed upward.

The operation of the two-stage compressor begins by pulling the larger piston downward while the intake valve is open. The larger cylinder fills with air from the atmosphere. As the crankshaft continues to turn, the larger piston is pushed upward and the air

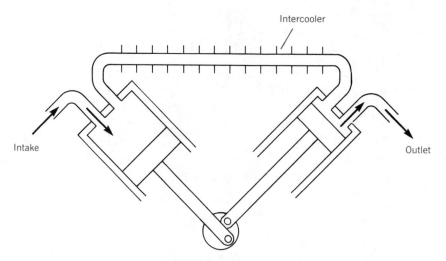

Figure 3.14 Two-stage compressor

in the cylinder is compressed. As the piston approaches the top of
the cylinder, the outlet valve opens and the compressed air exits. As
noted earlier, the compressed air is much hotter than the **ambient
temperature.** (Ambient temperature is defined as the temperature
of the air outside the compressor.) The air leaving the larger cylinder is directed through an **intercooler.** The purpose of the intercooler is to cool the air before it is drawn into the smaller cylinder.

Cooling the air before it enters the smaller cylinder reduces the
pressure. Cool air is more easily compressed than hot air, and thus
the efficiency of the compressor is increased. Since cooling the air
reduces the pressure and since the purpose of compressing the air is
to raise the pressure, you might wonder why we go to the trouble of
cooling the air between the two stages. In order to understand why
this is done, recall that the air being drawn into the first stage of the
compressor is at ambient temperature. When the air is compressed,
the spaces between the air molecules are reduced, and the molecules collide with each other and with the walls of the cylinder more
frequently. These collisions cause the temperature of the air to rise
and the pressure to be further increased. The increase in temperature is a result of compression but is not the purpose of compression. If the compressed air is held in the first stage of the
compressor, the temperature of the air will eventually return to that
of the ambient air and the increase in pressure that was caused by
the increase in temperature will be lost, but the air will still be compressed because the molecules are closer together.

After the air has cooled, it can be further compressed more easily by the second stage of the compressor. Although this process of compressing air in the first stage and allowing it to return to ambient temperature while in the first stage has the advantage of making further compression easier, it is obviously too time-consuming. The intercooler is used to accelerate the process of cooling the air before it enters the second stage for further compression. An intercooler may be as simple as a copper tube with fins. The tube and fins radiate the heat to the atmosphere.

If the system requires more cooling than can be accomplished by a simple intercooler, a chilled water system can be employed (Figure 3.15). In such a system, a copper **fin tube** is mounted inside a housing, and chilled water is pumped into the housing. As the water passes over the fin tube, the heat from the compressed air is transferred to the water. The heated water is pumped back to a **water chiller** to be cooled again, and the cycle is repeated.

After the compressed air from the larger cylinder has been cooled, it flows into the smaller cylinder as the piston is pulled downward. When the piston in the smaller cylinder is pushed upward, the air in the cylinder is further compressed.

Multistage compressors are more efficient than single-stage compressors. Less energy is required to turn a multistage compressor because the heat from compression is removed between each stage of compression.

Vane Compressor.   A vane compressor is similar in design to the vane pump used in hydraulic systems (Figure 3.16). Atmospheric air rushes in to fill the vacuum created by the rotor. As the rotor turns, the air is carried into chambers which reduce its volume and thus compress it.

The major advantage of the vane compressor over the reciprocating piston compressor is its constant delivery of compressed air. (Recall that the output of a reciprocating compressor is a pulsating output.)

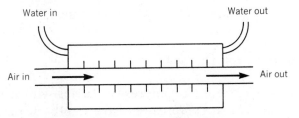

Figure 3.15   Chilled water cooling system

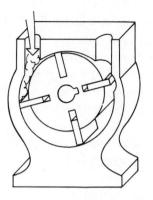

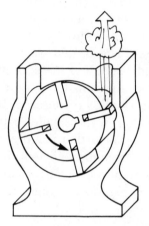

Figure 3.16 Operation of a vane compressor

The major disadvantage of the vane compressor is its limited output pressure. It is possible to get much higher output pressure from a multistage reciprocating compressor.

Helical Compressor.   A helical compressor (Figure 3.17) compresses air through the action of two meshing rotors resembling screws (this type of compressor is often called a "screw compressor"). Atmospheric air enters at one end of the compressor, flows past the turning rotors, and exits as compressed air at the other end.

INTAKE FILTERS

Air entering a compressor carries with it dirt from the atmosphere. Most compressors have tightly fitting parts in a housing. Airborne dirt drawn into the compressor will **score** (scratch) cylin-

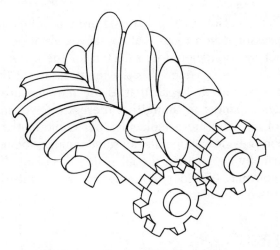

Figure 3.17 Helical (screw) compressor

der walls, destroy bearings, and significantly reduce the life of the compressor.

An **intake filter** will trap much of the dirt before it enters the compressor. The air filter on an automobile engine serves the same purpose as the intake filter on a compressor. The air filter on an automobile engine traps dirt before it can enter the engine, and the intake filter on a compressor traps dirt before it can enter the compressor (Figure 3.18).

The intake filter does not have to be mounted on the compressor. It can be mounted in another room or even outside. The filter is simply plumbed to the compressor with pipe.

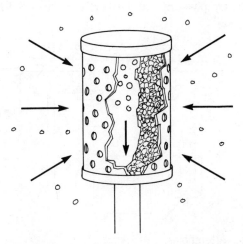

Figure 3.18 Intake filter

The location of the filter should be chosen carefully. If the compressor is in a room that contains a great deal of airborne dirt, a cleaner location should be selected for the filter.

Also to be considered is the temperature of the room where the filter is mounted. Recall that as the temperature of a gas increases the gas expands, meaning that there are fewer molecules per cubic foot of air. The more molecules per cubic foot the more efficiently the compressor will operate. In some regions, the outside temperature is normally cooler than the air inside the plant. If this is your situation you may consider mounting the filter outside and plumbing it into the compressor.

AFTERCOOLERS

An **aftercooler** is often used in conjunction with a compressor. The major purpose of an aftercooler is to do as its name implies: to cool the gas after it has been compressed.

An aftercooler also removes much of the entrapped water vapor in the compressed air. The water carried in hot compressed air can be substantial — as much as 1.4 quarts of water in 1000 ft^3 of air. An industrial compressor of moderate size can produce as much as 50 gallons of water in a 24-hour period.

The construction of an aftercooler is similar to that of the intercooler described earlier. The compressed gas is passed through a fin tube. The gas may be cooled by radiation of heat from the fin tube to the surrounding air, or the tube may be encased in a housing and cooled by passing water over it (Figure 3.19).

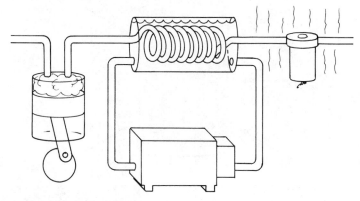

Figure 3.19 Aftercooler

If more water vapor must be removed than can be removed with a simple water-type aftercooler, a refrigeration unit can be added. The refrigeration unit cools the water that is passing over the fin tube, thus lowering the temperature of the compressed air. More water vapor turns into water and can be drained from the system.

RECEIVER TANK

A **receiver tank** is always included in a pneumatic system. The receiver tank performs the same function as the accumulator in a hydraulic system: it stores the compressed air for use by the system (Figure 3.20).

When the components in a pneumatic system are turned on, they demand a constant supply of high-pressure air to operate. Because the compressor alone may not be able to meet the demand, the receiver tank takes up the slack, supplying air as needed.

Figure 3.20 Receiver tank

The receiver tank also performs another function. As we said earlier, many compressors do not supply an even flow of compressed air. The piston compressor supplies air only while the piston is moving upward. When the piston is moving downward the compressor is drawing air in from the atmosphere, preparing for the next compression stroke. The uneven air flow from the compressor can damage pneumatic components and cause the system to operate erratically. The receiver tank absorbs the pulses from the compressor and supplies the system with a constant flow of high-pressure air.

PRESSURE SWITCH

The receiver tank stores the compressed air from the compressor until it is needed by the system. If the system does not need air, the receiver tank stores the compressed air, and the pressure in the tank rises. For safety, the pressure in the tank cannot be allowed to continue to rise. A **pressure switch** is used to turn off the compressor when the tank has reached a preset pressure. When the system begins to operate, the compressed air flows out of the receiver tank and the pressure in the tank goes down. When the pressure gets low enough, the pressure switch turns the compressor back on to refill the tank.

Figure 3.21 shows a pressure switch. The pressure switch is normally mounted on the receiver tank. As the pressure in the tank rises, the piston is pushed up. When the pressure is high enough the piston overcomes the spring tension and turns the power switch off, thus cutting the power to the compressor. As the compressed air in the tank is used by the system, the pressure in the tank begins to drop. After the pressure in the tank drops far enough so that the spring pushes the piston down, the compressor switch turns on. (Additional information on pressure switches can be found in Chapters 9 and 10.)

The spring can be adjusted with a bolt. As the bolt is turned the spring is compressed further, and the pressure in the tank must be higher before it can overcome the spring tension and turn the compressor off.

SAFETY RELIEF VALVE

The pressure switch should keep the pneumatic system operating within safe limits; however, should the pressure switch fail and

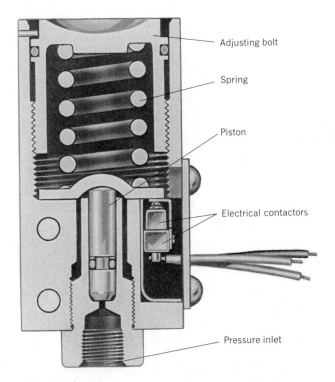

Figure 3.21 Pressure switch

not turn the compressor off, the pressure in the tank will continue to rise. If the pressure should become too high the receiver tank or other component in the system may explode. For this reason a **safety relief valve** is always included in the system.

Figure 3.22 shows a safety relief valve along with its schematic symbol. As the pressure in the system rises, the **poppet** is pushed upward. If the poppet is pushed up far enough, the exhaust port is opened and the pressure in the tank is vented to the atmosphere.

DESICCANT DRYERS

Not all systems require an aftercooler for cooling the air and removing moisture. Often the simpler **desiccant dryer** will give the desired results. Even systems that utilize an aftercooler may include a desiccant dryer as a final attempt to trap all of the moisture in the system.

The filter contains a chemical that converts the water vapor into water. Three types of desiccant dryers are used in industry: the **deli-**

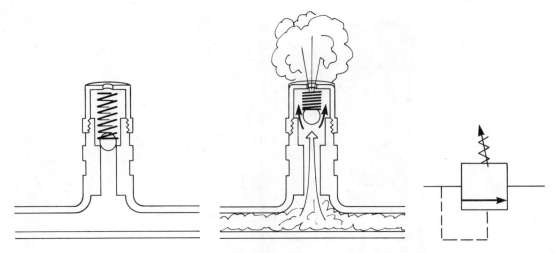

Figure 3.22   Safety relief valve

quescent dryer, the **chemically regenerative dryer,** and the **heat regenerative dryer.**

Deliquescent Dryer

In a **deliquescent dryer** (Figure 3.23), water vapor in the compressed air is passed through a chemical called a **deliquescent drying agent.** As water vapor passes through the chemical it is absorbed. Deliquescent chemicals include lithium chloride and calcium chloride.

The deliquescent agent in the dryer reacts chemically with the water vapor. The chemical can absorb only a certain amount of water

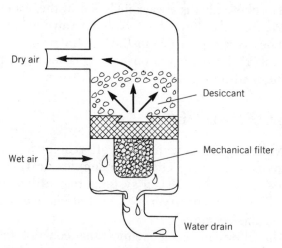

Figure 3.23   Deliquescent dryer

vapor before it begins to liquefy. When this happens, the chemical must be replaced.

There are some problems with this type of dryer. The deliquescent agents are corrosive. As the compressed air is passed through the corrosive chemical it picks up some of the chemical and carries it throughout the system. In the long term this can cause maintenance problems; however, this type of dryer has the lowest initial cost and operating cost and is a popular choice in industry.

Chemically Regenerative Dryer

In a **chemically regenerative dryer,** two desiccant dryers are used. As one of the dryers becomes saturated, the incoming air is diverted to the other canister while the first canister is being renewed.

To renew the desiccant in the saturated canister, a small portion of the dried air exiting the second canister is diverted to the saturated canister (Figure 3.24). The dry air coming from the second canister quickly dries the desiccant in the saturated canister, which then can be used again. The switching from one dryer to the other is controlled by a timer and may occur several times each minute.

Heat Regenerative Dryer

The **heat regenerative dryer** is similar to the chemically regenerative dryer. There are two canisters as in the chemically regenerative dryer. When one of the canisters becomes saturated, the system automatically switches to the other canister. The difference between the two systems is that in the heat regenerative dryer the desiccant is dried by passing heated air through it (Figure 3.25), rather than by simply diverting dried air from the other canister as in the chemically regenerative dryer.

PNEUMATIC ACTUATORS

At this point we have looked at the major components that prepare the air for use in producing power. The actuators that produce work in pneumatic systems are similar to those used in hydraulic systems. Often it is not possible to tell the difference between hydraulic and pneumatic actuators by simply looking at them. Although the actuators look similar from the outside, they are not identical. Normally, hydraulic and pneumatic actuators cannot be interchanged. They differ in the types of seals and glands used.

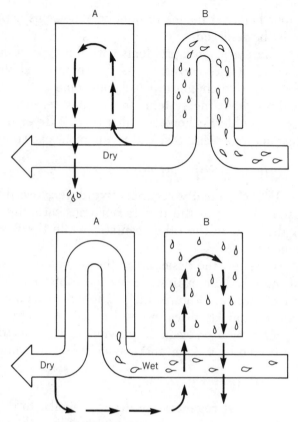

Figure 3.24 Chemically regenerative dryer

Linear Actuators

The same types of **linear actuators** (cylinders) that were described for hydraulic systems (Chapter 2) are also available for pneumatic systems. These cylinders are the **single acting cylinder,** the **double acting cylinder,** and the **double acting, double rod cylinder.** Another type of cylinder that is often used in pneumatic systems but rarely found in hydraulic systems is the **single acting spring return cylinder** (Figure 3.26).

The single acting spring return cylinder has a single input just as do other single acting cylinders. When air is pumped into the blind end of the cylinder until the pressure exceeds the spring tension, the rod extends (Figure 3.26). As long as the pressure behind the piston is higher than the spring tension, the rod remains extended. When the pressure is reduced behind the piston, the piston

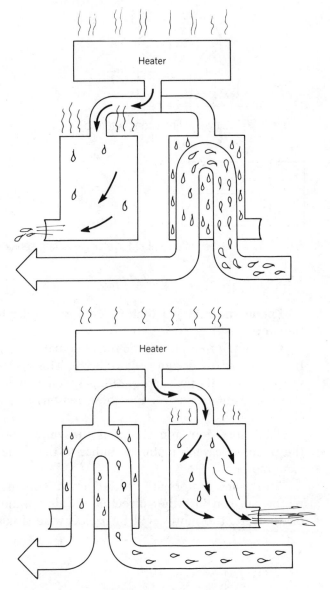

Figure 3.25 Heat regenerative dryer

and rod retract. The schematic symbol for a single acting spring return cylinder is shown in Figure 3.27.

Cushioned pneumatic cylinders are also available. They operate the same as the hydraulic cushioned cylinders described in Chapter 2.

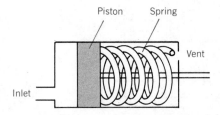

Figure 3.26   Single acting spring return cylinder

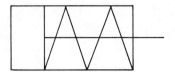

Figure 3.27   Schematic symbol for a single acting spring return cylinder

Rotary Actuators

Pneumatic rotary actuators are very similar to hydraulic rotary actuators.

One of the first pneumatic rotary actuators was the **piston motor.** It is of a "radial" design (Figure 3.28). This type of motor is a low-speed motor. Its normal operating speed is below 1000 rpm. It is also relatively expensive. For these reasons it is not a very popular choice.

The more common choice is the **vane motor.** The design of the pneumatic vane motor is similar to the design of the hydraulic vane motor.

Most vane motors are bidirectional. This means that the vane motor can run in either direction. If air is pumped into the port on the left, the motor will turn clockwise (Figure 3.29). If air is

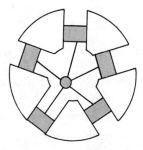

Figure 3.28   Piston motor

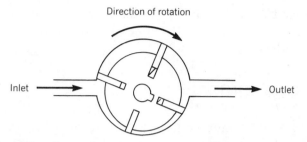

Figure 3.29 Vane motor (clockwise rotation)

pumped into the port on the right, the motor will turn counterclockwise (Figure 3.30).

The speed at which a vane motor turns is controlled by the volume of air pumped to the motor. The torque developed by a vane motor is determined by the pressure and the area of the vanes (F = PA). The higher the pressure and the larger the motor, the more torque the motor delivers.

The torque developed by a pneumatic rotary actuator is not a function of speed as it is with electric motors. A pneumatic motor develops the same torque when the shaft is rotating and when it is stalled due to a heavy load. This is an important advantage. Another advantage is the safety of operating a pneumatic motor around damp areas. Electric motors can produce dangerous electric shocks.

The vane motor is the most popular of the rotary actuators because of its relatively low cost, its power, its variable speed, and its safety.

Another type of rotary actuator is the **turbine motor.** High-pressure air is directed through a nozzle. As the air exits the nozzle it expands rapidly. This expanding air is directed across a turbine (fan). The turbine spins at a very high speed. Because of the high

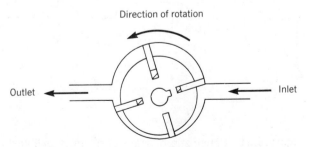

Figure 3.30 Vane motor (counterclockwise rotation)

speed and the difficulty of gearing it down to a usable speed, turbine motors are limited to special applications such as high-speed grinders.

FLOW CONTROL

At this point we have looked at the air make-up system, including the compressor, the receiver tank, the aftercooler, and dryers. We have also taken a brief look at pneumatic actuators. A simple system using these components is shown in Figure 3.31. In addition to these elements, the system requires some means of regulating the flow rate to the components to control their speed.

Pneumatic flow controls are similar to the flow controls found in hydraulic systems. Flow control can be as simple as a needle valve or as complex as a pressure flow compensator. If you are not certain of the operation of flow control valves, review the section on flow control valves in Chapter 2.

PRESSURE REGULATORS

Controlling the flow rate in a pneumatic system is only half the battle. Different actuators in a system require different pressures. **Pressure regulators** are used throughout the system to regulate the pressure delivered to the various actuators.

Figure 3.32 shows a cutaway view of a pressure regulator. A spring pushing against a diaphragm pushes a poppet (spool) down. Air entering the regulator passes through the orifice created by the open poppet. As back pressure begins to build due to the load on

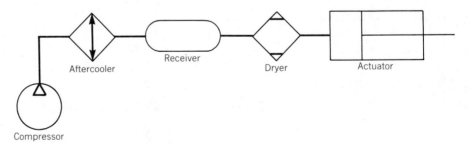

Figure 3.31 Schematic diagram of a simple pneumatic system (without flow control)

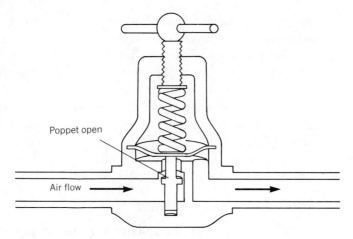

Figure 3.32 Pressure regulator (with poppet open)

the actuator, the diaphragm is pushed up, partially closing the poppet and reducing the flow of air to the actuator. As back pressure builds, the poppet is pushed further closed. Finally, when the back pressure becomes high enough, the poppet is fully closed and no more air flows to the actuator. Pressure at the actuator is at a maximum (Figure 3.33).

If the actuator should move the load, the volume in the actuator increases and the pressure drops (Figure 3.32). As the pressure drops, the spring pushes the diaphragm and the poppet downward,

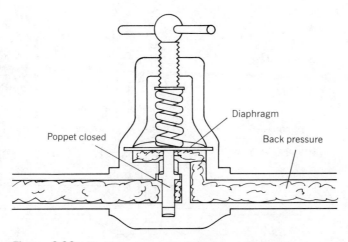

Figure 3.33 Pressure regulator (with poppet closed)

opening a passage for air. The pressure again begins to build in the actuator and the cycle repeats itself.

The pressure being supplied to the actuator is controlled by the setting of the spring. As the screw on the top of the regulator is turned in, the tension on the spring increases. The higher the tension on the spring, the higher the back pressure must be for the diaphragm to overcome the spring tension and close off the valve (Figure 3.34).

Some regulators use a piston rather than a diaphragm. Their operation is identical to that described for diaphragm pressure regulators.

Pilot Operated Regulator

The **pilot operated regulator** is made up of two separate regulators. One of them is called the **slave regulator** and the other is called the **pilot regulator** (Figure 3.35).

The slave regulator operates the same as the standard regulator: the only difference is the absence of the main spring. Air is directed to the top of the piston or the diaphragm from a "pilot line." There is still a small spring that opens the poppet. Pilot pressure from another regulator is directed to the top of the diaphragm, and back pressure from the actuator pushes upward to close the valve. The balance between pilot pressure and back pressure controls the final pressure. The pilot regulator is identical to the regulator that was described earlier.

The advantage of the pilot operated regulator is that the slave regulator can be located wherever it is needed, even if it is in an

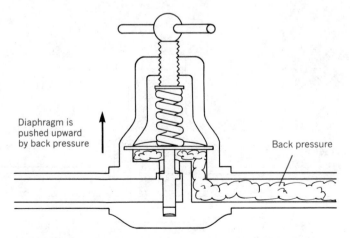

Diaphragm is pushed upward by back pressure

Back pressure

Figure 3.34   Back pressure in a pressure regulator

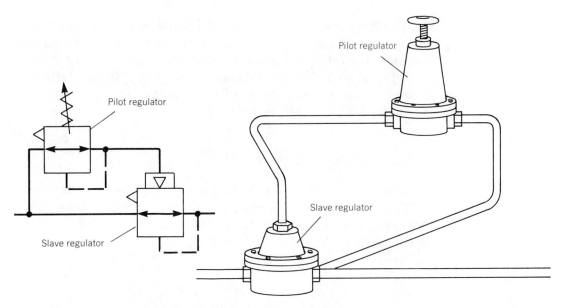

Figure 3.35 Pilot operated regulator

inconvenient location for adjustment. The pilot regulator can be placed in a location where it can be conveniently adjusted. Adjustment of the pilot regulator adjusts the slave regulator.

THE FRL

Often a regulator is combined with other elements and installed as a package. The package consists of a **filter**, a **regulator**, and a **lubricator** (**FRL**) (Figure 3.36).

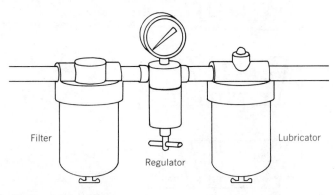

Figure 3.36 FRL

The filter removes rust, dirt, scale, and water that still remain in the system. The regulator controls the operating pressure of the actuator. The lubricator introduces a mist of oil into the air to lubricate the valves and actuators downstream. The FRL also includes a pressure gauge to show the regulated pressure. The schematic symbol for an FRL is shown in Figure 3.37.

DIRECTIONAL CONTROL VALVES

The **directional control valves** used in pneumatic systems are similar to those used in hydraulic systems. The major difference in their design is the use of **O rings.** The schematic symbols for directional control valves are shown in Figure 3.38.

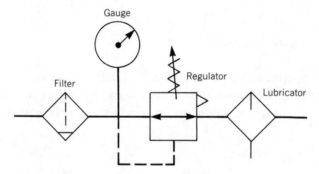

Figure 3.37   Schematic symbol for an FRL

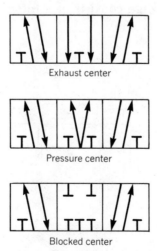

Exhaust center

Pressure center

Blocked center

Figure 3.38   Schematic symbols for directional control valves

We will analyze the operation of the blocked center directional control valve shown in Figure 3.38. All directional control valves can be analyzed in the same way.

The center block shows the inlet and exhaust ports on one side and the connections to the actuator on the other side. The center port on the inlet side is connected to system pressure (Figure 3.39). The other two connections are for exhaust. In a pneumatic system, fluid does not have to be returned to the tank as it does in a hydraulic system. The air is simply exhausted to the atmosphere.

In this valve all ports are blocked in neutral. When the spool is shifted to the right, air is directed to the blind end of the cylinder and the air in the rod end of the cylinder is exhausted to the atmosphere (Figure 3.40). For illustration, the symbol has been shifted.

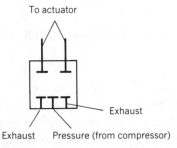

Figure 3.39 Center block of symbol for a blocked center directional control valve, showing inlet and exhaust ports and connection to actuator

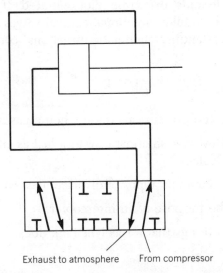

Figure 3.40 Blocked center directional control valve with spool shifted to the right

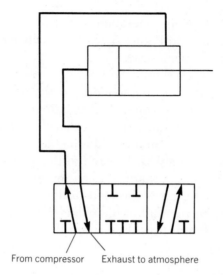

From compressor   Exhaust to atmosphere

Figure 3.41   Blocked center directional control valve with spool shifted to the left

In schematic drawings the symbol is never shifted. You have to imagine the shift of the symbol when analyzing a schematic drawing.

When the spool is shifted to the left, the pressure line is connected to the rod end of the cylinder and the blind end is connected to the exhaust port (Figure 3.41).

Pneumatic directional control valves can be operated by levers, pedals, or solenoids, or can be pilot operated. If you understand the operation of hydraulic directional control valves you should have no trouble understanding pneumatic directional control valves.

QUESTIONS FOR CHAPTER 3

1. Why does the pressure increase when a gas is compressed?

2. Does air flow in a pneumatic system just as oil does in a hydraulic system? Explain.

3. Name two common types of displacement compressors.

4. What is the purpose of an intercooler?

5. Name two things that a receiver tank does.

6. Describe the operation of a safety relief valve and draw its schematic symbol.

7. Describe the operation of a heat regenerative dryer.

8. Describe the operation of a pressure regulator.

9. Draw a directional control valve with all ports blocked in neutral.

CHAPTER FOUR

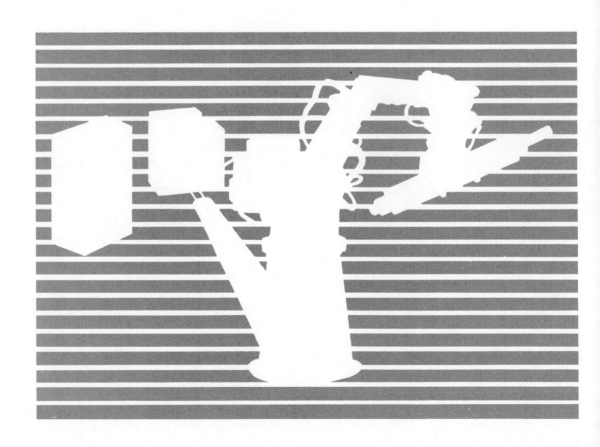

Electric Motors and Mechanical Drives

Many robot manufacturers are now using electric motors as drives for their robots (Figure 4.1). Previously, electric motors were not able to supply the power necessary to carry the desired loads, but in recent years electric motors have been developed that can power robots (Figure 4.2).

Electric-motor drives for robots have certain advantages over hydraulic and pneumatic drives. One advantage is the simplicity of an electric drive in comparison with hydraulic and pneumatic drives. Both hydraulic and pneumatic drives require plumbing and storage of fluids (air or oil) under pressure, and develop leaks which require service.

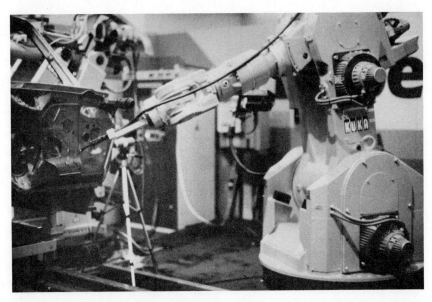

Figure 4.1 Robot with electric drive

Figure 4.2   Electric servo motor

Electric-drive robots are also more repeatable. The position of the arm of a hydraulic or pneumatic robot will vary depending on the load that the robot is carrying. The heavier the load, the more the fluid compresses and the less repeatable the robot. This problem can be overcome by using complex electronic circuitry to monitor the robot's position and to correct for the variation due to load; however, this also adds to the complexity of the robot.

MAGNETISM AND ELECTRIC MOTORS

Electricity passing through a wire creates a magnetic field (Figure 4.3). If the wire is wound to form a coil, the magnetic

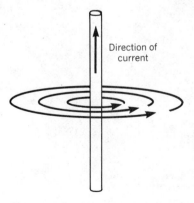

Direction of
current

Figure 4.3   Magnetic field created by passage of an electric current through a wire

lines of force are concentrated into a smaller area, which increases the strength of the magnet.

The direction in which the current flows through the coil and the direction in which the wire is wrapped determine which end of the coil becomes the north pole and which end becomes the south pole (Figure 4.4). A magnet formed by passing current through a coil of wire is called an "electromagnet."

Like magnetic poles repel each other and unlike magnetic poles attract each other. Placing an electromagnet between two permanent magnets makes a basic motor. If current is passed through

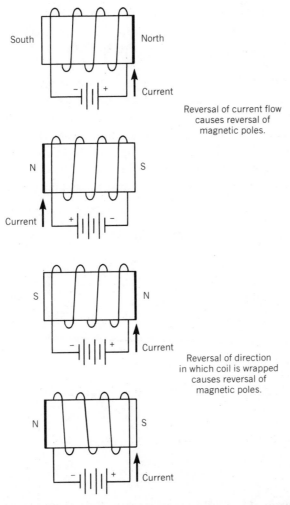

Reversal of current flow
causes reversal of
magnetic poles.

Reversal of direction
in which coil is wrapped
causes reversal of
magnetic poles.

Figure 4.4 Illustration of how magnetic poles are reversed by reversal of direction of current flow or by reversal of direction in which coil is wrapped

the electromagnet so that the north pole of the permanent magnet is lined up with the north pole of the electromagnet and the south pole of the permanent magnet is lined up with the south pole of the electromagnet, the electromagnet will attempt to turn. If the electromagnet is allowed to turn it will rotate until the north pole of the permanent magnet is lined up with the south pole of the electromagnet and the south pole of the permanent magnet is lined up with the north pole of the electromagnet (Figure 4.5).

If the current is reversed in the electromagnet, the electromagnet will again attempt to turn so that the north pole of the permanent magnet and the south pole of the electromagnet again align.

The switching of current flow in the electromagnet is done with a device called a **commutator**. A commutator is made by splitting a ring and mounting it on the shaft that passes through the electromagnet. One end of the electromagnet's wire is hooked to one half of the split ring and the other end of the wire is hooked to the other half of the split ring (Figure 4.6). **Brushes**, normally made of carbon, are mounted so that they rub against the commutator and are connected to the power source (battery or other dc power source) (Figure 4.7). When power is applied to the electromagnet (also called the armature) through the brushes and commutator, a magnetic field is set up in the armature, and the armature begins to turn. Just as

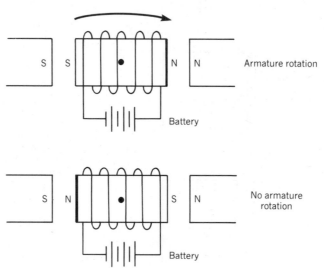

Figure 4.5   Unlike magnetic poles attract; like poles repel

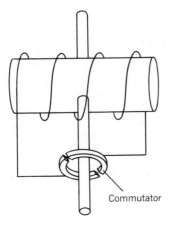

Figure 4.6   Commutator mounted on the shaft of a coil

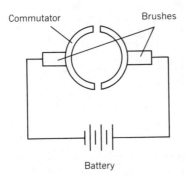

Figure 4.7   Commutator brushes

the unlike poles of the armature and the magnets align, the brushes make contact with the other half of the commutator. The poles of the armature change. We again have alignment between the north pole of the permanent magnet and the north pole of the armature as well as between the south pole of the permanent magnet and the south pole of the armature. The poles repel each other and the armature continues to turn.

    This simple motor cannot develop much torque (torque is the rotational force created by the armature). If a second electromagnet is added to the armature the motor will have higher torque because one of the electromagnets will always be in close proximity to the permanent magnets. Only one of the electromagnets in the armature is turned on at a time. Figure 4.8 shows the armature electromagnets and the sequence in which they are energized.

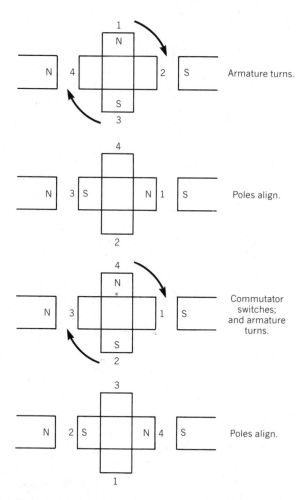

Figure 4.8 Operation of a dual-coil armature

SHUNT WOUND MOTORS

Up to this point we have been using permanent magnets for the field and a group of electromagnets for the armature as an example of a simple motor. Most dc motors use electromagnets rather than permanent magnets for the field. By application of direct current to the field coils (electromagnets), a magnetic field is created.

Figure 4.9 is a schematic diagram of a **shunt wound motor.** In a shunt wound motor, the field and the armature are wired in parallel. The field winding is connected directly across the applied voltage

Battery

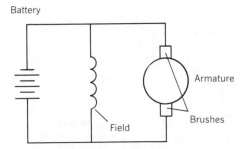

Armature

Brushes

Field

Figure 4.9 Schematic illustration of a shunt wound motor

(battery or other dc source). It is common practice to use fine wire with many turns to make up the field winding. The armature, on the other hand, has far fewer turns but a wire of sufficient size to carry the armature current.

When the armature begins to turn in the magnetic field created by the field winding, a voltage is created in the armature. The motor is now acting like a generator. Recall that we said earlier that current passing through a wire creates a magnetic field. The opposite is also true. A voltage is created when a wire is moved in a magnetic field. As the motor's armature turns in the magnetic field created by the field winding, a voltage is "induced" in the armature (Figure 4.10). The induced voltage is in opposition to the applied voltage on the armature and is called a **counter electromotive force,** or **CEMF**.

When the armature is standing still and a voltage is applied to the armature, the armature draws a high current. The resistance of the armature is low. For example, assume that the armature resistance is 2 ohms (Ω) and that the applied voltage is 24 volts (V). Then, using Ohm's law, it can be determined that the current drawn by the armature is 12 amperes (A) (Figure 4.11):

$$I = V/R$$

$$I = 24/2$$

$$I = 12 \text{ A}$$

As the armature begins to turn, the coils in the armature cut through the magnetic field created by the field winding, and CEMF is generated. The CEMF opposes the applied voltage. It is similar to connecting two batteries in series opposing (positive terminal to

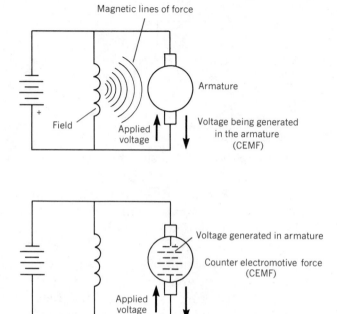

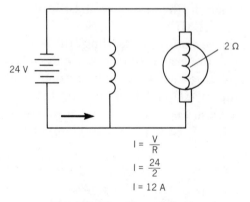

Figure 4.10 Illustration of induced voltage (CEMF) in the armature of a shunt wound motor

$$I = \frac{V}{R}$$

$$I = \frac{24}{2}$$

$$I = 12 \text{ A}$$

Figure 4.11 Voltage/current relationship for a stationary armature (no CEMF)

positive terminal) (Figure 4.12). The effective voltage then is the difference between the applied voltage and the CEMF. Referring to the previous example, the armature resistance is 2 Ω and the applied voltage is 24 V. With the armature turning, it is generating 20 V of CEMF. The effective voltage across the armature is only

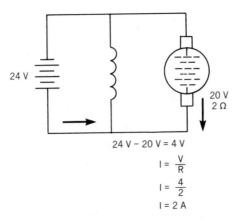

24 V

20 V
2 Ω

24 V – 20 V = 4 V

$$I = \frac{V}{R}$$

$$I = \frac{4}{2}$$

$$I = 2\ A$$

Figure 4.12   Voltage/current relationship for a turning armature (CEMF)

4 V (24 V − 20 V = 4 V). Again, substituting into the Ohm's law formula, we have:

$$I = V/R$$

$$I = 4/2$$

$$I = 2\ A$$

As you can see, the armature is drawing very little current when it is rotating because of the CEMF.

The rotational force that the armature generates when it is turning is called torque. The torque of a shunt wound motor is controlled, in part, by the armature current. The higher the armature current, the higher the torque. When a voltage is applied to the armature of a shunt wound motor while the armature is at rest, the armature current is high because there is no CEMF to limit the current. As the armature begins to turn in the strong magnetic field generated by the field winding, CEMF is generated, the armature current goes down quickly, and the torque also goes down.

A shunt wound motor will accelerate to a constant speed because of the constant field strength. The motor will continue to accelerate until the CEMF is almost equal to the applied voltage. The torque drops as the motor accelerates. If a load is applied to the motor, the speed will drop, the CEMF will drop, the armature current will go up, and the torque of the motor will increase, accelerating the load. When the torque and the load are in balance, the speed will again stabilize—often close to the original unloaded speed. This characteristic is referred to as "speed regulation." The shunt wound motor

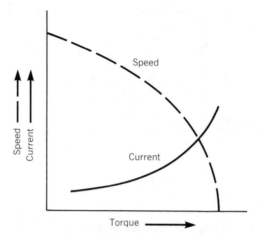

Figure 4.13   Speed-torque curve for a shunt wound motor

has good speed regulation. Figure 4.13 shows the speed-torque curve for a typical shunt wound motor.

The speed of a shunt wound motor can be controlled by controlling the field current. Figure 4.14 is a schematic diagram of a shunt wound motor with a **rheostat** (variable resistor) in series with the field coil. With the rheostat adjusted so that it is not adding any resistance to the field circuit and with a voltage applied to the armature, the motor will accelerate to a maximum speed. If resistance is added to the field circuit by adjusting the rheostat, the motor speed will increase.

The motor speed increases because the field strength is reduced, which reduces the CEMF in the armature. As the CEMF goes down, the armature current goes up and the speed increases. The more resistance that is added to the field circuit, the higher the speed. It should be noted that if too much resistance is added to the field circuit or if field current is completely lost, the motor may accelerate to such a high speed that the armature will fly apart.

The speed of a shunt wound motor can be reduced to less than its maximum speed by adding a rheostat to the armature circuit (Figure 4.15). As resistance is added to the armature circuit, the armature current is reduced, which reduces the torque, and the motor slows down.

SERIES WOUND MOTORS

At this point we have examined the operation of the shunt wound motor. The **series wound motor** is shown in Figure 4.16.

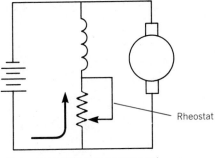

Motor accelerates to constant speed.

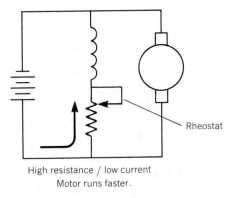

High resistance / low current
Motor runs faster.

Figure 4.14 Shunt wound motor with a rheostat in series with the field coil

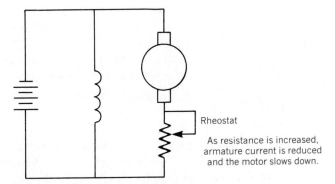

As resistance is increased,
armature current is reduced
and the motor slows down.

Figure 4.15 Shunt wound motor with a rheostat in series with the armature circuit

The construction of a series wound motor is similar but not identical to that of a shunt wound motor. The major difference is in the winding of the field. Recall that we said that the field of a shunt wound motor has many turns of thin wire. The field of a series

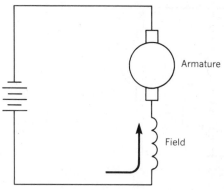

Full motor current through field

Figure 4.16 Schematic illustration of a series wound motor

wound motor is wound with a few turns of heavy wire. Heavy wire is used for the field because the field carries the full motor current.

When voltage is applied to a series wound motor, high current is drawn by the field and the armature. With a high current being drawn by the field and the armature, the torque is very high. As the armature begins to rotate, CEMF is generated. The current in the armature begins to drop. Since the armature is in series with the field, the field current also goes down.

Recall that reducing the field current in a shunt wound motor makes the motor run faster. The same is true of the series wound motor. The difference is that in a series wound motor the field current drops automatically as the motor accelerates. As the armature turns faster, the CEMF goes up and the current goes down.

As the field current goes down, the motor accelerates, the CEMF goes up, the current goes down, and the motor runs faster. If there is no load to limit the speed of the motor it will accelerate to a very high speed. This condition is called a **runaway**. Figure 4.17 shows the speed-torque curve for a series wound motor.

Series wound motors are always connected directly to a load to prevent runaways. Never connect a series wound motor to machinery using pulleys and belts. If the belts should break or slip off the pulleys the motor will run away, destroying the motor and possibly other equipment.

With this kind of danger you may wonder why anyone would choose to use a series wound motor. The series wound motor is an excellent choice when it is necessary to have very high torque at low speed. For example, series wound motors are often used for cranes and hoists, because the series wound motor has the starting torque

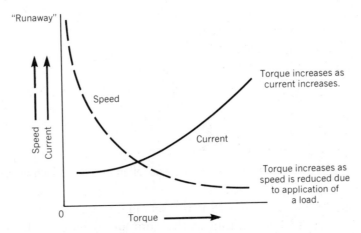

Figure 4.17 Speed-torque curve for a series wound motor

to move heavy loads. The more slowly the motor turns the higher its torque due to the low CEMF and the high armature current.

COMPOUND MOTORS

The speed regulation of a shunt wound motor is excellent. The shunt wound motor attempts to maintain a constant speed in spite of the load. With a light load or no load, the motor accelerates to a maximum speed. As the load is increased, the motor speed goes down, torque increases, and the motor attempts to return to its original speed. Because of the increased load, the motor will never return to its no-load speed, but the speed will not be substantially lower (Figure 4.13).

The speed regulation of a series wound motor is poor. If the motor is allowed to run with no load, it will continue to accelerate until it self-destructs. As a load is added to a series wound motor, its speed drops dramatically and the torque rises. The series wound motor, however, has much higher starting torque than the shunt wound motor.

As you can see, both motors have advantages and disadvantages. The ideal motor would combine the advantages of the series wound motor and the shunt wound motor. The ideal motor would have the high starting torque of a series wound motor and the speed regulation of a shunt wound motor. The **compound motor** has both a series winding and a shunt winding, and has operating characteristics that approach those of the ideal motor.

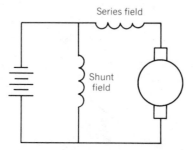

Figure 4.18   Schematic illustration of a compound motor

Figure 4.18 is the schematic drawing of a compound motor. The series winding of the compound motor can be wound to aid the magnetic field of the shunt winding. This type of motor is called a **cumulative compound motor.** The cumulative compound motor has excellent starting torque. Although its starting torque is not as good as that of a simple series wound motor, it is far superior to that of a shunt wound motor. The speed regulation of a cumulative compound motor also is acceptable, although not as good as that of a simple shunt wound motor. Figure 4.19 shows the speed-torque curve for a cumulative compound motor.

A compound motor can also be wound so that the series and shunt fields magnetically oppose each other. This type of motor is called a **differential compound motor.** A differential compound motor has lower starting torque than a cumulative compound motor, but better speed regulation. The characteristics of the differential compound motor are not much better than those of the simple shunt

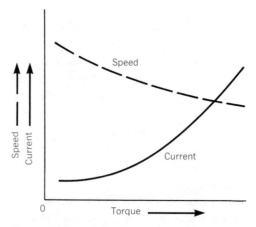

Figure 4.19   Speed-torque curve for a cumulative compound motor

wound motor. For this reason the differential compound motor has not found wide acceptance in industry.

PERMANENT MAGNET MOTORS

The motors described thus far have been constructed with an electromagnet to create the field magnetism. The **permanent magnet motor** eliminates the wound field and substitutes powerful permanent magnets to produce a magnetic field.

Permanent magnet (PM) motors are not new. The permanent magnet motor was developed years ago but did not find wide acceptance because the permanent magnets available at that time were weak and lost their magnetic properties quickly.

The development of new types of magnets such as **Alnico V magnets** and **ceramic magnets** have made the PM motor a popular choice in robots and other forms of automation.

Wound field motors have been used by industry for many years. In robot applications they have a significant disadvantage—the nonlinearity of the speed-torque curve. Figure 4.13 shows the speed-torque curve for a typical shunt wound motor. Notice that, as the speed drops due to an increase in the load, the torque increases proportionally until the speed arrives at a point where the torque no longer increases.

Figure 4.20 shows the speed-torque curve for a permanent magnet motor. Notice that the speed-torque curve is linear. As the speed of the motor decreases, the torque continues to increase.

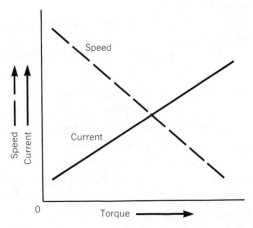

Figure 4.20   Speed-torque curve for a permanent magnet motor

Demagnetizing of PM Motors

The material used in making the magnets for a PM motor is placed in a very strong magnetic field during manufacture. The magnetizing field is created with very powerful electromagnets. Once magnetized, the material should maintain its magnetic properties for many years.

Although the magnets available today are much better, they can still be demagnetized. If the permanent magnets in a PM motor should become demagnetized, the motor cannot develop the designed torque and must be disassembled so that the magnets can be remagnetized.

Whenever current is applied to the armature of a permanent magnet motor, the permanent magnets are partially demagnetized. When the power to the armature is turned off, the magnets return to full strength. This effect can be observed when a permanent magnet motor is very heavily loaded. The motor will have an apparent loss of torque.

If the armature current is allowed to rise too high, the permanent magnets will not return to their original strength. When the magnetic field created by the armature becomes very strong due to high armature current, the armature permanently demagnetizes the field magnets. If this should occur, the only solution is to disassemble the motor and remagnetize the permanent magnets.

It is common to use magnets that are strong enough to withstand armature current at seven to eight times the rated armature current of the motor. If the rated current for a PM motor is 10 A, the motor can withstand armature currents up to 70 A before the permanent magnets begin to demagnetize.

The problem of high current and demagnetization of the permanent magnets normally occurs when the motor is operating in one direction and is suddenly reversed. To reverse a motor, the voltage applied to the armature is simply reversed. Recall that when a motor is running, a CEMF is generated. CEMF is a voltage in the direction opposite to that of the applied voltage (Figure 4.21). The CEMF limits the current. When the applied voltage is suddenly reversed, the CEMF is no longer opposing the applied voltage. The CEMF and the applied voltage are in the same direction, and the CEMF adds to the applied voltage (Figure 4.22). As a result, the armature current being drawn by the motor becomes very high. The very high armature current creates a very strong armature field which can demagnetize the permanent magnets.

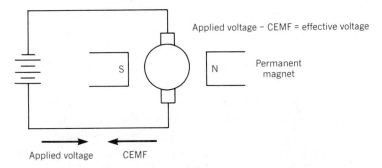

Figure 4.21 Relationship between applied voltage and CEMF in a permanent magnet motor (applied voltage − CEMF = effective voltage)

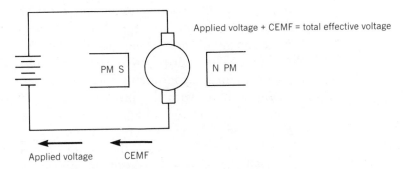

Figure 4.22 Results when motor is reversed (applied voltage + CEMF = effective voltage)

To protect the permanent magnets from being demagnetized when reversing the motor, the current must be limited. One method used by engineers is to initially apply a small reverse voltage. Since the voltage is small, the effect of the applied voltage plus the CEMF does not allow the current to reach levels at which the permanent magnets may be demagnetized. As the armature begins to slow down, the voltage is increased. When the armature comes to a stop and begins to run in reverse, the CEMF will again limit the armature current.

PRINTED ARMATURE MOTORS

One major problem with all the motors described up to this point is the construction of the armature. The armatures of these motors are constructed by wrapping wire around an armature made of a soft steel. This type of armature is heavy. When the armature is at rest (not turning), it takes time to accelerate the armature. Once the armature is turning, its mass attempts to keep it turning. This is sometimes called the "flywheel effect."

There are many applications in robotics and other forms of automation where it is necessary to bring the motor up to full speed almost instantly, then stop the motor and reverse it. In some applications this process of starting, stopping, and reversing may occur hundreds of times per second. The **printed armature motor** (Figure 4.23), also called the "disc armature motor," was designed to fill this need. This type of motor was originally made by etching copper away from a printed circuit board to form the wires of the armature, but manufacturing methods have changed. Today, the common method of producing a printed armature motor is to stamp out the wires for the armature from flat sheets of copper and then laminate many of these stampings together with insulation between the sheets. The "wires" of the copper laminations are connected together to form a continuous wire.

One problem with the printed armature motor is the difficulty of building a motor with high torque while keeping the mass of the armature low. The **inertia** of a disc increases with the fourth power of its diameter. To increase the torque, the diameter of the disc is increased. Since the inertia of the disc increases with the fourth power of its diameter, increasing the diameter of the disc rapidly increases the difficulty of quickly reversing the direction of the motor. This problem has been overcome by some printed armature motor manufacturers.

SHELL ARMATURE MOTORS

The **shell armature motor** was developed as an alternative solution to the problem of making a high-torque motor with low inertia. Figure 4.24 shows a shell armature.

The shell armature consists of an armature shaft, a commutator, and copper or aluminum wire. The wire is wound to form a hollow shell or cup. The size of the wire is determined by the current requirement of the motor. The wire is held together with epoxy and fiberglass. This forms an armature of light weight.

Production of a high-torque motor requires a larger armature. By making the armature longer rather than increasing its diameter (as was required for the printed armature motor), the inertia of the motor can be kept to a minimum.

Figure 4.23 Printed armature motor

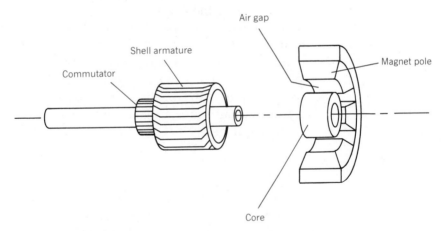

Figure 4.24 Shell armature

BRUSHLESS DIRECT-CURRENT MOTORS

The dc motors described up to this point have one common feature: they all have brushes and a commutator to transmit power to the armature.

As the armature rotates and the brushes move from one segment of the commutator to the next, the brushes arc. This can cause electrical disturbances, often called "noise." Electrical noise can cause problems in the computer that controls the robot. The arcing of the brushes also pits the commutator. As pits develop in the commutator, the brushes wear quickly.

A **brushless dc motor** has been developed to overcome this problem. The brushless dc motor is a type of permanent magnet motor. The permanent magnets are mounted on the armature shaft rather than being mounted as field magnets, and the field is wound.

Figure 4.25 is an illustration of a brushless dc motor. When the field is energized the armature rotates until the north pole of the armature is in line with the south pole of the field and the south pole of the armature is in line with the north pole of the field. At this point the armature stops rotating. Before the armature will rotate again, the magnetic field of the field winding must be reversed. This process of reversing the field is done electronically and is called **electronic commutation**.

Electronic Commutation

To control the instant at which the voltage is switched in the field, signals must be sent from the armature to the electronic control. One common method is shown in Figure 4.26. The disc mounted on the armature spins between a light and a receiver. When light strikes the receiver a signal is sent to the electronic control. When the beam of light is broken by the disc, the receiver's output sends a signal to the electronic control to switch the applied voltage.

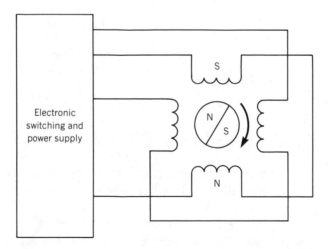

Figure 4.25   Schematic illustration of a brushless dc motor

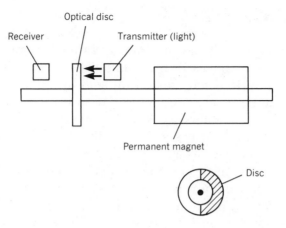

Figure 4.26 Optical-disc method of electronic commutation

Another method utilizes the **Hall effect** to sense the position of the armature. When a magnetic field passes near a semiconductor (transistor), the semiconductor's resistance goes down and more current flows. A Hall-effect device is mounted near the armature, and the magnets of the armature switch the Hall-effect device on and off. The signal from the Hall-effect device is used to switch the electronic control circuit which applies the voltage to the field winding (Figure 4.27).

The advantage of the brushless dc motor is its long life. If the electronics used for commutating the motor are properly designed, the motor life will be almost indefinite. If the motor does fail, the most likely cause would be the bearings that support the armature.

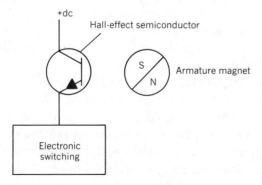

Figure 4.27 Hall-effect method of electronic commutation

TACHOMETERS

Robots respond to commands from a computer. The computer sends out signals to the manipulator drive. The electrical signals are converted into motion through hydraulic actuators, pneumatic actuators, or electric motors. After sending out such a signal, the computer monitors the resulting progress of the manipulator. When the manipulator has arrived at the desired location, the computer stops sending the signal and the manipulator stops.

The command for the manipulator to move is both a **positional command** and a **velocity command.** The manipulator must move to the correct position and at the correct speed. The device used to monitor the speed of the arm is called a **tachometer**.

A tachometer is a generator. A generator converts mechanical energy into electrical energy, whereas a motor converts electrical energy into mechanical energy.

Recall that when we began describing motors we said that current passing through a wire creates a magnetic field. During the description of motor characteristics we also said that when the armature turns in a magnetic field a voltage is produced. Suppose that the armature of a motor is turned by a crank and that a field magnetism exists, created by the permanent magnets. As the armature is turned in the magnetic field, the armature windings cut through the magnetic field, and a voltage is produced.

The tachometer (generator) is similar to a small dc motor. Recall that when we were describing the PM motor we said that its speed is directly proportional to the applied armature voltage whereas the speed of a wound field motor is not. A tachometer built with permanent magnets for the field will produce an output voltage that is directly proportional to the speed at which the armature is being turned, whereas a wound field tachometer will not produce a voltage that is directly proportional to the armature speed. For this reason, tachometers are normally built with permanent magnet fields.

Many tachometers consist of an armature, a permanent magnet field, a commutator, and brushes. The output voltage of the tachometer at a specified speed is dependent on the strength of the field and the number of turns on the armature.

Tachometers are rated by the number of volts they produce per one thousand revolutions per minute (V/krpm). Some common output voltage ratings for tachometers are 1, 2.5, 3, 5, 7, 13, 18.9, 31.5, and 50 V/krpm. This rating is called a **voltage constant.**

Now suppose that we choose a tachometer with a voltage constant of 5 V/krpm. If the armature of this tachometer is turned at 3600 rpm, what is the output voltage? Because the voltage constant is expressed in V/krpm and the motor speed is expressed simply in rpm, we must convert either the voltage constant into rpm or the speed into krpm. Although both methods will work, it is often easier to convert the motor speed into krpm. To make this conversion, we simply divide the speed in rpm by 1000. A speed of 3600 rpm equals 3.6 krpm.

Now that we have both numbers stated in the same units, all that is necessary is to multiply the speed in krpm times the voltage constant: 3.6 krpm × 5 V/krpm = 18 V. The output voltage of the tachometer at 3600 rpm is 18 V.

It is also possible to convert the output voltage of a tachometer into a speed if we are given the output voltage and the voltage constant. For example, consider a tachometer that is being turned by a motor. The output voltage of the tachometer is 9.5 V and the voltage constant of the tachometer is 7 V/krpm. All that needs to be done is to divide the output voltage by the voltage constant:

$$\frac{\text{Output voltage}}{\text{Voltage constant}} = \text{Speed in krpm}$$

$$\frac{9.5\ \text{V}}{7\ \text{V/krpm}} = 1.357\ \text{krpm} \quad \text{or} \quad 1357\ \text{rpm}$$

Since the output voltage of a tachometer is directly proportional to its speed and the output voltage can be computed for a given speed, the output voltage of a tachometer can be used by a robot's controller to monitor the speed of the robot's motors.

Figure 4.28 is a block diagram of such a control. The robot's computer outputs a number. The number is converted to a specific voltage by an electronic device called a **digital-to-analog converter (D/A).** The output voltage of the D/A is fed to another electronic device called a **comparator.** The output of the comparator is sent to an amplifier, and the output of the amplifier is sent to the motor. The armature of a tachometer is connected to the armature of

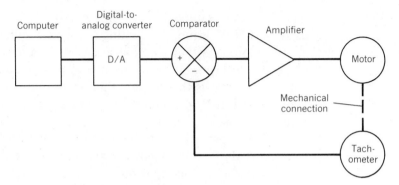

Figure 4.28   Block diagram of a robot motor speed control system

the motor so that when the motor's armature turns the tachometer's armature also turns.

The output terminals of the tachometer are wired to another input of the comparator. The comparator compares the command voltage of the D/A with the voltage of the tachometer. If the motor is turning too slowly, the voltage of the tachometer will be smaller than that of the D/A. The difference between the command voltage and the voltage of the tachometer is outputted by the comparator, amplified by the amplifier, and fed to the motor. The voltage being applied to the motor is higher, and the motor speeds up.

If the output of the tachometer is greater than the command voltage from the D/A, the comparator will send out a lower voltage to the amplifier. The output of the amplifier will be lower, and the motor will turn more slowly.

You can see that this type of system ensures that the robot's motors will always turn at the command speed. If the load that the robot is carrying increases, the robot's motor will slow down. The tachometer in conjunction with the electronics will sense this slowing and correct the speed automatically. If, on the other hand, the robot is not carrying a load, the robot's motor will run faster. Again, the tachometer in conjunction with the electronics will sense the increase in speed and automatically slow the robot's motor to the command speed.

A more complete description of the electronics and control systems used in robots can be found in subsequent chapters.

A tachometer is often built on the armature shaft of a servo motor (Figure 4.29). If the armature of the motor is constructed using a soft steel core wound with wire, the construction of the

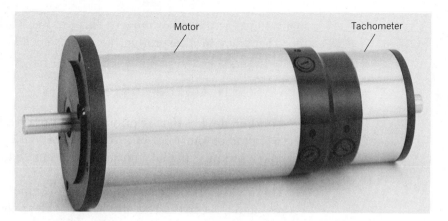

Figure 4.29 Tachometer built on the armature shaft of a servo motor

tachometer's armature will be the same. If, on the other hand, the motor's armature is a shell-type armature, the armature of the tachometer will also be of a shell-type construction.

Tachometers are also made in their own housings (Figure 4.30). These tachometers are often mounted at the joints of robots powered by hydraulics. Recall that tachometers are rated in V/krpm. If the shaft of the tachometer were connected directly to the joint of the robot, its rotational speed might be less than 1 rpm, and because it would be rotating so slowly its output voltage would be too low to use effectively. The common method of solving this problem is to mount a gear train between the joint of the robot and the shaft of the tachometer. By attaching a gear train between the joint and the tachometer shaft, the tachometer's armature is made to rotate

Figure 4.30 Tachometer

many times as the robot moves. This produces a higher voltage and more accuracy in interpreting the robot's speed.

OPTICAL ENCODERS

Another device that is often used to determine the speed of a servo motor or the joint of a robot is the **optical encoder.** Optical encoders produce digital information rather than a voltage. An optical encoder is shown in Figure 4.31.

The optical encoder has three major components: a disc, a light source, and a receiver. The disc has both clear and opaque areas and is mounted on a shaft (Figure 4.32). The light source is mounted on one side of the disc, and a receiver is mounted on the other side of the disc.

As the shaft is turned the disc turns. The clear portion of the disc allows light to pass through it and to strike the receiver, and the opaque portion of the disc stops the light from striking the receiver. When light strikes the receiver the receiver outputs a voltage, and when light is blocked from striking the receiver there is no output from the receiver.

The output from the receiver is a series of pulses (Figure 4.33). The pulses are fed to an electronic counter. The counter counts the number of pulses it receives during a short period of time. The robot designer can choose the period of time during which pulses are counted.

Figure 4.31   Optical encoder

Figure 4.32 Incremental encoder disc

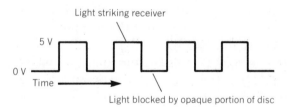

Figure 4.33 Output from an optical encoder when rotating

For example, if a robot designer chooses a disc that produces 500 pulses per revolution, the counter counts pulses for 1/10 of a second, and the counter counts 3000 pulses, what is the rotational speed of the motor? If the counter counts 3000 pulses in 1/10 of a second, then the encoder would produce 30,000 pulses in 1 second (3000 × 10 = 30,000) and the encoder would produce 1,800,000 pulses in one minute (30,000 × 60 = 1,800,000). We know that the encoder produces 500 pulses per revolution; thus, 1,800,000 pulses/minute divided by 500 pulses/revolution = 3600 rpm (1,800,000/500 = 3600).

Figure 4.34 is a block diagram of a speed control system that includes an optical encoder. The computer outputs a desired rpm,

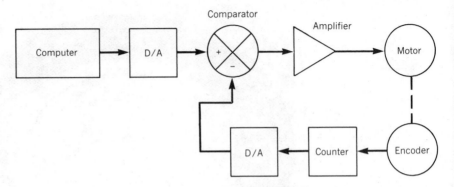

Figure 4.34   Block diagram of a speed control system incorporating an optical encoder

which the D/A converts into a voltage. This voltage is sent to a comparator. The comparator in turn feeds an amplifier, and the amplifier delivers the voltage to the motor. Connected to the motor is an optical encoder. The encoder is connected to an electronic counter. The count from the counter is fed to a D/A, which converts the count into a voltage. The output from the second D/A is connected to the other input of the comparator. If the output from the second D/A matches the output of the computer's D/A, then the motor is turning at the commanded speed. If, however, the output voltage of the second D/A is higher than the output voltage of the computer's D/A, then the motor is running too fast. The output voltage of the comparator goes down and the motor slows. If, on the other hand, the output voltage of the second D/A is lower than the output voltage of the computer's D/A, then the motor is turning too slowly. The output voltage of the comparator goes up and the motor accelerates.

For more information on optical encoders and control systems, see Chapter 9.

Hall-effect Tachometers

The Hall effect was described in the section on brushless dc motors. A Hall-effect device is mounted near the armature. As the magnetic field from the armature sweeps past the Hall-effect device, it outputs a pulse that is counted by a counting circuit. The counting circuit is similar to that described for the optical encoder.

STEPPING MOTORS

All of the motors described in this chapter have been continuous rotation motors. When a voltage is applied to the motor, the armature turns. The armature will continue to rotate until the applied power is removed.

The armature of a **stepping motor** does not continue to turn when power is applied; rather, the armature turns through a specific number of degrees and then stops. The advantage of a stepping motor over a standard motor is the ability to predict how far the shaft has rotated, whereas it is not possible to predict how far the shaft of a standard motor has rotated. The stepping motor can be connected to a digital computer. When the computer outputs a given number of pulses, the stepping motor responds by turning a specific number of degrees. Stepping motors cannot develop much torque, and for this reason they are not normally used on robots. However, they are used in many other automated machines such as NC milling machines, disk drives, and printers on computers. To understand the operation of the stepping motor we will briefly review basic motor principles.

Armature Rotation

Figure 4.35 shows a simple motor. The motor has a permanent magnet armature and a wound field. When the field is energized, the armature will rotate until the north pole of the armature is aligned with the south pole of the field and the south pole of the armature is aligned with the north pole of the field.

At this point the armature is held by the magnetic field. For the armature to continue rotating, the applied voltage to the field must be reversed. When the applied voltage is reversed, the north pole of the field becomes the south pole and the south pole of the field becomes the north pole. Because like fields are now aligned, the magnetic fields repel each other and the armature again rotates until the unlike poles are aligned.

The major problem with this design is that it is impossible to predict the direction of armature rotation. The armature may turn clockwise or counterclockwise. This problem can be eliminated by adding another set of field windings (Figure 4.36).

Figure 4.37 illustrates the operation of a four-pole stepping motor. Notice that all of the fields are energized. The south pole

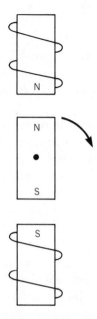

Figure 4.35 Simple stepping motor

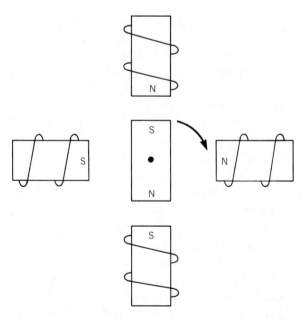

Figure 4.36 Addition of a second set of field windings ensures armature rotation in the desired direction

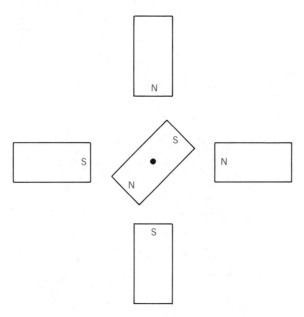

Figure 4.37   Operation of a four-pole stepping motor

of the armature is located between the two north field poles, and the north pole of the armature is equidistant between the two south field poles.

In Figure 4.38 the fields have been switched and the armature follows, turning counterclockwise. Figure 4.39 shows the fields switched again, and the armature again follows. Notice that the pair of north field poles are rotating counterclockwise and the armature is following.

Figure 4.40 shows the schematic diagram and the switch sequence for control of the motor. Notice that one switch is changed at a time. When one switch is changed, one pair of field magnets reverses and the armature rotates. Changing the switches in the order shown in this figure will cause the motor to rotate clockwise.

Figure 4.41 shows the switch sequence used to rotate the armature in a counterclockwise direction. Notice that the switching sequence is the same as that for clockwise rotation if the clockwise chart is read from the bottom up.

This type of stepping motor is called a **bipolar motor.** The bipolar motor requires a dual-voltage power supply to reverse the magnetic fields of the poles.

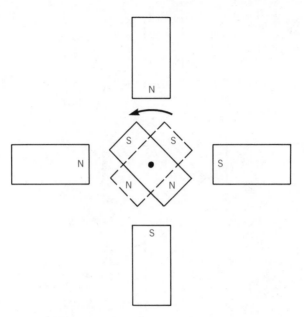

Figure 4.38   Motor shown in Figure 4.37, with poles switched (armature turns counterclockwise)

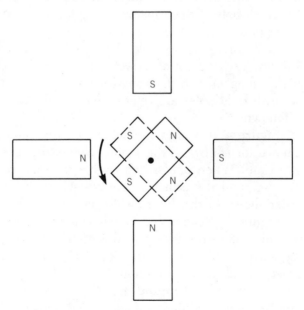

Figure 4.39   Motor shown in Figure 4.38, with poles switched again (armature continues to turn counterclockwise)

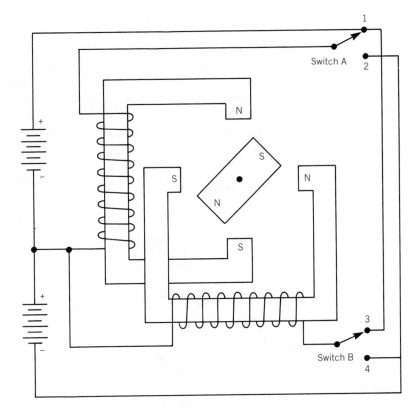

	Clockwise rotation	
Step #	Switch A	Switch B
1	1	3
2	2	3
3	2	4
4	1	4
1	1	3

Same switch positions: armature has made one complete revolution.

Figure 4.40 Schematic diagram and switch sequence (clockwise rotation) for a four-pole stepping motor

	Counterclockwise rotation	
Step #	Switch A	Switch B
1	1	3
2	1	4
3	2	4
4	2	3
1	1	3

Figure 4.41 Switch sequence for counterclockwise rotation of a four-pole stepping motor

Bifilar Stepping Motors

The **bifilar stepping motor** is shown in Figure 4.42. It is similar to the bipolar stepping motor. The difference is the center tap of the field winding, which allows the use of a single-voltage power supply. When a positive voltage is applied to the top of the vertical coil, the current flows through the coil to the center tap and to ground. Notice, however, that only one-half of the coil is active for any given switch position. When the voltage is applied to the bottom of the coil, the windings of the coil appear to be in the opposite direction. Voltage applied to the bottom half of the coil passes

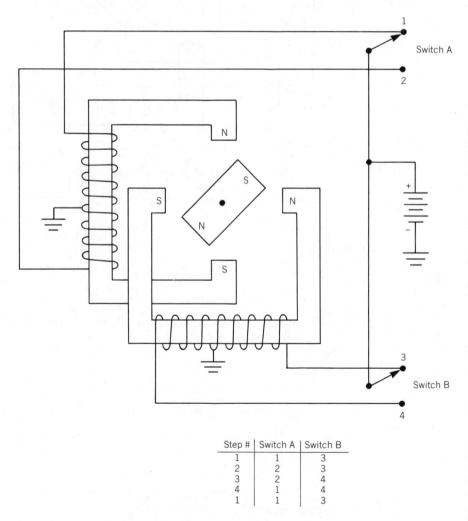

Step #	Switch A	Switch B
1	1	3
2	2	3
3	2	4
4	1	4
1	1	3

Figure 4.42   Schematic diagram and switch sequence for a bifilar stepping motor

through the coil to ground. The magnetic poles are reversed. The switching sequence of the bifilar motor is simple and follows a definite pattern. This is the same output sequence that is produced by a digital counting circuit called a **shift register.** Shift registers are described in detail in Chapter 8.

The bifilar stepping motor illustrated in Figure 4.42 is a four-pole motor. There are also eight-pole bifilar stepping motors.

Step Angle

In each of the stepping motors discussed up to this point, a single permanent magnet is used for the armature. In reality the armature of a stepping motor is made up of many poles. The more poles in the armature and the more poles in the field (also called a stator), the less the shaft rotates with each pulse from the ring counter. The amount of armature rotation with each pulse is called the **step angle.** The step angle is measured in degrees. Stepping motors are available with step angles of less than one degree.

Figure 4.43 is an illustration of the field and armature of a stepping motor. Notice that there are teeth machined into the field and into the armature. Each tooth forms a magnetic pole. The more teeth machined into the armature, the smaller the step angle.

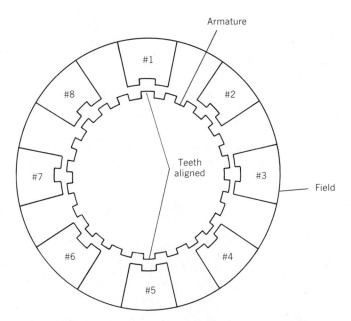

Figure 4.43 Stepping motor with teeth machined into the armature and field

There are more teeth in the armature than in the field (stator). Notice that teeth are aligned only between the poles at fields 1 and 5. When a pulse is sent to the motor, the armature advances so that teeth align at fields 2 and 6. The next pulse advances the armature so that teeth align at fields 3 and 7. The next pulse advances the armature so that teeth align at fields 4 and 8. Finally, the next pulse aligns teeth at fields 1 and 5. Each pulse moves the armature the equivalent of ¼ of a tooth. The sequence described above advances the armature counterclockwise.

The motor illustrated in Figure 4.44 has 50 teeth. Since the armature has 50 teeth and it takes 4 steps to advance one tooth, it takes 200 steps to complete one full revolution. The step angle can be computed by dividing 360 degrees by 200 steps: 360/200 = 1.8 degrees.

Variable-reluctance Stepping Motors

The stepping motors described up to this point have all had permanent magnet armatures. Another type of stepping motor is the **variable-reluctance stepping motor.**

The variable-reluctance stepping motor has a soft iron armature rather than a permanent magnet armature. Figure 4.45 is a diagram

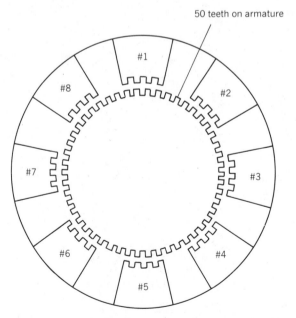

Figure 4.44   Stepping motor with 50 teeth machined into the armature

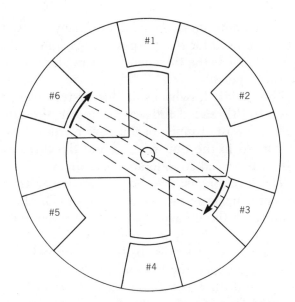

Figure 4.45   Variable-reluctance stepping motor

of a variable-reluctance stepping motor. The dotted lines illustrate the concentration of the magnetic lines of force. When poles 3 and 6 are turned on, the magnetic field builds at the poles. The soft iron armature will be pulled into alignment to complete the path for the magnetic field.

If the fields are turned on in the order shown in Figure 4.46, the motor will rotate clockwise. If the fields are turned on in the reverse order, the motor will rotate counterclockwise.

BALL BEARING SCREW DRIVES

When servo motors and stepping motors are used in robots and automated systems it is often necessary to convert the rotary motion

Step	Fields
1	3 and 6
2	2 and 5
3	1 and 4
4	3 and 6

Figure 4.46   Order in which the fields of a variable-reluctance stepping motor are turned on for clockwise rotation

of a motor into linear motion. For many years, threaded rods with nuts were used for this purpose in machine tools. If the nut is kept from turning as the threaded rod is rotated, the nut will move along the threaded rod.

The major drawbacks of this system were friction and wear. When the threaded rod was turned in the nut, the high friction produced wasted much of the power created by the motor. As the rod turned in the nut, the nut and the threaded rod would wear. This wear caused considerable inaccuracy.

These problems have been overcome with the development of the **ball bearing screw drive** (Figure 4.47). The groove in the screw is similar to the groove in a standard screw but is cut to allow ball bearings to roll freely in the groove.

The nut is cut so that balls can roll freely through the nut. The balls roll between the nut and the screw. As the balls roll between the nut and the screw they reach the end of their travel. A cage is provided for recirculation of the balls to the other end of the nut.

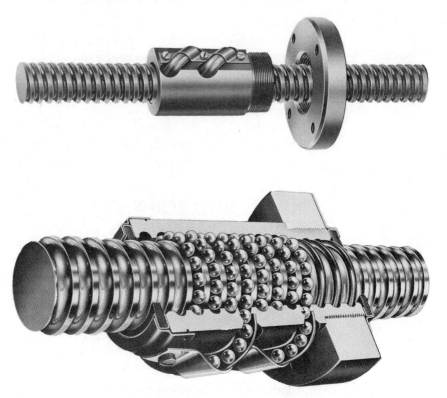

Figure 4.47   Ball bearing screw drive

The rolling friction of the ball bearing screw drive (also known as a ball screw drive) is substantially lower than that of the standard screw-and-nut construction.

Because of its obvious advantages, the ball screw drive has become popular in robots and other automated systems.

TRANSMISSIONS

The ball screw drive converts the rotary motion of a servo motor into linear motion. In addition to linear motion, many robots require rotary motion. For example, all the joints of a jointed arm robot can be driven with a rotary drive.

Most servo motors cannot be connected directly to the joint of a robot to produce rotary motion. The power developed by most servo motors is not high enough to move the arm of the robot if the shaft of the servo motor is connected directly to the robot's joint. If, on the other hand, we connect the armature of the servo motor to a **transmission** and the output of the transmission to the robot's joint, there will be sufficient power to move the robot's arm and the load that the robot is carrying.

When a transmission is used, the servo motor turns at high speed. The servo motor turns the input shaft of the transmission at high speed. Gears in the transmission reduce the speed and increase the torque. The output shaft of the transmission turns at low speed and high torque, driving the robot's arm.

The problem with a standard gear transmission is the torque developed in the gears. Since only one tooth of each gear is driving one tooth of another gear, the gears wear quickly (Figure 4.48). As the gears wear, the output shaft is driven less and less evenly. The robot's arm no longer moves smoothly, and the positioning of the arm becomes inaccurate.

Figure 4.49 shows the problem of gear wear. When the drive gear is turning, one tooth of the drive gear is putting pressure on

Figure 4.48   Gears in a standard gear transmission

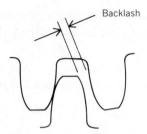

Figure 4.49 Backlash

one tooth of the driven gear. When the drive motor reverses direction, the drive gear will turn several degrees before a tooth on the drive gear comes in contact with the driven gear. This looseness in the gears is called **backlash**.

As the teeth wear, the backlash in the gears becomes worse, and as the backlash becomes worse the teeth wear even more quickly. If this situation is allowed to continue, the gears in the transmission will be destroyed. The backlash in the gears can be removed by bringing the gears closer together (Figure 4.50). As the backlash is removed, the friction between the gears is increased. The increased friction between the gears wastes power and generates heat and noise in the transmission.

Harmonic Drives

Until recently, machine designers had no alternative to the standard gear transmission when it was necessary to reduce the speed and increase the torque output of a motor. The new transmission that overcomes many of the problems of the standard transmission is called a **harmonic drive.**

The harmonic drive was developed and patented by the Harmonic Drive Division of the Emhart Machinery Group, Wakefield, Massachusetts. Figure 4.51 shows the harmonic drive. The advantages of the harmonic drive are a virtual absence of backlash, reduced gear wear, and reduced friction.

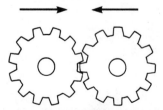

Figure 4.50   Gears pushed together to eliminate backlash

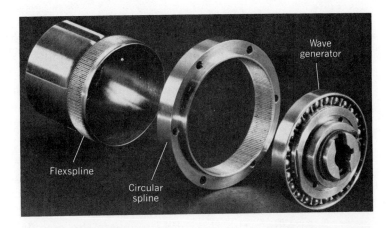

Figure 4.51 Harmonic drive

The harmonic drive consists of three major elements: the **circular spline,** the **wave generator,** and the **flexspline**.

The wave generator is an ellipse. The flexspline is a flexible cup with teeth cut on the outside diameter. The wave generator is slid into the flexspline, distorting it into an ellipse. The circular spline is a nonflexible ring with gear teeth cut on the inside diameter. The wave generator and flexspline assembly is slid into the circular spline.

The outside teeth on the flexspline engage the teeth of the circular spline. Since the flexspline is distorted into an ellipse by the wave generator, the teeth of the flexspline engage the circular spline at only two points which are 180 degrees apart.

The number of teeth on the flexspline is always less than the number of teeth on the circular spline. For our first example we will assume that the circular spline has 200 teeth and the flexspline has 198 teeth. As the wave generator is driven by a motor, the point at which the flexspline engages the circular spline shifts. For example, if the wave generator is being turned clockwise, the teeth of the flexspline will follow the wave generator, and its teeth will be engaged in a clockwise direction.

After one complete revolution of the wave generator the flexspline will have engaged 198 teeth of the circular spline. Since there are 200 teeth on the circular spline and only 198 teeth have been engaged in one full revolution, the circular spline's position has been shifted by two teeth. After 100 revolutions of the wave generator, the circular spline will have made one revolution.

The ratio of a harmonic drive is controlled by the number of teeth on the circular spline and the difference in number of teeth between the circular spline and the flexspline. For example, if the circular spline has 300 teeth and the flexspline has 298 teeth, the ratio is 300:2, or 150:1. If the circular spline has 200 teeth and the flexspline has 197, the ratio is 200:3, or 66.67:1.

By choosing different numbers of teeth for the circular spline and the flexspline, virtually any ratio is possible. The total gear reduction possible with a harmonic drive is limited only by how many teeth can be machined into the circular spline and the flexspline.

One major difference between the harmonic drive and the standard gear transmission is the lack of backlash in the harmonic drive. When the wave generator rotates, the teeth of the flexspline are forced into the teeth of the circular spline. This linear engagement of the flexspline teeth with the circular spline teeth (as opposed to the sliding friction found in a standard gear transmission) greatly reduces tooth wear, and the tightness of the engagement between the circular spline and the flexspline eliminates backlash.

The advantages of the harmonic drive have been recognized by robot designers, and it is becoming a common method of reducing the speed and increasing the torque of servo motors.

In the next chapter we will begin the study of the electronic control circuits that are used for control of robots.

QUESTIONS FOR CHAPTER 4

1. Describe the operating characteristics of a shunt wound motor.

2. What does "CEMF" mean, and what causes it in a motor?

3. Describe the operating characteristics of a series wound motor.

4. What is the advantage of a compound motor?

5. Describe the construction and the advantages of a printed armature motor.

6. Describe the construction and the advantages of a shell armature motor.

7. If a tachometer is rated at 3 V/krpm and is turning at 1800 rpm, what is its output voltage?

8. If the output voltage of a tachometer is 11 V and the tachometer is rated at 5 V/krpm, what is its speed?

9. What is the Hall effect?

10. Describe the differences between bipolar and bifilar stepping motors.

11. What is the advantage of a ball bearing screw drive?

12. What is the advantage of a harmonic drive over a standard gear transmission?

13. What is the ratio of a harmonic drive in which the circular spline has 300 teeth and the flexspline has 297 teeth?

CHAPTER FIVE

Digital Logic

The electronic circuits in robots and other automated machines must supply the commands which control the electric motors, hydraulic systems, and pneumatic systems. In addition, the electronics must store and execute the programs that control the robot and must make decisions based on information received by the robot's sensors.

The **logic circuits** in the control perform the decisionmaking functions of computer numeric control machines and robots. Logic circuits process the information stored in the memory and compare it with information received through the robot's sensors. When specific conditions are met, the control outputs a command.

The goal of the game called "twenty questions" is to guess what object your opponent is thinking of by asking questions that can be answered with either a "yes" or a "no." After asking a series of yes/no questions it is easy to identify the object that your opponent has in mind.

Logic circuits operate in much the same way. The input to the system is in the form of a "yes" or a "no." If the input is turned "on" the logic circuit is receiving a "yes" answer to its question, and if the input is turned "off" the logic circuit is receiving a "no" answer to its question. Through these on and off signals the logic circuit makes a decision.

An example of the use of a logic circuit in industry is a machine used for automatic filling of soap boxes. There are three questions that are asked before the soap dispenser fills the box: first, "Is a properly oriented box located under the soap dispenser?"; second, "Is there enough soap to fill the box?"; and third, "Is the box empty?". If the answers to all three questions are "yes," the dispenser opens and fills the waiting soap box.

Figure 5.1 is a schematic diagram of the soap box filling machine. The logic circuit consists of three switches connected together. The

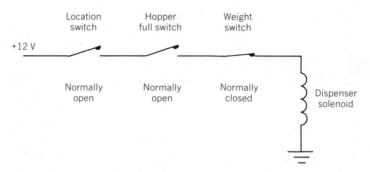

Figure 5.1   Schematic diagram of the soap box filling machine

first switch closes when a box is properly located under the soap dispenser. The second switch closes when the soap dispenser is full and opens when the level of soap in the dispenser is low. The final switch is a switch that determines whether the soap box is full or empty. If the box is empty the switch closes. When the box is filled with soap, the weight of the box and the soap forces the switch open.

The switches are wired one to the other. This is called a **series circuit.** The output of one switch is connected to the input of the following switch. The last switch is connected to a **solenoid.** A solenoid is a coil of wire wound around a hollow tube. The end of a steel plunger is placed inside the tube. When voltage is applied to the coil, a magnetic field is formed and the steel plunger is pulled into the coil of wire. One end of the steel plunger is connected to a spring loaded trap door in the dispenser. When voltage is applied to the solenoid, the plunger is pulled into the coil of wire and the door in the dispenser opens. When the voltage supply to the solenoid is cut off, the spring closes the door in the dispenser.

When a box is properly located under the dispenser, the "box in position" switch is closed. If the soap hopper is full of soap, the "hopper full" switch is also closed. Finally, if the soap box is empty, the "box empty" switch is closed. When all three switches are closed, voltage is applied to the solenoid. The solenoid opens the hopper door and soap pours into the soap box. The box will continue to fill until the weight of the box and the soap is great enough to force the "box empty" switch open. When the "box empty" switch opens, the solenoid is turned off and the dispenser door closes.

All logic circuits can be thought of as open and closed switches. However, most logic circuits used in robots and other automated machines are built using diodes, transistors, or integrated circuits. Before studying logic circuits in detail we will briefly review diodes and transistors.

DIODES

A diode is the electrical equivalent of a hydraulic check valve. The diode allows current to flow in only one direction.

Diodes are made of silicon or germanium. The silicon or germanium is chemically modified through a process called **doping**. The doping process changes the silicon or germanium from an insulator into a "semiconductor." If the doped silicon or germanium has an excess of electrons it is called an "N" type crystal, and if it has a shortage of electrons it is called a "P" type crystal. A diode is formed by fusing together a "P" type crystal and an "N" type crystal (Figure 5.2).

Forward Bias

When the diode is connected as in Figure 5.3, current flows from the positive battery terminal through the "P" material, through the "N" material, through the resistor, and back to the negative battery terminal. The diode is said to be **forward biased.** When a diode is forward biased it is a low-resistance path and allows current to flow.

Reverse Bias

If the connections to the battery are reversed as in Figure 5.4, very little current flows. The diode is said to be **reverse biased.**

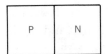

Figure 5.2 Diode

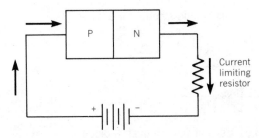

Figure 5.3 Diode connected for forward bias (full current flow)

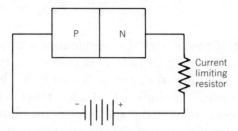

Figure 5.4 Diode connected for reverse bias (very little current flow)

Conventional Current Flow

The schematic symbol for a diode is shown in Figure 5.5. Two terms that you should remember are **anode** and **cathode**. The anode is the positive ("P") material of the diode and the cathode is the negative ("N") material of the diode.

Electrons flow from negative to positive and thus flow from the cathode to the anode. Look again at Figure 5.5: electrons are flowing in the direction opposite to that of the arrow.

The conventional theory of electricity says that electrons flow from negative to positive but that current flows from positive to negative. The diode symbol was designed with the conventional theory in mind. Current flows in the direction of the arrow. The arrow is pointing from positive to negative (Figure 5.6). It is important to remember that electrons flow from negative to positive but that the conventional theory of electricity says that current flows from positive to negative.

When analyzing electronic circuits in this book we will use this convention. This convention has been chosen because it is the most widely used theory in industry and engineering practice.

Figure 5.5 Schematic symbol for a diode

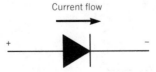

Figure 5.6 Current flow through a diode

TRANSISTORS

Transistors are similar to diodes except that they include one additional element. The **NPN transistor** (Figure 5.7) is made up of one piece of positive ("P") material "fused" between two pieces of negative ("N") material.

Another type of transistor is the **PNP transistor,** which consists of one piece of negative ("N") material "fused" between two pieces of positive ("P") material (Figure 5.8).

A transistor has a wire attached to each of its three segments (Figure 5.9). One of the outside segments is called the **emitter.** The other outside segment is called the **collector.** The middle segment is called the **base.** The degree of doping determines which outside segment is the emitter and which is the collector.

Figure 5.10 shows the schematic symbols for NPN and PNP transistors. The emitter, collector, and base are identified for each type of transistor. Notice that for each transistor there is an arrow on the emitter. The arrow on the emitter of the NPN transistor is pointing out of the transistor, and the arrow on the emitter of the PNP transistor is pointing into the transistor.

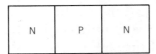

Figure 5.7   NPN transistor

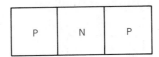

Figure 5.8   PNP transistor

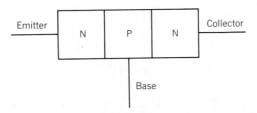

Figure 5.9   Transistor segments

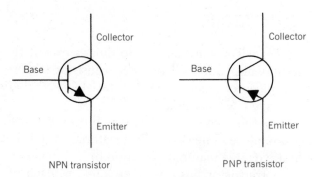

Figure 5.10   Schematic symbols for NPN and PNP transistors

Figures 5.11 and 5.12 show a circuit comprised of two batteries, a light bulb, an NPN transistor, and a switch. The positive terminal of battery #2 is connected to the bulb. The other lead from the bulb is connected to the collector of the NPN transistor. The negative terminal of battery #2 is connected to the emitter. There is another battery (#1) with its negative terminal connected to the emitter of the transistor. The positive terminal of battery #1 is connected to a resistor, the resistor is connected to the switch, and the other lead of the switch is connected to the base of the transistor. When the switch is open (off), as in Figure 5.11, there is very high resistance between the collector and the emitter of the transistor. The resistance is so high that very little current flows from battery #2 through the bulb and through the transistor. Since very little current is flowing, the bulb does not light.

In Figure 5.12 the switch is closed. With the switch closed, a current flows from the positive terminal of battery #1 through the

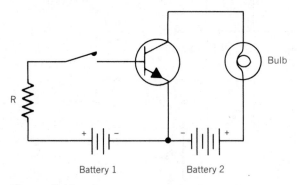

Figure 5.11   Circuit comprising two batteries, a light bulb, an NPN transistor, and a switch, with switch open (off)

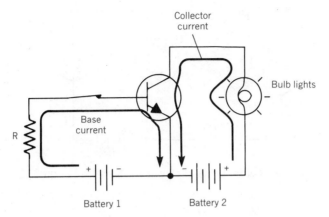

Figure 5.12 Same circuit as in Figure 5.11, but with switch closed (on)

resistor and the switch, through the base and the emitter, and back to the battery. If the resistor is of the correct value, the current flowing into the base of the transistor reduces the resistance between the collector and the emitter. With a low resistance between the collector and the emitter, current flows from the positive terminal of battery #2 through the bulb, through the collector and the emitter, and back to the negative terminal of battery #2. Since current is now flowing through the bulb, the bulb lights. If the switch is turned off, the resistance between the collector and the emitter again becomes very high, and current stops flowing into the base of the transistor. Very little current flows from battery #2 through the bulb, and the bulb goes out.

Notice that the current flowing through the base-emitter circuit and through the collector-base-emitter circuit is flowing in the direction of the arrow, just as it did in the diode.

Figure 5.13 shows the same circuit as that shown in Figures 5.11 and 5.12 except that the NPN transistor has been replaced with a PNP transistor. Notice that the positive terminals of the two batteries are both connected to the emitter. The emitter must be positive with respect to the base and the collector on a PNP transistor before current can flow in the emitter-base-collector circuit. The current flowing in the emitter-base circuit and in the emitter-base-collector circuit is flowing in the direction of the arrow.

Figure 5.14 shows another transistor circuit — the circuit of a one-transistor record player. When the needle in the phonograph cartridge rides in the groove of a record, the needle vibrates. These small vibrations generate a small voltage in the cartridge, and this

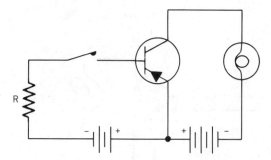

Figure 5.13   Same circuit as in Figures 5.11 and 5.12, but with PNP transistor

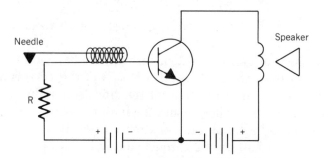

Figure 5.14   Circuit of a one-transistor record player

varying voltage is applied to the base of the transistor. The varying voltage on the base of the transistor controls the resistance between the collector and the emitter. The higher the voltage on the base of the transistor, the lower the resistance. If a speaker is connected between the battery and the collector, the vibrations of the speaker will duplicate the vibrations of the needle in the record groove. This is called "analog amplification."

When the transistor is operating as shown in Figure 5.14, the resistance of the transistor varies but does not drop to the smallest possible value. The transistor is operating as an **analog amplifier.** Analog amplifiers are used in robots and NC machines to control servo valves or motors.

On the other hand, the transistors shown in Figures 5.11 and 5.12 are switched from the highest to the lowest possible resistance between the collector and the emitter. When this resistance is as low as possible, the transistor is said to be **saturated**. When transistors are used for digital circuits they are either off or saturated.

Transistor Cut-off and Saturation

When transistors are used in digital circuits the transistors should be **cut off** (highest possible collector-emitter resistance) or **saturated** (lowest possible collector-emitter resistance).

When the voltage between the base and the emitter of a silicon transistor is zero, the transistor is cut off. As the base-emitter voltage is raised, the transistor remains cut off until a specific voltage is reached. Silicon transistors begin to conduct when the base voltage is approximately 0.5 V. As the base voltage continues to rise, the collector-emitter resistance goes down. When the base voltage reaches approximately 0.8 V, the collector-emitter resistance can go no lower, and the transistor is saturated (Figure 5.15).

When a transistor is used in a digital circuit, the base of the transistor should be kept at 0 V. Unfortunately, in the real world a small voltage from another circuit may be imposed on the base (Figure 5.16). If this voltage is 0.5 V, the transistor will start to conduct.

One method of eliminating this problem is to add a diode to the base circuit. The diode, like the transistor, does not begin to conduct until the voltage reaches a certain minimum value. For silicon diodes, this minimum voltage is approximately 0.5 V. This cut-in voltage is added to the cut-in voltage of the transistor. The transistor will not begin to conduct until the input voltage is approximately 1.0 V. This helps ensure that the transistor will not begin conducting when a small voltage is imposed on the base (Figure 5.17).

When a diode is used in the base-emitter circuit of a transistor to raise the cut-in voltage, a total of approximately 1.5 V must be applied to the diode. This will ensure that the transistor is saturated.

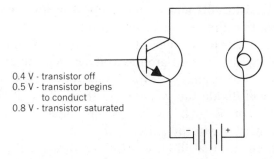

0.4 V - transistor off
0.5 V - transistor begins
       to conduct
0.8 V - transistor saturated

Figure 5.15 Transistor circuit with various input voltages

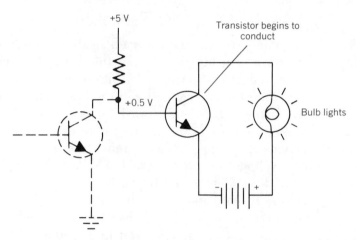

Figure 5.16   Transistor circuit with unwanted voltage on base of transistor

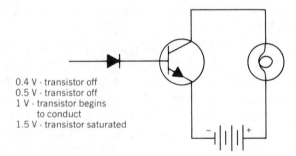

Figure 5.17   Same circuit as in Figure 5.15 with diode input

Another method of raising the input voltage of a transistor to ensure that the transistor does not begin to conduct in response to a small voltage imposed on the base is to add resistors to the input circuit. Resistors R1 and R2 in Figure 5.18 form a voltage divider. Assuming that the resistance of resistor R1 is equal to the resistance of resistor R2, the voltage at the junction of the two resistors, and therefore the base-to-ground voltage, will be one-half of the input voltage. For example, if the input voltage is 0.5 V, the voltage divider will divide the voltage and the voltage at the junction will be 0.25 V. This is far below the cut-in voltage of the transistor, and so the transistor will remain cut off.

To ensure that the transistor is saturated, the input voltage must also be much higher. If the input voltage is 1.2 V, the voltage at the junction will be 0.6 V. The transistor will be partially conducting

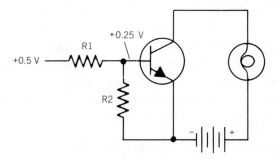

Figure 5.18   Same circuit as in Figure 5.15 with voltage divider input

and will not be saturated. To ensure that the transistor is saturated, the input voltage should be at least 1.8 V. The voltage at the junction will be 0.8 V, which is the maximum voltage that the base-emitter junction can achieve. The remaining voltage will be dropped by resistor R1. The transistor will be saturated (Figure 5.19).

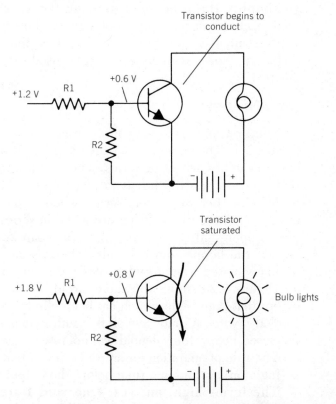

Figure 5.19   Same circuit as in Figure 5.18, but with higher input voltage

LOGIC GATES

Logic gates answer "yes/no" questions. Recall the operation of the soap box filling machine. The circuit was made up of three switches wired in series. If the soap box was in place, the hopper was full, and the soap box was empty, the soap hopper would fill the soap box. This circuit is an example of a logic gate.

Truth Tables.   A truth table is used to describe the output of a logic gate or combination of gates under every possible input condition. You should develop a facility for reading and writing truth tables, because they are one of the primary tools used by logic circuit designers and technicians. Truth tables will be used extensively in this text to help explain the operation of logic circuits.

Logic Symbols.   Logic gates are used in computers, robots, and NC machines. In fact, logic gates are used in most automated machines. To simplify the process of drawing the schematic diagrams of such machines, a standard set of logic symbols has been developed. These symbols eliminate the need to draw the complete circuit for each gate. Logic symbols will also be used extensively throughout this text.

The AND Gate

The first logic gate we will examine is the **AND gate.** Refer to Figure 5.1. The schematic of the soap box filling machine shows three switches wired in series. There is a normally open switch that senses when the box is in place under the soap hopper, and there is another switch that senses whether the soap hopper is full or empty. These two switches are wired in series. For the soap hopper to open and fill the soap box, the soap box must be in place "AND" the hopper must be full. (There is one other switch — the normally closed weight switch. We will ignore it for now.)

Figure 5.20 shows the same circuit as the one in Figure 5.1 except that the weight switch has been eliminated. The switches are labeled "Switch A" and "Switch B." Both switches (A "AND" B) must be closed before the solenoid will operate.

The rule of operation for an AND gate is that both A and B must be high (on) at the same time before the output is high (on).

The terms **high, on, yes, true,** and **1** are used to describe when an output or an input is turned "on." The terms **low, off, no,**

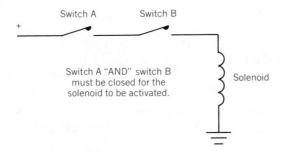

Figure 5.20 AND gate made with two switches wired in series

false, and **0** are used to describe when an output or an input is turned "off."

Figure 5.21 shows the truth table and logic symbol for a two-input AND gate. In the truth table (at left), the first two columns represent the inputs of the gate and are labeled "A" and "B." The right-hand column shows the output of the gate and is labeled "A · B = Q." This should be read as A "and" B, and not as A "times" B. In the first condition shown in the table, both the A and B inputs are off (0), and the output Q is also off (0). In the second condition, A is off (0) and B is on (1), and the output Q is off (0). In the third condition, A is on (1) and B is off (0), and the output Q is off (0). In the final condition, both A and B are on (1), and the output Q is on (1). As can be seen from this truth table, the output is turned on only when both inputs are on.

In the logic symbol shown at right in Figure 5.21, notice that only the inputs and output are shown. There are no lines indicating the supply voltage and ground connection wires necessary for powering the gate. Although the supply voltage and ground connection wires are not shown, the gate must have the proper supply voltage and must also be grounded. These connections are assumed and are not normally shown.

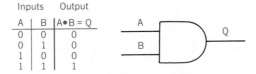

Figure 5.21 Truth table and logic symbol for a two-input AND gate

Totem Pole AND Gate

Figure 5.22 is the schematic diagram of a two-input AND gate. This circuit conforms to the truth table in Figure 5.21: the bulb is turned on only when both switches (A and B) are on.

The two NPN transistors are wired in series—that is, the emitter of transistor 1 is connected to the collector of transistor 2. With both switches off, both transistors are off and no current flows. The bulb is off.

When switch A is closed and switch B is open (Figure 5.23), transistor Q1 remains off. Since transistor Q2 is off, the resistance of the collector-emitter circuit of transistor Q2 is very high. Current does not flow through the base-emitter junction of transistor Q1 and the collector-emitter circuit of transistor Q2 because of the high resistance. Both transistors are cut off, and the bulb remains off.

When switch A is open and switch B is closed (Figure 5.24), transistor Q2 is forward biased and transistor Q1 is cut off. Since there is still a high resistance at the collector-emitter circuit of transistor Q1, the bulb does not light.

The bulb will be turned on only when both switches are closed (Figure 5.25). When switch B is closed, transistor Q2 saturates. When switch A is closed, current can flow through the base-emitter junction of transistor Q1 and through the collector-emitter circuit

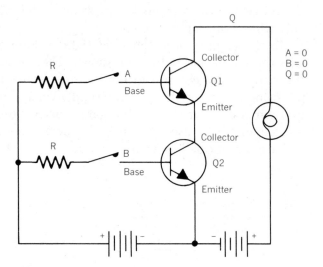

Figure 5.22 Schematic diagram of a "totem pole" AND gate with both switches open

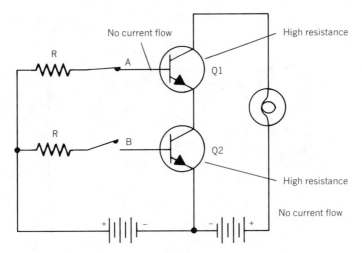

Figure 5.23   Totem pole AND gate with switch A closed and switch B open

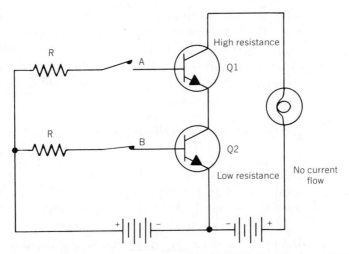

Figure 5.24   Totem pole AND gate with switch A open and switch B closed

of transistor Q2, and transistor Q1 saturates. There are now low-resistance paths in the collector-emitter circuits of both transistors. Current flows through the bulb, and it lights.

The AND gate illustrated in Figures 5.22 to 5.25 is called a **totem pole AND gate** because the transistors are stacked like the segments of a totem pole. When transistors are wired in series there is a small voltage drop through each of them even when they are

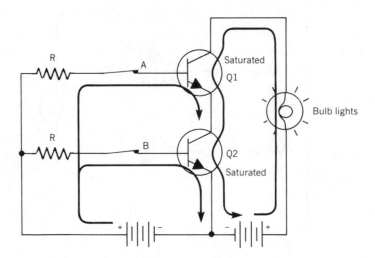

Figure 5.25   Totem pole AND gate with both switches closed

saturated. As more and more transistors are stacked in the "totem pole" the total voltage drop can become significant. For this reason many designers choose not to use the totem pole design. If, however, the totem pole design is the easiest and most efficient design for a particular application, then totem poles are normally limited to only two transistors in series.

Diode Transistor AND Gate

Figure 5.26 shows another design for an AND gate — the **diode transistor AND gate.** In this gate the inputs must be grounded in order to obtain low input or zero input. If the inputs are not connected to ground, the AND gate sees them as high (1).

In our previous examples (Figures 5.22 to 5.25), the inputs required a voltage to be applied to the input of the gate before the gate saw the input as high. A low was simply an open circuit. A low in that circuit could also be a ground.

In Figure 5.26, grounding the input makes the input low. If the input is allowed to be open it will be seen by the gate as a high. If a positive voltage is applied to the input, the gate will also see the input as a high. The chart in Figure 5.27 shows that both gates can be operated using the same type of inputs — ground is a low and a positive voltage is a high.

In analyzing the operation of an electronic circuit it is necessary to assume some set of input conditions. Given those conditions, we carefully trace through the circuit to determine the output.

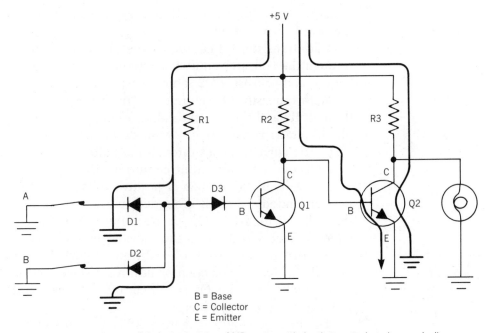

Figure 5.26   Diode transistor AND gate with both inputs low (grounded)

	Totem pole AND	Diode transistor AND	
Low	Ground or open switch	Ground or closed switch to ground	Same
High	Positive voltage or closed switch to positive voltage	Positive voltage or open switch	

Figure 5.27   Chart comparing operation of totem pole and diode transistor AND gates

The first input condition that we will discuss (Figure 5.26) is the condition wherein both inputs are low (grounded). With both inputs grounded, the current flows from the positive 5-V supply through resistor R1, through diodes D1 and D2, and to ground. Since diodes D1 and D2 are forward biased, they have 0.8 V across them. This 0.8 V appears across diode D3 and the base-emitter junction of transistor Q1. The voltage at the base-emitter junction is 0.4 V, and the transistor is cut off.

With transistor Q1 cut off, current from the supply does not flow through the collector-emitter circuit of Q1. Transistor Q2 is forward biased, and transistor Q2 saturates.

With transistor Q2 saturated, current flows from the 5-V supply through R3 and through the collector-emitter circuit of Q2. The applied voltage is essentially being dropped across resistor R3, and very little current passes through the bulb. The bulb remains off.

A second possible input condition (Figure 5.28) occurs when input A is high (either open or positive voltage) and input B remains low (grounded). With B grounded, current from the supply flows through resistor R1 and diode D2. Transistor Q1 is cut off, and the circuit operates as it did in the previous example.

A third possible input condition (Figure 5.29) occurs when input B is high (either open or positive voltage) and input A is low (grounded). Transistor Q1 is still cut off, and the circuit operates as it did in the previous examples.

The final possibility (Figure 5.30) is when both inputs are high at the same time (either open or positive voltage). With both inputs high, transistor Q1 is forward biased. Current flows from the supply through resistor R1, diode D3, and the base-emitter junction of transistor Q1. Transistor Q1 saturates.

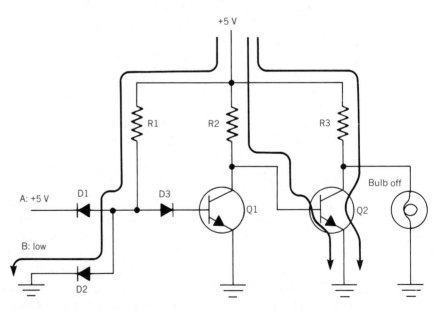

Figure 5.28 Diode transistor AND gate with input A high and input B low

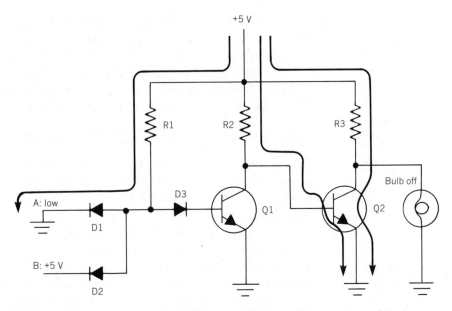

Figure 5.29 Diode transistor AND gate with input A low and input B high

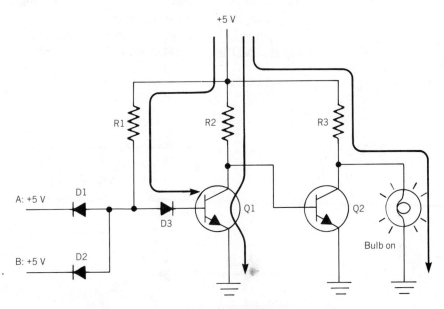

Figure 5.30 Diode transistor AND gate with both inputs high

Current also flows from the supply through resistor R2, through the collector-emitter circuit of Q1, and to ground. The supply voltage is dropped across resistor R2. Transistor Q2 is no longer forward biased and is cut off.

With transistor Q2 cut off, the collector-emitter circuit of transistor Q2 is a high-resistance path. Current flows from the supply through resistor R3 and through the bulb. The bulb lights.

The only input condition that lights the bulb is the condition wherein both inputs are high. This meets the requirements of an AND gate, and the circuit is operating properly.

AND Gates With More Than Two Inputs

All the AND gates described so far have had only two inputs. It is possible to build AND gates with many inputs. The schematic diagram of a three-input AND gate is shown in Figure 5.31. Notice that the only difference between this circuit and the one described earlier is a greater number of input diodes.

In order for the output to be high, all of the inputs must be high. If any of the diodes is grounded (low), there will be a current path through resistor R1, through the grounded diode, to ground. This will hold transistor Q1 at cut-off, and the output will be low.

In spite of the number of inputs of an AND gate, the rule of operation is the same. All inputs must be high at the same time in order for the output to be high.

Figure 5.32 shows the logic symbol and truth table for a four-input AND gate. Notice that the truth table uses 1's for highs and

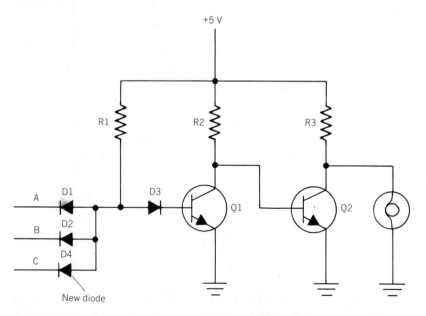

Figure 5.31 Schematic diagram of a diode transistor three-input AND gate

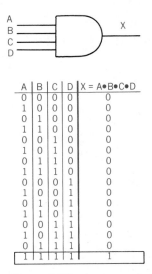

A	B	C	D	X = A•B•C•D
0	0	0	0	0
1	0	0	0	0
0	1	0	0	0
1	1	0	0	0
0	0	1	0	0
1	0	1	0	0
0	1	1	0	0
1	1	1	0	0
0	0	0	1	0
1	0	0	1	0
0	1	0	1	0
1	1	0	1	0
0	0	1	1	0
1	0	1	1	0
0	1	1	1	0
1	1	1	1	1

Figure 5.32   Logic symbol and truth table for a four-input AND gate

0's for lows. This is common practice when writing truth tables. We will use this convention when showing truth tables, but will normally refer to the inputs and outputs as highs and lows in the text.

Automated Soap Box Filler With a Three-input AND Gate

Before an automated soap box filler can fill a soap box, three conditions must be met: a box must be in place under the hopper, the soap hopper must be full, and the soap box must be empty. A three-input AND gate can be used to test for all three conditions.

The AND gate that we will choose for the automated soap box filler will recognize an input as "high" when it is open or when a positive voltage is applied. When the input is grounded, the AND gate will recognize the input as "low." Figure 5.32 is an illustration of such an AND gate.

Figure 5.33 shows the schematic diagram and truth table for an automated soap box filler. If a box is properly located under the hopper, switch A will be open. There is a positive voltage on the input, and the AND gate sees this as a high input. If a box is not properly located under the hopper, the switch will close, grounding the input (the input will be low).

The "hopper full" switch is labeled "B." When the hopper is full the switch is open, and the AND gate sees the input as high. When the hopper empties, the switch closes, grounding the input, and the AND gate sees the input as low.

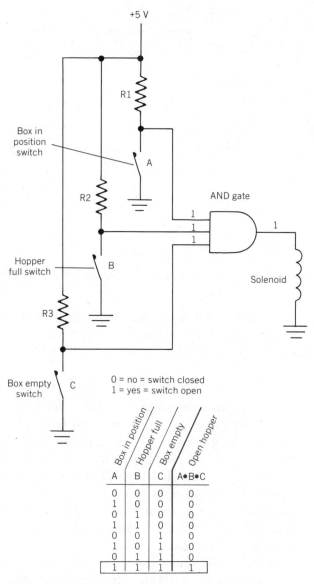

Figure 5.33   Schematic diagram and truth table for a soap box filling machine

The third condition that must be met is that the soap box must be empty. If the soap box is empty, the "box empty" (weight) switch, labeled "C," will be open. The AND gate sees this as a high input. When the box fills with soap, the weight switch closes, grounding the input, and the AND gate sees this as a low condition.

With inputs A, B, and C all high (all switches open), the AND gate's output is high. The output of the AND gate is connected to the hopper solenoid. When the output of the AND gate is high, the solenoid opens the hopper and soap fills the box.

When the soap box is full, the weight switch closes, grounding the input of the AND gate (Figure 5.34). The output of the AND gate goes low, the solenoid is turned off, and the hopper stops filling the box.

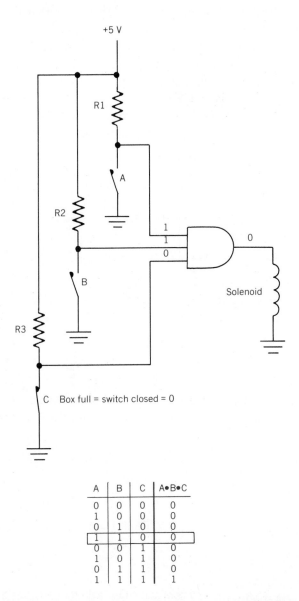

Figure 5.34 Soap box filling machine with "box full" switch closed (low)

The OR Gate

Another logic gate is the **OR gate.** A schematic diagram of a simple OR gate is shown in Figure 5.35. If either switch A "OR" switch B is closed, the bulb will light.

The logic symbol and truth table for the OR gate are shown in Figure 5.36.

The rule of operation of an OR gate is that if any or all inputs are high, the output will be high.

A transistor OR gate is shown in Figure 5.37. When both inputs A and B are low (grounded or no input voltage), transistor Q1 is cut off. The supply voltage, through resistor R1, forward biases transistor Q2. Transistor Q2 is saturated. With transistor Q2 saturated, the supply voltage is dropped across resistor R2 and the output is low.

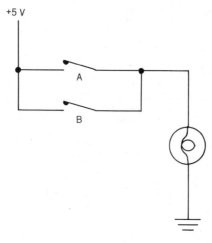

Figure 5.35   Schematic diagram of a simple OR gate made with two switches

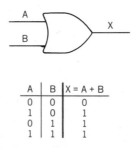

A	B	X = A + B
0	0	0
1	0	1
0	1	1
1	1	1

Figure 5.36   Logic symbol and truth table for an OR gate

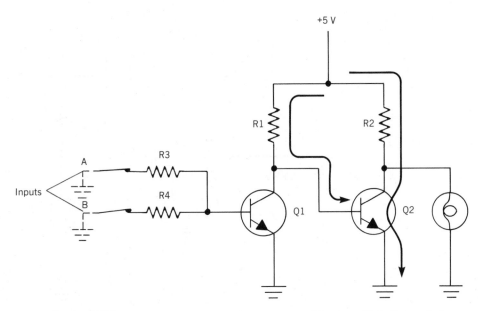

Figure 5.37 Schematic diagram of a transistor OR gate with both inputs low

If either input is high (positive voltage), transistor Q1 saturates (Figure 5.38). With transistor Q1 saturated, the supply voltage is dropped across resistor R1, removing the forward bias on transistor Q2. Transistor Q2 cuts off. With transistor Q2 cut off, the output approaches the supply voltage, turning on the bulb.

OR Gates With More Than Two Inputs

OR gates can have any number of inputs. Figure 5.39 shows a schematic diagram of a four-input OR gate. The rule for operation is the same for all OR gates regardless of the number of inputs. If any or all of the inputs are high, the output will be high.

Figure 5.40 shows the logic symbols for four- and eight-input OR gates. Notice that the basic symbol is the same. Inputs are simply added to the symbol by drawing additional lines.

Modified Automated Soap Box Filler (With OR Gate)

The automated soap box filler can be modified by adding a second soap hopper. The second soap hopper increases the capacity of the machine. The second soap hopper also includes a "hopper full" switch. If either or both hoppers are full, the soap box filler will continue to operate.

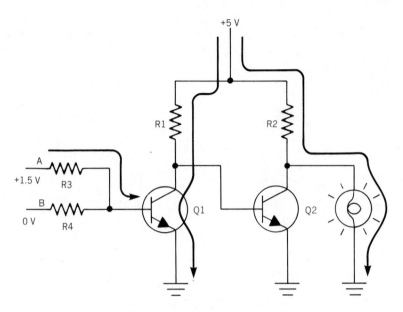

Figure 5.38 Transistor OR gate with one input (A) high

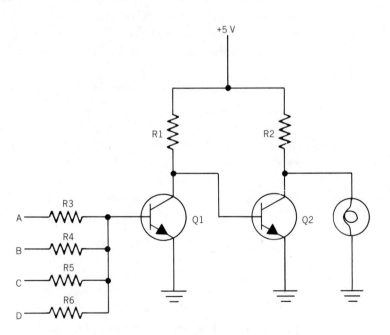

Figure 5.39 Schematic diagram of a four-input OR gate

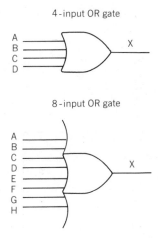

4-input OR gate

8-input OR gate

Figure 5.40 Logic symbols for four- and eight-input OR gates

The schematic diagram of the automated soap box filler with two hoppers is shown in Figure 5.41. If a soap box is in position under the hoppers, the normally closed "box in position" switch is open and input A of the AND gate is high.

The "hopper full" switch of hopper 1 is connected to one input of the two-input OR gate, and the "hopper full" switch of hopper 2 is connected to the other input of the OR gate. If hopper 1, hopper 2, or both hoppers are full, the OR gate's output will be high. The OR gate's output is connected to input B of the three-input AND gate.

The third input of the AND gate is connected to the "box full" switch, as it was in the previous examples.

If a box is in position under the hoppers, the box is empty, and either hopper 1 or hopper 2 or both are full, the hoppers will open, filling the soap box. When both hoppers are empty the output of the OR gate will be low. The B input of the AND gate will be low, and the machine will not operate.

If either hopper 1 or hopper 2 is refilled, or if both are refilled, the machine will again operate when all other conditions are met.

The NOT Gate

A simple illustration of the **NOT gate** is shown in Figure 5.42. When the switch is off (low), the bulb is on (high). When the switch is on, the supply voltage is dropped across the resistor, current flows from the supply through the resistor to ground, and the bulb is off.

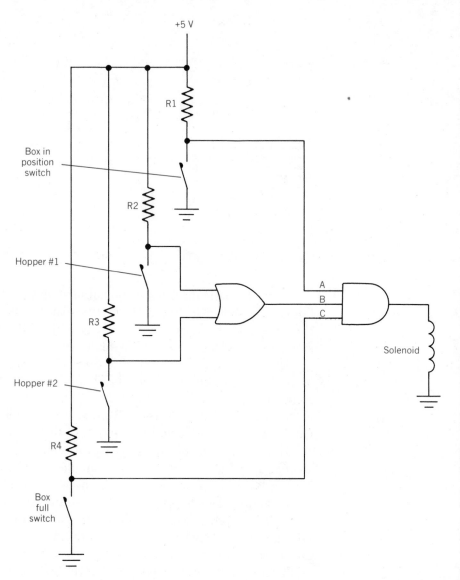

Figure 5.41   Soap box filling machine with two hoppers

Figure 5.43 is a schematic diagram of a transistor NOT gate. When the input is low, transistor Q1 is cut off, and the output is high.

Figure 5.44 is the same schematic as that shown in Figure 5.43 but with the input high. Transistor Q1 saturates. Current flows from the supply through resistor R1, through the collector-emitter circuit of Q1, and to ground. The output is low.

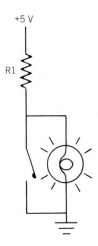

+5 V

R1

Figure 5.42 Schematic diagram of a NOT gate

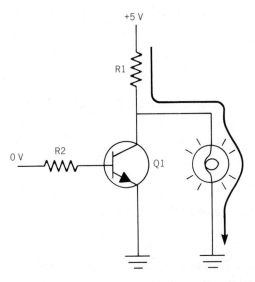

+5 V

R1

0 V   R2

Q1

Figure 5.43 Transistor NOT gate with input low

Figure 5.45 shows the logic symbol for a NOT gate. A triangle is the symbol for an amplifier (Figure 5.46). The addition of the small circle at the output indicates that the output is "inverted." If the output of an amplifier is inverted, the output is the opposite of the input: when the input is negative the output is positive, and when the input is positive the output is negative. This is exactly what a NOT gate does. If the input is low the output is high, and if the input is high the output is low. The output is "inverted."

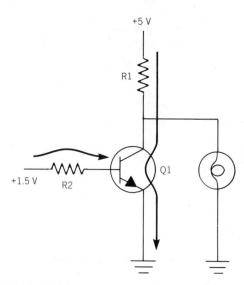

Figure 5.44   Transistor NOT gate with input high

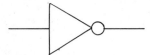

Figure 5.45   Logic symbol for a NOT gate

Figure 5.46   Logic symbol for an amplifier

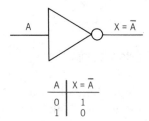

Figure 5.47   Logic symbol and truth table for a NOT gate

Notice that in Figure 5.47 the input is labeled "A" and the output is labeled "$\overline{A}$." The bar over the A means that the output is the opposite of the input (inverted). If the input of a NOT gate is labeled "A," the output is read as "NOT A."

The rule for the operation of a NOT gate is that the output is always the opposite of the input. If the input is low (0) the output is high (1), and if the input is high (1) the output is low (0).

Combination Gates

The three gates that have been discussed up to this point are the AND gate, the OR gate, and the NOT gate. All other gates are combinations of these three basic gates.

NAND Gate

One combination gate is the **NAND gate.** A NAND gate is an AND gate connected to a NOT gate (Figure 5.48). The rule of operation of a NAND gate is that all inputs must be high for the output to be low, or that if any input is low the output will be high.

Figure 5.49 shows the truth table for an AND gate and a NAND gate. Notice that the output of the NAND gate is simply the inverse of the output of the AND gate. A and B are the inputs of the AND gate. The output of the AND gate is labeled "X." The output of the AND gate is the same as the input of the NOT gate. The output of the NOT gate is the inverse of its input and is labeled "$\overline{X}$."

The logic symbol for a NAND gate is shown in Figure 5.50. Notice that the symbol is identical to that of the AND gate except for the small circle at the output. The circle indicates that the output of the AND gate is inverted.

If you have trouble remembering the truth table for the NAND gate, write the truth table for the AND gate first and then simply

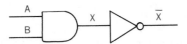

Figure 5.48   NAND gate (an AND gate followed by a NOT gate)

Inputs		AND	NAND
A	B	X	$\overline{X}$
0	0	0	1
1	0	0	1
0	1	0	1
1	1	1	0

Figure 5.49   Truth table for an AND gate and a NAND gate

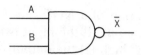

Figure 5.50   Logic symbol for a NAND gate

invert the AND gate's output. If the output is 0, write a 1; if the output is 1, write a 0.

NOR Gate

The **NOR gate** is another combination gate. The NOR gate is an OR gate followed by a NOT gate (Figure 5.51). If either input A "OR" input B is high, the output of the OR gate will be high. The NOT gate inverts the output of the OR gate, making it low. The truth table for an OR gate and a NOR gate is shown in Figure 5.52.

The rule of operation of a NOR gate is that if any input is high the output will be low. The output of the NOR gate is high only when all of its inputs are low.

The logic symbol for a NOR gate is shown in Figure 5.53. The symbol for a NOR gate is the same as the symbol for the OR gate except for the circle on the output, which indicates that the output of the OR gate is inverted.

EXCLUSIVE OR Gate

Another combination gate is the **EXCLUSIVE OR gate,** normally written as **XOR**. We will use this designation throughout this text.

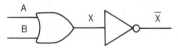

Figure 5.51   NOR gate (an OR gate followed by a NOT gate)

Inputs		OR output	NOR output
A	B	X	$\overline{\text{X}}$
0	0	0	1
1	0	1	0
0	1	1	0
1	1	1	0

Figure 5.52   Truth table for an OR gate and a NOR gate

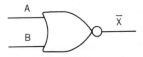

Figure 5.53 Logic symbol for a NOR gate

The XOR gate is similar to the OR gate. It follows the rule of operation of the standard OR gate except when an even number of inputs are high. Recall that when both inputs of an OR gate are high, the output is high. However, when an even number of inputs are high the output of the XOR gate is low. Figure 5.54 shows the truth table for a two-input XOR gate.

The logic symbol for an XOR gate is shown in Figure 5.55. It is the same as the symbol for the standard OR gate except for the curved line across the inputs.

The XOR gate is not a simple combination gate, but rather is a combination of several gates. One possible way of building an XOR gate is shown in Figure 5.56. The XOR gate is made by using an OR, a NAND, and an AND gate.

We will examine the operation of the XOR gate by tracing through every possible combination of inputs. First (Figure 5.56), if

A	B	X
0	0	0
1	0	1
0	1	1
1	1	0

Figure 5.54 Truth table for a two-input XOR gate

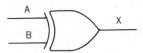

Figure 5.55 Logic symbol for an XOR gate

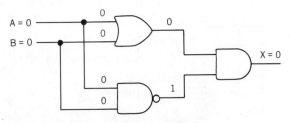

Figure 5.56 XOR gate with both inputs low

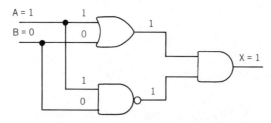

Figure 5.57   XOR gate with input A high and input B low

both input A and input B are low (0), the output of the OR gate will be low (0). Since the NAND gate is connected to the same input lines as the OR gate, its inputs will also be low (0), and since both inputs to the NAND gate are low, its output will be high (1). The inputs to the AND gate are low (0) and high (1), and its output is low. This conforms to the first condition of the truth table.

The second possibility (Figure 5.57) is for input A to be high and input B to be low. The output of the OR gate is high. With a high and a low as inputs to the NAND gate, its output is also high. Therefore, the inputs to the AND gate are both high, and its output is high.

The third possibility (Figure 5.58) is for input A to be low and input B to be high. With input A low and input B high, the output of the OR gate is high and the output of the NAND gate is also high. Both inputs of the AND gate are high, and its output is also high.

The final possibility (Figure 5.59) is for both input A and input B to be high. With both inputs to the OR gate high, its output is high. With both inputs of the NAND gate high, its output is low. Since one input to the AND gate is low and the other input is high, the output of the AND gate is low. This conforms to the final possibility of the XOR truth table.

Negated Input Gates

The AND, OR, and NOT gates are the basic gates. The NAND, NOR, and XOR gates are combination gates. There are other combi-

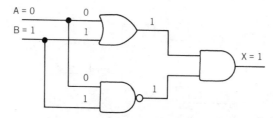

Figure 5.58   XOR gate with input A low and input B high

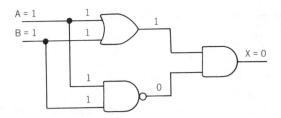

Figure 5.59 XOR gate with both inputs high

nation gates that are used by designers to simplify circuit design. Although these gates are not used as frequently as the first six gates that have been described, you will occasionally find them in circuits and should be familiar with them.

These specialized gates have some or all of their inputs inverted. A gate with inverted inputs is called a **negated input gate.**

Negated Input AND Gate

A **negated input AND gate** is an AND gate with its inputs inverted. Figure 5.60 shows two NOT gates connected to the inputs of an AND gate. The standard symbol for a negated input AND gate is shown in Figure 5.61. The symbol for a negated input AND gate is the same as the symbol for the standard AND gate except for the small circles on the inputs.

The truth tables for a negated input AND gate and for a NAND gate are shown in Figure 5.62. Many students make the mistake of thinking that inverting all the inputs of an AND gate is the same as inverting the output. Comparing the truth tables for a NAND gate

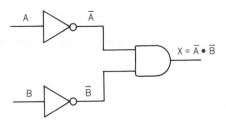

Figure 5.60 Negated input AND gate (two NOT gates followed by an AND gate)

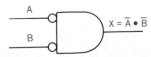

Figure 5.61 Logic symbol for a negated input AND gate

Negated input AND gate   NAND gate

A	B	X = $\overline{A} \cdot \overline{B}$
0	0	1
1	0	0
0	1	0
1	1	0

$\neq$

A	B	X = $\overline{A \cdot B}$
0	0	1
1	0	1
0	1	1
1	1	0

Figure 5.62   Truth tables for a negated input AND gate and a NAND gate

and a negated input AND gate demonstrates that in fact this is not true.

Now compare the truth table of the negated input AND gate with that of the NOR gate (Figure 5.63). The truth tables are identical. The negated input AND gate is the logic equivalent of the NOR gate.

Negated Input OR Gate

The **negated input OR gate** has inverted inputs. The logic symbol and the truth table for a negated input OR gate are shown in Figure 5.64. Compare this truth table with that of the NOR gate (Figure 5.65). The truth tables are not the same. Negating the inputs of an OR gate is not equivalent to inverting the output of an OR gate.

Now compare the truth table of the negated input OR gate with that of the NAND gate (Figure 5.66). The truth tables are

Negated input AND gate   NOR gate

A	B	X = $\overline{A} \cdot \overline{B}$
0	0	1
1	0	0
0	1	0
1	1	0

$=$

A	B	X = $\overline{A + B}$
0	0	1
1	0	0
0	1	0
1	1	0

Figure 5.63   Truth tables for a negated input AND gate and a NOR gate

$X = \overline{A} + \overline{B}$

A	B	X = $\overline{A} + \overline{B}$
0	0	1
1	0	1
0	1	1
1	1	0

Figure 5.64   Logic symbol and truth table for a negated input OR gate

Negated input OR gate   NOR gate

A	B	X = $\overline{A} + \overline{B}$
0	0	1
1	0	1
0	1	1
1	1	0

$\neq$

A	B	X = $\overline{A + B}$
0	0	1
1	0	0
0	1	0
1	1	0

Figure 5.65 Truth tables for a negated input OR gate and a NOR gate

Negated input OR gate   NAND gate

A	B	X = $\overline{A} + \overline{B}$
0	0	1
1	0	1
0	1	1
1	1	0

$=$

A	B	X = $\overline{A \bullet B}$
0	0	1
1	0	1
0	1	1
1	1	0

Figure 5.66 Truth tables for a negated input OR gate and a NAND gate

identical. The negated input OR gate is the logic equivalent of the NAND gate.

Special Gates

In addition to gates that have all of their inputs negated there are some gates that have some normal inputs and some negated inputs. Figure 5.67 shows such a gate along with its truth table. The gate's output is high when inputs A and B are high and input C is low.

Developing Truth Tables for Special Gates.   There are many special gates with some negated inputs and some normal inputs. When you find a gate like this in a circuit it is a good idea to write a truth table for the gate to eliminate confusion.

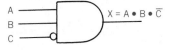

$$X = A \bullet B \bullet \overline{C}$$

A	B	C	X = $A \bullet B \bullet \overline{C}$
0	0	0	0
1	0	0	0
0	1	0	0
1	1	0	1
0	0	1	0
1	0	1	0
0	1	1	0
1	1	1	0

Figure 5.67 Logic symbol and truth table for a special gate with both negated and normal inputs

If you are having trouble writing the truth table for the special gate, draw the schematic for the gate by adding the NOT gate to the circuit (Figure 5.68). After completing the drawing, write a truth table for each gate and then combine the truth tables (Figure 5.69).

LOGIC CIRCUIT ANALYSIS

In troubleshooting of logic circuits it is often necessary to predict the output given certain inputs. If the output of the logic circuit does not correspond with the predicted output, the bad circuit has been found.

There are two methods of analyzing logic circuits. The first method is called simply **circuit analysis.** Use of this method requires that the logic levels (1's or 0's) be written next to the inputs on the logic diagram and that the output of each subsequent gate be determined.

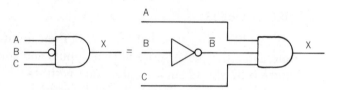

Figure 5.68   Schematic diagram of a special gate

B	$\overline{B}$
0	1
1	0

A	B	$\overline{B}$	C
0	0		0
1	0		0
0	1		0
1	1		0
0	0		1
1	0		1
0	1		1
1	1		1

Input A	Input B	Output of NOT gate	Input C	
A	B	$\overline{B}$	C	$X = A \bullet \overline{B} \bullet C$
0	0 → 1	0	0
1	0 → 1	0	0
0	1 → 0	0	0
1	1 → 0	0	0
0	0 → 1	1	0
1	0 → 1	1	1
0	1 → 0	1	0
1	1 → 0	1	0

When input A = 1, input B = 0, and input C = 1, output X = 1, because A = 1, $\overline{B}$ = 1, and C = 1 are combined in an AND gate.

Figure 5.69   Separate and combined truth tables for the special gate shown in Figure 5.68

The second method of analyzing logic circuits is a special algebraic system called **Boolean algebra.** Figure 5.70 shows a logic circuit. We will analyze this circuit using the circuit analysis method. Subsequently we will analyze several circuits using Boolean algebra.

Circuit Analysis

If we check the circuit shown in Figure 5.70 we find that the inputs have been assigned the following values: A = 1, B = 1, C = 0, D = 1, E = 1, and F = 0. With A = 1 and B = 1, the output of the AND gate G1 is 1. With C = 0 and D = 1, the output of the OR gate G2 is 1. With E = 1 and F = 0, the output of the AND gate G3 is 0.

Because the outputs of gates G1 and G2 are both 1's, the output of AND gate G4 is also 1. Since the output of gate G3 is 0, the output of the NOT gate G5 is 1. Because the output of AND gate G4 is 1 and the output of the NOT gate G5 is 1, the output of AND gate G6 is also 1.

Any logic circuit can be analyzed using this method. It is a simple matter of determining the output of each gate based on the input conditions of the gate.

Boolean Algebra

Boolean algebra is another method of predicting the output of a circuit given any combination of input conditions. Boolean algebra is a special form of algebra developed by George Boole. George Boole

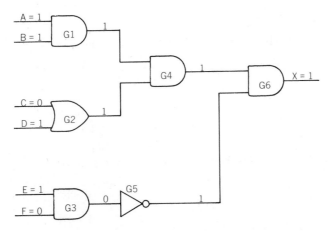

Figure 5.70 Logic circuit to be analyzed by circuit analysis (see text)

was a mathematician who lived from 1815 to 1864. He developed this system of algebra to aid the philosophers and logicians of his day in determining the most likely result given certain conditions. It was many years later that electronic engineers recognized the value of Boole's system and applied it to electronics.

AND Gate.   Figure 5.71 shows the Boolean equation for an AND gate. The inputs to the gate are designated with capital letters such as A, B, and C. The letter used to designate the output of a gate is X or Q. The output of the AND gate specified by the letter X is the result of "ANDing" the inputs. The formula is $X = A \cdot B$.

To use the equation, substitute logic values into the equation and complete the operation. For example, the first condition in the truth table for an AND gate is $A = 0$ and $B = 0$. Substitute these numbers into the equation:

$$X = A \cdot B$$
$$X = 0 \cdot 0$$
$$X = 0$$

The second possibility is $A = 1$ and $B = 0$. Substituting these values into the equation gives:

$$X = A \cdot B$$
$$X = 1 \cdot 0$$
$$X = 0$$

The third possibility is $A = 0$ and $B = 1$:

$$X = A \cdot B$$
$$X = 0 \cdot 1$$
$$X = 0$$

The final possibility is $A = 1$ and $B = 1$:

$$X = A \cdot B$$
$$X = 1 \cdot 1$$
$$X = 1$$

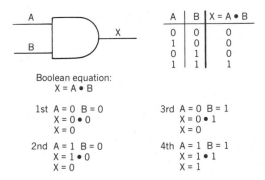

A	B	X = A • B
0	0	0
1	0	0
0	1	0
1	1	1

Boolean equation:
X = A • B

1st A = 0 B = 0
X = 0 • 0
X = 0

3rd A = 0 B = 1
X = 0 • 1
X = 0

2nd A = 1 B = 0
X = 1 • 0
X = 0

4th A = 1 B = 1
X = 1 • 1
X = 1

Figure 5.71 Boolean equation for an AND gate

All possible combinations of inputs have been considered, and the results found by substituting into the Boolean equation conform to the truth table of the AND gate.

If there are more than two inputs to an AND gate, the Boolean formula can be applied by simply adding more letters to designate the inputs. For example, the Boolean equation for a four-input AND gate is $X = A \cdot B \cdot C \cdot D$.

OR Gate. The Boolean equation for an OR gate is $X = A + B$. The equation and the truth table for an OR gate are shown in Figure 5.72. We will prove the equation just as we did earlier:

If A = 0 and B = 0, then

X = A + B

X = 0 + 0

X = 0

If A = 1 and B = 0, then

X = A + B

X = 1 + 0

X = 1

If A = 0 and B = 1, then

X = A + B

X = 0 + 1

X = 1

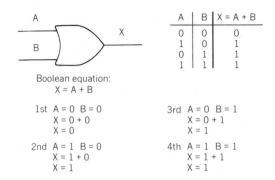

Figure 5.72 Boolean equation for an OR gate

If $A = 1$ and $B = 1$, then

$X = A + B$

$X = 1 + 1$

$X = 1$

Notice that in Boolean algebra $1 + 1 = 1$, and not 2 as in normal addition. As you know, there are only two possible input or output conditions—"on" or "off." The number 1 is used to represent "on" and the number 0 is used to represent "off." A logic gate does not have "2" as a possible output condition, and therefore the only possible answer is "1." Since this can lead to confusion, most engineers read the Boolean equation $X = A + B$ as $X = A$ "OR" B rather than as $X = A$ "plus" B.

As you know, OR gates can have many inputs. The equation for a four-input OR gate is $X = A + B + C + D$. If all inputs are high (1's), the result is $X = 1$. Remember that the "+" in Boolean algebra tells you to perform an "OR" operation and not to add the inputs together.

NOT Gate. The Boolean equation for the NOT gate is $X = \overline{A}$, often written as $A = \overline{A}$. For the NOT gate the input is inverted: if $A = 0$ then $\overline{A} = 1$, and if $A = 1$ then $\overline{A} = 0$ (Figure 5.73).

NAND Gate. The Boolean equation for a NAND gate is $X = \overline{A \cdot B}$. The bar over the $A \cdot B$ means that the output is inverted from that of a normal AND gate. Recall that a NAND gate is simply an AND gate followed by a NOT gate. The Boolean equation for an AND gate is $X = A \cdot B$. If $A \cdot B$ is the input to a NOT gate, then the output of the NOT gate will be $\overline{A \cdot B}$ (Figure 5.74).

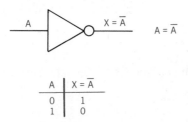

Figure 5.73   Boolean equation for a NOT gate

Figure 5.74   Boolean equation for a NAND gate

Note that the bar spans the entire expression and not simply each individual letter. The bar over the entire expression means that the inversion occurs after the "AND" operation. If the equation were written with a separate bar over each letter $(\overline{A} \cdot \overline{B})$, the inputs would be inverted before the algebraic operation is executed (Figure 5.75). This equation means that the inputs are inverted before they are ANDed. A negated input AND gate is the same as a NOR gate and is not equivalent to a NAND gate: $\overline{A} \cdot \overline{B} = \overline{A + B}$.

NOR Gate.   The Boolean equation for a NOR gate is $X = \overline{A + B}$. This equation tells us that the output of the NOR gate is simply the inversion of the output of the standard OR gate. Again it is important to recognize that the bar spans the entire "A + B" expression (Figure 5.76).

If the bar were only over the A and the B $(\overline{A} + \overline{B})$ this would mean that the inputs of the OR gate are inverted before the "OR" operation (Figure 5.77). If the inputs are inverted before the "OR" operation, the gate is a negated input OR gate and performs the operation of a NAND gate: $\overline{A} + \overline{B} = \overline{A \cdot B}$.

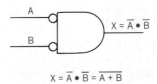

$X = \overline{A} \cdot \overline{B} = \overline{A + B}$

Figure 5.75   Boolean equation for a negated input AND gate

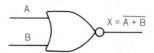

Figure 5.76 Boolean equation for a NOR gate. See Figure 5.63.

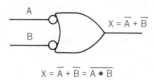

$$X = \overline{A} + \overline{B} = \overline{A \cdot B}$$

Figure 5.77 Boolean equation for a negated input OR gate

Writing Complex Boolean Equations. Now that you know the basic Boolean equations you can begin to write complex equations that describe the operations of multiple-gate logic circuits.

We will write a Boolean equation which describes the operation of the circuit shown in Figure 5.78. The output of the AND gate labeled G1 is $A \cdot B$, and the output of the AND gate labeled G2 is $C \cdot D$.

The two AND gates are connected to an OR gate (G3). The output of AND gate G1 and the output of AND gate G2 are ORed. The equation for the complete circuit is $X = (A \cdot B) + (C \cdot D)$. The formula is read "X = A AND B OR C AND D."

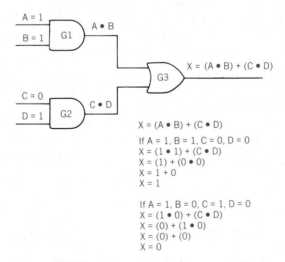

Figure 5.78 Schematic diagram and Boolean equation for a three-gate logic circuit

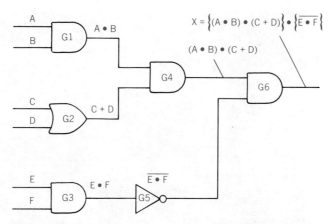

Figure 5.79 Schematic diagram and Boolean equation for a six-gate logic circuit

Examine Figure 5.79. We will use this circuit as a second example for writing Boolean equations. Inputs A and B are connected to an AND gate: the equation is A · B. Inputs C and D are ORed: the equation is C + D. The results of A · B and C + D are ANDed by gate G4: the result is $(A \cdot B) \cdot (C + D)$.

Inputs E and F are ANDed by gate G3 and then inverted by gate G5. The equation for this operation is $\overline{E \cdot F}$. When this result is combined with the equations for inputs A, B, C, and D the result is $X = \{(A \cdot B) \cdot (C + D)\} \cdot \{\overline{E \cdot F}\}$.

The use of Boolean algebra may seem difficult, but with a little practice you will find that writing a Boolean equation is easy and is a convenient way of determining the output of a complex logic circuit.

INTEGRATED CIRCUITS AND THEIR FABRICATION

The logic gates that have been described in this chapter can be built using individual transistors, resistors, and diodes. Today, however, the circuits can be purchased as complete units called **integrated circuits,** or "IC's" for short.

The integrated circuit has revolutionized the electronics industry. Thousands of transistors can now be fabricated and connected on one chip.

When IBM introduced the first computer in the early 1950's (Figure 5.80), the price was about one million dollars. The same computing power can now be fabricated on a single integrated circuit that sells for about ten dollars (Figure 5.81).

Figure 5.80   The first computer (IBM)

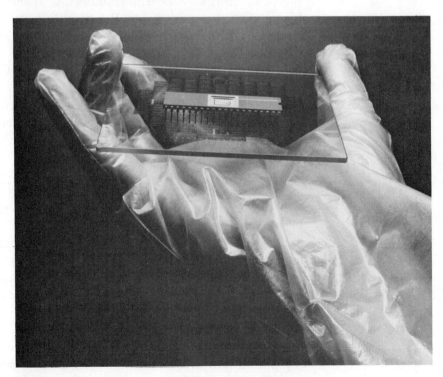

Figure 5.81   IC chip with computing power equivalent to that of the computer shown in Figure 5.80

Fabrication of IC's

The process of manufacturing an IC begins with design. The design of an IC has been simplified through the use of specialized computer programs. Previously the process of designing a chip required the designer to draw each transistor and wire to scale. Today the designer simply pushes buttons on a **CAD** (Computer Aided Design) computer terminal, and the computer makes the drawing.

After design, the next step is manufacturing. The IC chips are manufactured in very clean rooms (Figure 5.82). The room must be as dust free as possible, because one speck of dust will destroy a chip. The air in the room must contain less than 100 particles per cubic foot. This is more than 100 times cleaner than the operating room of a modern hospital.

The manufacturing process begins with a **silicon wafer** (Figure 5.83). The wafer shown is only a few inches in diameter. However, after the manufacturing process is complete the wafer can contain as many as 200 separate IC chips, and each chip in turn can contain thousands of transistors.

The first step in converting the silicon wafer into IC chips is to form a layer of **silicon dioxide** on the surface of the wafer (Figure 5.84). After the silicon dioxide has been deposited on the wafer, a layer of light-sensitive material is applied.

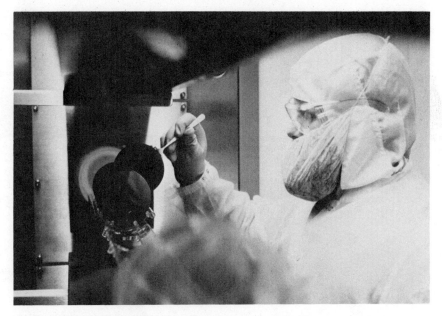

Figure 5.82 Clean room used for manufacture of IC chips

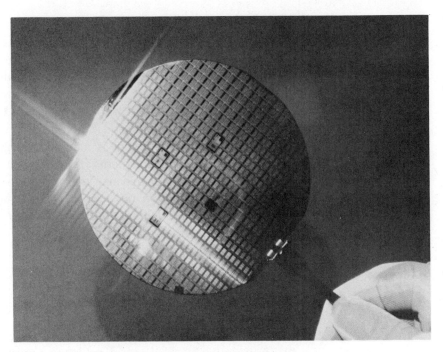

Figure 5.83   Silicon wafer used for making IC chips

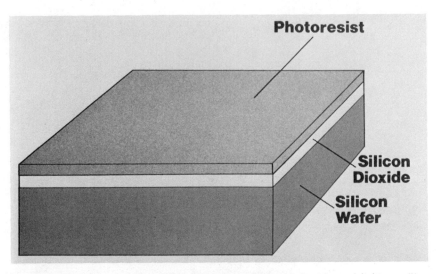

Figure 5.84   Silicon wafer after deposition of silicon dioxide and light-sensitive material (photoresist)

Next, a **photo mask** made from the designers' drawing is placed over the wafer and the wafer is exposed to ultraviolet light. The photo mask allows the light to strike the wafer in some areas but not in others.

The wafer is then "developed" in much the same way as a photograph is developed in a darkroom (Figure 5.85). The develop-

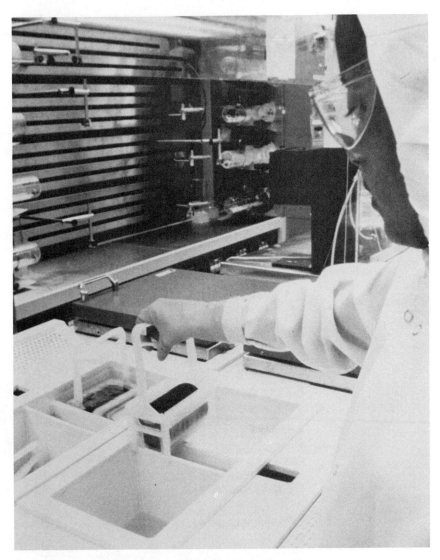

Figure 5.85 "Development" of IC wafers

ment chemically changes the areas that were exposed to the ultraviolet light. After being developed, the wafer is dipped in an acid bath. The acid etches the unexposed areas of the wafer.

Figure 5.86 is an electron microscope photograph of a wafer that has been etched with acid. The lines on the wafer are only one micron wide (a piece of notebook paper is approximately 40 microns thick).

Now a P or an N area can be formed by putting the wafer into a special oven that contains a particular type of impurity in the atmosphere. The wafer is baked in the oven at over 1000 °C. Under this extreme temperature the impurity in the atmosphere diffuses into the etched portion of the wafer, forming a P or an N area. Resistors and capacitors are formed at the same time.

By repeating the process of applying photosensitive material, exposing the wafer to ultraviolet light, developing the wafer, and baking the wafer, the transistors are formed (Figure 5.87).

The next step is to connect the various NPN and PNP transistors together with wires. First, a thin layer of aluminum is formed on the wafer (the wafer is exposed to an aluminum atmosphere, and the aluminum coats the entire wafer). The wafer is then coated with another layer of photosensitive material, exposed to ultraviolet light, and developed again.

The wafer is again placed in an acid bath, and the area that was not exposed to the ultraviolet light is eaten away, forming the wires.

Figure 5.86   Electron microscope photograph of an IC wafer after acid etching

Figure 5.88 shows the aluminum wires that are left after etching of the wafer.

The wafer is now complete. It contains hundreds of complete chips. Unfortunately, not all of the chips are good. The next step is to test the chips (Figure 5.89). Dozens of probes contact the various chips automatically, running hundreds of tests. Any circuits that fail the test are marked with a pen and discarded later.

Figure 5.87   Baking of IC wafers

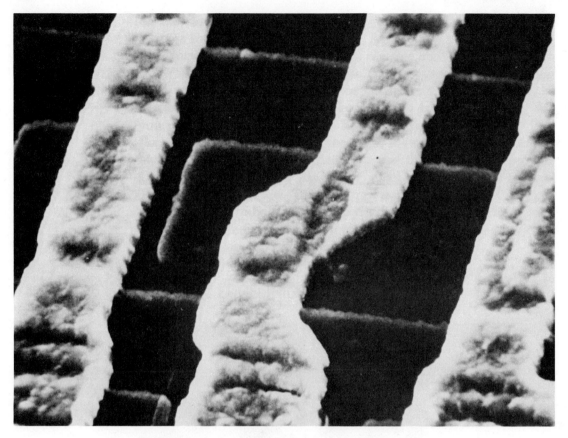

Figure 5.88 IC wafer after formation of wires

The next steps in the manufacturing process are scribing of the wafer and breaking of the wafer into individual chips. The chips that passed the test are then mounted in a case, and very fine wires are attached to these chips.

Finally, the chip is sealed in the case and given one final test to ensure that it survived the delicate processes of mounting and connection.

IC Packages

There are three different styles of cases used for mounting IC chips (Figure 5.90). These three styles are the **flat pack** (surface mounted), the **can**, and the **dual in-line package.** The dual in-line package, often called the "DIP," is the most popular.

Figure 5.89 Testing of an IC wafer

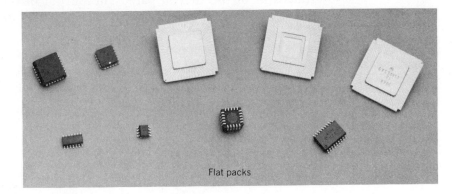

Flat packs

Cans

Dual in-line package

Figure 5.90 The three styles of IC packages

OPERATING CHARACTERISTICS OF IC'S

IC chips come in a variety of types or "families," but before studying the individual IC families we will discuss the general operating characteristics of IC's. These characteristics are **current sourcing** and **current sinking, fan-out number, wire-ANDing, hanging inputs,** and **noise immunity.**

Current Sourcing

Some families of IC's are referred to as **current sourcing** families. In current sourcing IC's, current flows out of the output when the output is logic high. The "Resistor Transistor Logic" or "RTL" family (see next section) is a current sourcing family.

Figure 5.91 shows the output of one RTL gate connected to the input of another RTL gate. When the output of the "driving gate" is high, the driving gate supplies current to the "driven gate." The current that is supplied by the driving gate forward biases the input transistor of the driven gate. The output of the driving gate is supplying the current to the driven gate. The driving gate is the "current source."

Current Sinking

Other IC families (such as the "Diode Transistor Logic" or "DTL" family: see next section) are called **current sinking** families.

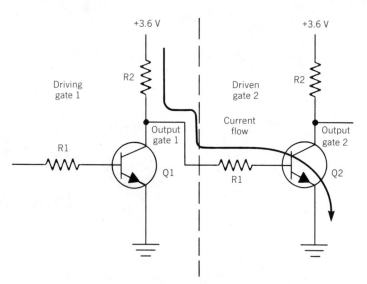

Figure 5.91   Current sourcing circuit (RTL)

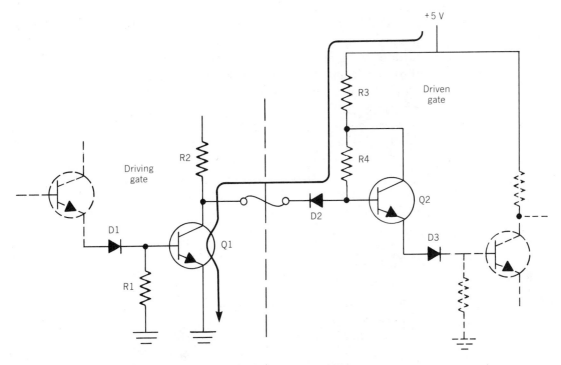

Figure 5.92 Current sinking circuit (DTL)

When the driving gate of a current sinking IC (Figure 5.92) is low, current flows from the driven gate through the collector-emitter circuit of the driving gate and to ground. This differs from current sourcing. Recall that in current sourcing IC's current flows from the driving gate to the driven gate when the driving gate is high.

Figure 5.92 shows current flowing from the driven gate to the driving gate. In this situation the driving gate is said to be a current "sink." Since a DTL IC does not supply current through its output, but rather sinks current, it is called "current sinking."

Fan-out Number

Another characteristic of IC families is the **fan-out number.** The fan-out number for an IC family is the number of inputs that a driving gate can supply. For example, if the gate is current sourcing and the fan-out number is 4, the driving gate can be connected to four different inputs and supply sufficient current for all of them to operate normally.

If the driving gate is connected to more gates than the fan-out number specifies, the circuit will operate erratically. When the driving gate is connected to more inputs than it can supply, the output voltage will drop. If the output voltage drops far enough we can no longer be assured that the input transistor will saturate. When the input transistor is operating in the middle range, it may switch on or it may be off. This situation obviously cannot be tolerated.

Figure 5.93 shows the output of one gate connected to the inputs of three other gates. The driving gate can supply sufficient current for the three inputs, and the circuit will operate properly.

The term "fan-out" also applies to current sinking logic. The fan-out number for current sinking logic specifies the number of driven gates that can be "sinked." If more gates are connected to current sinking logic than the fan-out number specifies, the driving gate may be damaged and the circuit will operate erratically.

Wire-ANDing

At times circuit designers want to connect two or more driving gates to one driven gate. This is called **wire-ANDing.** Some designers also refer to this as **wire-ORing.** Wire-ANDing is an acceptable practice for some logic families, such as RTL (Figure 5.94), but there are many logic families for which wire-ANDing is not permissible (the data sheets for the logic family in question should be checked before wire-ANDing is attempted). When an IC family cannot be wire-ANDed, an OR gate can be used to connect the driving gates to the driven gate (Figure 5.95).

Hanging Inputs

Often there are inputs to a gate that are not needed for a particular logic circuit. If these inputs are left unconnected they are called **hanging inputs.**

Hanging inputs can cause erratic operation of the circuit and cannot be tolerated. One problem with hanging inputs is their susceptibility to pickup of electrical noise. This electrical noise may be interpreted by the gate as a high input, causing the output to go high.

Another problem with hanging inputs is the way in which the logic gate may interpret the hanging input even if electrical noise is not present. If the logic gate interprets a hanging input as a high and the hanging input is on an OR gate, the OR gate's output will be locked high and the gate will be rendered useless (Figure 5.96).

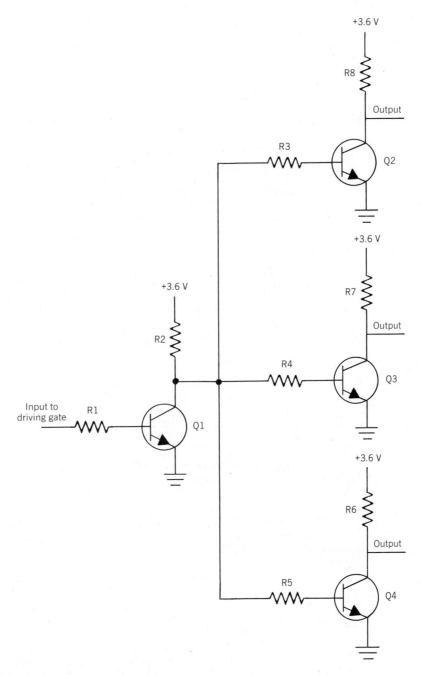

Figure 5.93 Logic circuit (RTL) with sufficient fan-out for proper operation

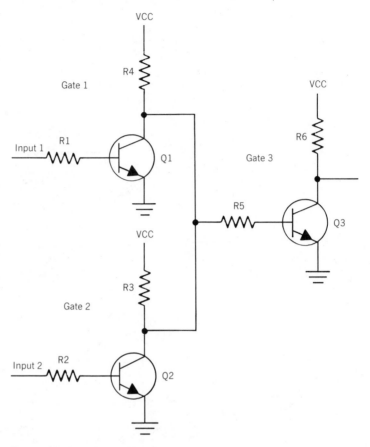

Figure 5.94   Wire-ANDing circuit (RTL)

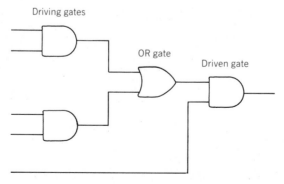

Figure 5.95   Use of an OR gate when wire-ANDing is not permissible

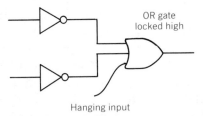

Figure 5.96 Logic circuit with OR gate output locked high due to a hanging input

In some logic families, hanging inputs are interpreted as being low. If a hanging input is interpreted as being low and is part of an AND circuit, the output of the AND gate will never go high. Even if all the used inputs go high, the AND gate will remain low if the hanging input is interpreted as being low (Figure 5.97).

Although there are instances in which hanging inputs will not affect the operation of the logic circuit, it is not good practice to leave them "hanging." In the example of the OR gate with a hanging input, the input should be connected to ground (logic low) (Figure 5.98). When the unused input is connected to ground, the OR

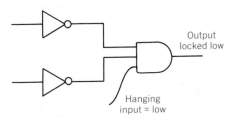

Figure 5.97 Logic circuit with AND gate output locked low due to a hanging input

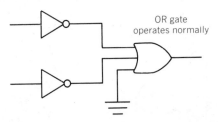

Figure 5.98 Logic circuit with OR gate hanging input connected to ground

gate recognizes it as a low input. If any one of the used inputs goes high, the output of the OR gate will go high, as intended.

In the example of an AND gate with a hanging input, good practice dictates that the hanging input be connected to the power supply (Figure 5.99). When the unused input is connected to the power supply, the gate will see it as logic high. If all of the used inputs are also logic high, the output of the AND gate will go high, and the gate will operate properly.

Another possible solution to the problem of hanging inputs is simply to connect any unused input to a used input. When a used input goes high, the unused input connected to it will also go high, and when the used input goes low the unused input will also go low. The gate will operate normally (Figure 5.100).

Noise Immunity

Another important characteristic of IC's is their **noise immunity**, which means the ability of the IC to ignore electrical noise.

Electrical noise is unwanted fluctuation of voltage or current. You have probably seen examples of electrical noise in your home. If you operate a hair dryer or other small appliance near a TV, the picture or sound will become distorted. This disturbance is caused by electrical noise.

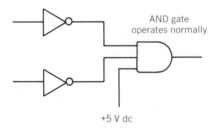

Figure 5.99   Logic circuit with AND gate hanging input connected to power supply

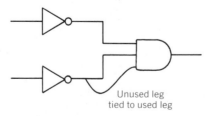

Figure 5.100   Logic circuit with hanging input connected to used input

Electrical noise that gets into a logic circuit can cause input lines of gates to be read as high when they are actually low or as low when they are actually high.

Manufacturing plants contain many possible sources of electrical noise. In fact, there is so much electrical noise in most manufacturing plants that it wasn't until recently (with the advent of high-noise-immunity IC chips and special noise shielding) that computers, in the forms of programmable controllers, robots, and other automation, were actually moved out onto the plant floor. Much of the electrical noise in manufacturing plants is caused by motors. Direct-current (dc) motors are notorious for generating electrical noise. Other sources of electrical noise are electric arc welding and the starting and stopping of large machinery. All of these sources can add up to high levels of electrical noise. Thus, if IC's are to be used in manufacturing plants, their noise immunity can be an important consideration.

Electrical noise is generated not only by external sources but also by the logic circuits themselves. The fast switching of inputs from logic low to logic high can induce voltages in other nearby input lines. Although all logic families normally have enough noise immunity to ignore internally generated noise, such noise is additive to external noise and thus can increase the likelihood of incorrect outputs.

IC FAMILIES

There are many different families of IC chips. The names of the various families describe the major components used in the chips or the special operating characteristics of the chips. Some of the names that describe the major components are **Resistor Transistor Logic (RTL)**, **Diode Transistor Logic (DTL)**, **Transistor Transistor Logic (TTL)**, and **Complementary Metal-Oxide Semiconductor (CMOS)**. The family names that describe special operating characteristics are **High Threshold Logic (HTL)** and **Emitter-Coupled Logic (ECL)**.

We will cover all of these families of IC's even though some of them have already become obsolete. When troubleshooting automated equipment, you will find examples of all of these IC families, and you should become familiar with all of them.

RTL (Resistor Transistor Logic)

The **RTL** family is one of the oldest IC families. Its name describes the basic components used in its design. Figure 5.101

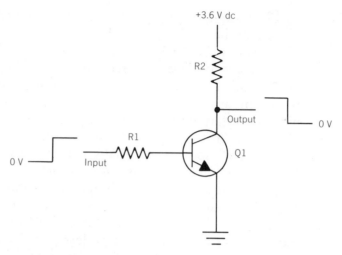

Figure 5.101   RTL inverter (NOT gate)

shows a simple RTL inverter (NOT gate). If the input is high the output will be low, and if the input is low the output will be high.

RTL is a current sourcing family. When the output of an RTL gate goes high it supplies current to bias the input transistor of the driven gate.

The fan-out numbers for most RTL's range from 4 to 10. This means than an RTL can supply current for 4 to 10 driven gates.

The noise immunity of RTL is considered to be about average. There are other IC families that are superior in noise immunity to the RTL family.

RTL IC's can be safely wire-ANDed.

DTL (Diode Transistor Logic)

The **DTL** family is another of the older IC families. The input circuits of DTL IC's contain diodes. The gate shown in Figure 5.102 is a three-input NAND gate. If all three input diodes are logic high at the same time, transistor Q1 will be forward biased and will saturate. When transistor Q1 is saturated, Q2 will also be forward biased and will also saturate. When Q2 is saturated, the output will be low.

If the inputs to a DTL gate are hanging, the input will be logic high. To bring the input to logic low, the hanging inputs must be grounded. Since DTL IC's see hanging inputs as logic high, it is possible to leave any unused input on a DTL AND gate hanging, but this is not good practice. Leaving the unused inputs hanging runs the risk of erratic operation due to electrical noise. It is much

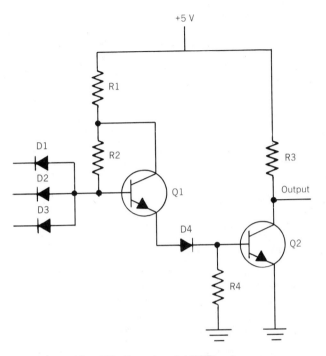

Figure 5.102 DTL three-input NAND gate

better practice to tie all unused inputs of the AND gate to the power supply or to a used input.

The unused inputs of a DTL OR gate cannot be left hanging. Since DTL IC's see hanging inputs as logic high, if any input is left hanging the output will be held high and the OR gate will be rendered useless. The unused inputs of DTL OR gates must be grounded.

DTL IC's are current sinking, have fair noise immunity, have fan-out numbers of 8 or more, and can be safely wire-ANDed.

HTL (High Threshold Logic)

The name **HTL** does not describe the components used in the IC's of this family, but rather describes an operating characteristic. In IC's of the HTL family a high input voltage is required in order for the gate to see the input as logic high. Since the input voltage needed for the input to be recognized as logic high is much higher than for other gates, HTL IC's have excellent noise immunity. The construction of HTL circuits is similar to that of DTL circuits. The major difference is the **zener diode** in the emitter leg of transistor Q1 (Figure 5.103).

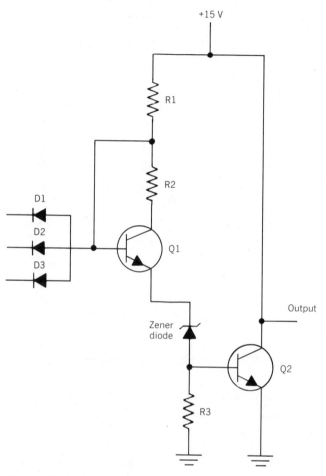

Figure 5.103   HTL circuit (note zener diode)

A zener diode is a special diode. When a zener diode is reverse biased it does not conduct. This is true of any diode. The difference is that at a specific reverse bias voltage the zener diode breaks down and begins to conduct. The breakdown voltage ("avalanche" voltage) of the zener diode used in HTL logic is 7 V. Before transistor Q1 begins to conduct, the forward bias voltage must reach 7.6 V (0.6 V for transistor Q1 and 7 V for the zener diode).

The excellent noise immunity of HTL circuits does not come free. Since the input voltage required to switch HTL gates is high, the power supply needed to drive the gates is larger than for other logic families.

The fan-out number for the HTL family of IC's is about 10.

ECL (Emitter-Coupled Logic)

The name of the **ECL** family describes the way in which the transistors are connected together. Emitter coupling of transistors and the operation of these gates in nonsaturated regions allow them to operate at very high speeds.

IC's of all the other logic families operate between cut-off and saturation. Recall the discussion of transistor operating characteristics near the beginning of this chapter. There is an operating region between the cut-in point and saturation. This operating region is used by amplifiers. Because ECL's are restricted to this operating region, they can be switched much faster than any other type of logic circuit.

Many IC families can operate at 5 to 10 million cycles per second (5 to 10 MHz). This may seem fast, but by computer standards it is quite slow. ECL circuits can operate at 120 MHz.

ECL IC's are often used for the input stages of high-speed counters. They are used to divide the high-speed incoming pulses down to a range that can be handled by other logic families.

As you may suspect, this speed does not come without some disadvantages. The major disadvantage of the ECL family is its low noise immunity.

TTL (Transistor Transistor Logic)

The **TTL** family is the most popular of all IC families. Figure 5.104 is a schematic diagram of a TTL inverter (NOT gate). Notice that the current is inputted to the gate through the emitter, rather than through the base as we have seen before. The output of TTL is of the "totem pole" type.

TTL is current sinking. When the output is logic low, current can flow from a driven gate through Q3 and to ground.

TTL can also be current sourcing. When the output is high, current flows from the supply through the collector-emitter junction of Q4 and through the diode to the output. Although TTL can supply some current, it is normally not recommended that it be operated as a current source. It is better practice to use TTL as a current sink.

The fan-out number of TTL is 10 (when used in the current sinking mode). The noise immunity of TTL is relatively high, but is not as good as that of HTL.

Generally, TTL cannot be wire-ANDed, but there is a special group of TTL gates, called "open collector" gates, that can be wire-

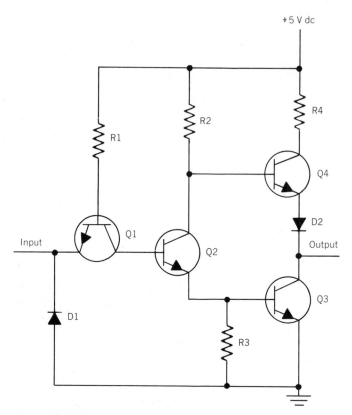

Figure 5.104 TTL inverter (NOT gate)

ANDed. Open collector gates can source relatively high current (compared to other IC families) and can be used to drive relays and other electromechanical devices without the need for additional transistor interfaces.

TTL has excellent speed. TTL can operate at about 25 million cycles per second (25 MHz). Although this is much slower than ECL, it is much faster than the other IC families. The combination of relatively high speed and relatively high noise immunity has made TTL a very popular choice.

The high speed and good noise immunity of TTL do not come without cost. The cost comes in the form of high power consumption. The current drawn by TTL is approximately 20 milliamperes (20 mA). This may not sound like very much current, but consider a circuit that has only 100 TTL's. The circuit will draw 2 A. A logic circuit that contains only 100 TTL's is not very large. There are circuits that contain thousands of TTL's, with power requirements of 30 A and more.

There is a special subgroup within the TTL family that requires far less power. The lower power group solves the problem of large power supplies, but what these IC's gain in power consumption they lose in speed. The low power group is very slow compared to the standard TTL.

Schottky TTL. The Schottky TTL is a subgroup of the TTL family. The Schottky TTL is the fastest of all TTL groups because of low current switching. Again, this speed does not come without cost. The Schottky TTL draws almost twice as much power as the standard TTL.

The fan-out number of Schottky TTL is 10. The specified fan-out is for gates in the same subfamily. The fan-out of Schottky TTL to other Schottky devices is 10. The fan-out of Schottky TTL to standard TTL is only about 5. The specification literature from TTL manufacturers specifies the fan-out within the same subfamily and often gives alternate fan-outs for other subfamilies.

CMOS (Complementary Metal-Oxide Semiconductor)

The **CMOS** family is one of the newest IC families. It has become established very quickly as a major family because of its special operating characteristics.

The first major difference between CMOS and all other IC families is that CMOS uses field-effect transistors rather than bipolar transistors. Many good books are available that describe the field-effect transistor, and we will not describe its special operating characteristics here. In this section we will discuss the advantages of CMOS and the special handling that it requires.

The power consumption of CMOS is only a few microamperes, whereas the power consumption of TTL is measured in milliamperes. The extremely low power consumption of CMOS reduces the size of the required power supplies.

Proper operation of TTL gates requires a power supply between 4.8 and 5.2 V. In fact, if the power supply voltage varies by more than a few tenths of a volt the TTL will operate erratically. CMOS, on the other hand, has a very wide operating range. CMOS will operate properly from 4.5 to 18 V. This greatly simplifies the design of power supplies.

The advantage of the wide operating voltage range of CMOS is partially offset by its low speed. CMOS is limited to approximately 5 MHz, whereas TTL can operate in the 25-MHz range.

Another concern is the special handling that CMOS requires. The CMOS family is very sensitive to static electricity. If a static electricity discharge is fed to the pins of a CMOS chip it will burn out.

To guard against damage due to static electricity, CMOS manufacturers ship the devices in special antistatic packages. CMOS devices should be left in their special packages until ready for use.

When handling CMOS devices, be careful not to touch the pins. The static electricity that builds up on your body can discharge through a CMOS chip and destroy it.

It is also important that the power be turned off before a CMOS chip is removed from its socket. If you remove the chip with the power still on, the chip will most likely burn out.

Even with all these special handling considerations, CMOS has become very popular. Most robots and other forms of automation use a combination of CMOS and TTL chips.

In the next chapter we will cover flip-flops, adders, counters, and displays. With these circuits and the logic gates that we have studied in this chapter we will design a basic pick-and-place robot.

QUESTIONS FOR CHAPTER 5

1. What is meant by the term "saturation"?

2. Draw the symbols and write the truth tables for the AND gate, the OR gate, and the NOT gate.

3. Draw the symbols and write the truth tables for the NAND gate, the NOR gate, and the XOR gate.

4. Draw the schematic described by the equation $X = (A \cdot B) + (C \cdot D)$.

5. Describe the process of making an IC chip.

6. What is meant by the term "current sinking"? Use schematics to illustrate your answer.

7. What is meant by the term "fan-out"?

8. How does a hanging input affect the operation of an AND gate?

9. What is meant by the term "noise immunity"?

10. What is the major advantage of ECL logic?

11. How does Schottky TTL differ from standard TTL?

12. What is the major advantage of CMOS?

CHAPTER SIX

Flip-Flops

In Chapter 5 we examined the operation of simple logic circuits. One characteristic that all of these logic gates have in common is a lack of memory: the inputs to these gates must be active to affect the output. Recall that all inputs to an AND gate must be high for the output to be high. When any of the inputs drops low, the output immediately drops low. Also, if one input of an AND gate goes high and then drops low, followed by the other input going high and then dropping low, the output of the AND gate will never go high. The AND gate does not have a memory. It does not remember that an input was high and does not keep the output high after the inputs have gone low.

This lack of memory also applies to the other simple logic gates, including NAND, NOR, OR, NOT, and XOR gates. There are many times when temporary memory is important in a logic circuit. The circuit that is used for temporary memory is called a **flip-flop** (it is also called a "bistable multivibrator").

RELAY FLIP-FLOP

Figure 6.1 shows a relay with its contact wired in parallel with a momentary contact switch. The momentary contact switch is pushed, closing the contact, and the relay coil is energized, which closes the relay contact. When the momentary contact switch is released, the relay remains energized.

Notice that the symbol for a relay contact is the same as the symbol for a capacitor. It is unfortunate that this ambiguity exists. When relays are included in a circuit, remember that these symbols on the schematic diagram stand for relay contacts and not for capacitors.

The relay "remembers," by remaining energized, that the momentary contact switch was pushed. To turn the relay off, the

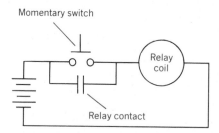

Figure 6.1   Simple relay circuit

main power must be interrupted. When the main power is interrupted, the relay contacts open. When the power is restored, the relay remains off until the momentary contact switch is pressed again.

An improvement in this circuit is shown in Figure 6.2. In this circuit there are two relays wired in parallel. Wired in series with the SET relay coil are the normally closed contact of the RESET relay (R1) and the normally open contact of the SET relay (S2). Wired in series with the RESET relay are the normally closed contact of the SET relay (S1) and the normally open contact of the RESET relay (R2).

When the SET button is pushed, the SET relay coil is energized. This closes the normally open contact S2 and opens the normally closed contact S1. When the SET button is released, the SET relay remains energized.

When the RESET button is pushed, the RESET relay is energized. The normally open contact R2 closes, and the normally closed contact R1, which is wired into the SET relay circuit, opens. This

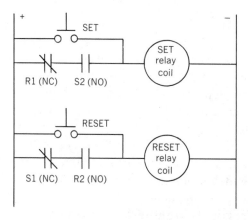

Figure 6.2   Relay flip-flop with SET and RESET relays

opens the SET relay circuit, and the SET relay contacts return to their normal positions (contact S1 returns to its normally closed position and contact S2 returns to its normally open position).

Recall that contact R2 is closed because the RESET relay coil is being energized by the RESET button, and that contact S1 has returned to its normally closed position. When the RESET button is released, the RESET relay remains energized.

If the SET button is pushed again, all the contacts reverse. The RESET relay turns off and the SET relay turns on. The SET relay remains on until the RESET button is pushed again, at which time all the contacts reverse. The SET relay turns off and the RESET relay turns on.

The relays turn on and off when the SET and RESET buttons are pushed. The relays are "flipping and flopping" back and forth, and this circuit is called a **relay flip-flop.** The SET relay remembers that the SET button was pushed and the RESET relay remembers that the RESET button was pushed.

TRANSISTOR R-S FLIP-FLOP

Figure 6.3 shows a transistor version of the relay flip-flop. The operation of the **transistor flip-flop** is identical to that of the relay flip-flop. Momentarily pressing the SET button forward biases the SET transistor, and the SET transistor becomes saturated. The 5-V supply voltage is essentially dropped across resistor S1. The RESET transistor is cut off. The forward bias for the SET transistor is supplied through resistors R1 and S2. The output Q is high.

When the RESET button is momentarily pushed, the RESET transistor is forward biased and saturates (Figure 6.4). The supply voltage is dropped across resistor R1. This removes the forward bias from the SET transistor, and the SET transistor is cut off. With the SET transistor off, current flows from the supply through resistors S1 and R2, through the base-emitter junction of the RESET transistor, and to ground. The RESET transistor is saturated and remains saturated even after the RESET button is released. The output (Q) is low.

The SET transistor will remain cut off and the RESET transistor will remain saturated until the SET button is pushed. When the SET button is pushed, the SET transistor is forward biased and saturates (Figure 6.3). The forward bias of the RESET transistor is removed, and the RESET transistor is cut off. When the SET

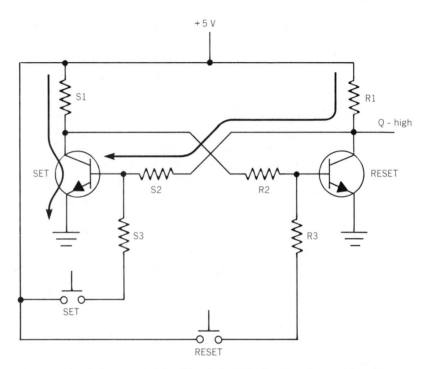

Figure 6.3   Transistor R-S flip-flop with SET transistor forward biased

button is released, the SET transistor will remain saturated until the RESET button is pushed. This type of flip-flop is called a **RESET-SET flip-flop** or an **R-S flip-flop.**

NOR R-S FLIP-FLOP

The R-S flip-flop can also be built using two NOR gates. A **NOR R-S flip-flop** is shown in Figure 6.5. Assume that the R (RESET) input is momentarily high. The output Q is low. Both inputs to the SET NOR gate are low, and therefore the output of this gate is high. This high output is connected to one input of the RESET NOR gate. The output of the RESET NOR gate is low. The flip-flop will remain in this state until the S (SET) input is brought high.

When the SET input is momentarily brought high (Figure 6.6), the output of the SET NOR gate goes low. The output of the SET NOR gate is connected to one input of the RESET NOR gate. With both inputs of the RESET NOR gate low, its output goes high. This holds the output of the SET NOR gate low. The flip-flop will remain in this state until the RESET input is brought high again.

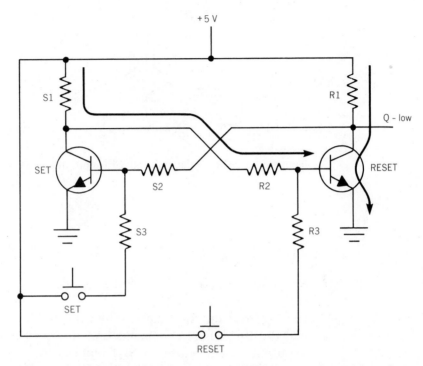

Figure 6.4 Transistor R-S flip-flop with RESET transistor forward biased

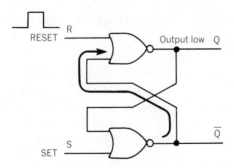

Figure 6.5 NOR R-S flip-flop with RESET input momentarily high

Notice that this circuit has both a Q and a $\overline{Q}$ output. The $\overline{Q}$ output is always the opposite of the Q output. If the Q output is 0 the $\overline{Q}$ output is 1, and if the Q output is 1 the $\overline{Q}$ output is 0. When there are both a Q and a $\overline{Q}$ output, the outputs are said to be **complements**. Most flip-flops have complementary outputs.

The truth table for a NOR R-S flip-flop is shown in Figure 6.7. When the SET and RESET inputs are both low, the output of the

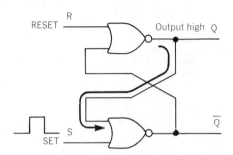

Figure 6.6   NOR R-S flip-flop with SET input momentarily high

R	S	Q	$\overline{Q}$
0	0	No change	
0	1	1	0
1	0	0	1
1	1	Disallowed	

Figure 6.7   Truth table for a NOR R-S flip-flop

flip-flop remains the same. When the SET input is brought high and the RESET input remains low, the Q output goes high. If the Q output is high before the SET input is brought high, the Q output simply remains high. Bringing the SET input high will "set" the output high. Once the output of the R-S flip-flop is "set" high it will remain high even after the SET input returns to a low condition. To switch output Q to the low condition the flip-flop must be "reset." When the RESET input is brought high while the SET input remains low, the output goes low.

There is one other possible condition: both SET and RESET inputs high at the same time. The SET input is attempting to drive the Q output high while the RESET input is attempting to drive the Q output low. This is an unstable condition and should not be allowed to occur in a digital circuit.

The logic symbol for a NOR R-S flip-flop is shown in Figure 6.8. This symbol shows the SET and RESET inputs as well as the Q and $\overline{Q}$ outputs.

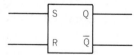

Figure 6.8   Logic symbol for a NOR R-S flip-flop

NAND R-S FLIP-FLOP

Another form of the R-S flip-flop is built using two NAND gates. The **NAND R-S flip-flop** is shown in Figure 6.9.

The operation of the NAND flip-flop differs from that of the NOR flip-flop. The truth table for the NAND flip-flop is shown in Figure 6.10. Notice that the disallowed condition for a NAND flip-flop is the condition wherein both the R and S inputs are low. The flip-flop does not change state when the R and S inputs are both high.

To set the Q output high, the S input is set low and the R input is set high. To reset the flip-flop (Q low), the R input is set low and the S input is set high. All of these conditions are the opposites of those for the NOR flip-flop.

The NAND flip-flop is normally written as $\overline{\text{R-S}}$. The logic symbol for an $\overline{\text{R-S}}$ flip-flop is shown in Figure 6.11. Notice that the inputs have bubbles on them. Recall that a logic-gate symbol with a bubble at the input indicates that the input is inverted. The same is true of a flip-flop. The bubbles on the inputs tell you that the inputs are active when they are low rather than when they are high.

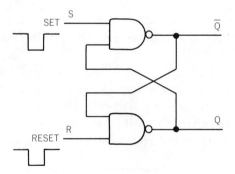

Figure 6.9 NAND R-S flip-flop

S	R	Q	$\overline{Q}$
0	0	Disallowed	
1	0	0	1
0	1	1	0
1	1	No change	

Figure 6.10 Truth table for a NAND R-S flip-flop

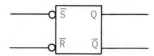

Figure 6.11 Logic symbol for a NAND R-S flip-flop

CLOCKED R-S FLIP-FLOP

The **clocked R-S flip-flop** is the same as the standard R-S flip-flop except for the addition of two AND gates (Figure 6.12). Recall that both inputs of an AND gate must be high for the output to be high. The SET and RESET inputs are not active until the input labeled "Clock" is also high. While the clock input is high, the clocked R-S flip-flop operates the same as a standard R-S flip-flop. At the instant the clock goes low, the flip-flop is frozen. If the Q output is high when the clock goes low the output will remain high, and if the Q output is low when the clock goes low the output will remain low. While the clock is low, the R and S inputs can continue to switch without any effect on the output of the flip-flop.

Clock Circuits

A **clock circuit** generates a regular string of pulses (Figure 6.13). The purpose of a clock circuit is to control the timing of a flip-flop or other digital circuit.

If an R-S flip-flop is connected to a clock circuit (the clock input of the flip-flop is connected to the output of the clock circuit), the output of the flip-flop will change only while the clock input is high. For many flip-flops it is important that the outputs switch at the same time and at regular intervals in spite of the fact that the

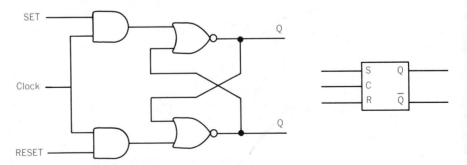

Figure 6.12 Clocked R-S flip-flop circuit and logic symbol

Figure 6.13 String of pulses generated by a clock circuit

inputs switch at random. This is easily accomplished by using clocked flip-flops.

Clock circuits will be studied in detail later in this chapter. For now, just remember that a clock circuit generates a regular string of pulses and that such pulses are used to control the timing of many digital circuits.

Timing Diagrams

A **timing diagram** shows when the output of a clocked digital circuit changes and provides an easy means of analyzing the operation of a logic circuit. Figure 6.14 is an example of a timing diagram. Timing diagrams will be used throughout this text to illustrate the operation of logic circuits and control circuits.

Examine the timing diagram in Figure 6.14. At a point in time that we have labeled T1, the clock goes high. The S and R inputs are both low, and Q is low also (these initial conditions are arbitrary). At a later point in time, labeled T2, the clock goes high again. At the same instant the clock goes high the S input also goes high. This brings the Q output high. Q remains high until time T3. At T3, the clock input is high and the R input goes high. Since the clock is high, at the instant that R goes high the flip-flop resets and Q goes low. At time T4, the S input again goes high; however, the clock is low. The flip-flop ignores the S input, and Q remains low.

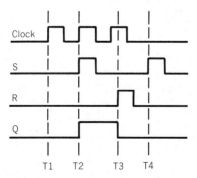

Figure 6.14 Timing diagram for a clocked R-S flip-flop

D LATCH

The clocked R-S flip-flop can be modified so that the output of the flip-flop will follow the input while the clock input is high. This circuit is called a **D latch** and is shown in Figure 6.15. There is only one input—the D (data) input. The D input is connected to the S input of an R-S flip-flop. The D input is also connected to the R input through an inverter. If the clock input is high and the D input goes high, the Q output goes high. If the clock input is high and the D input goes low, the inverter (NOT gate) brings the R input high, and the Q output goes low. When the clock input goes low the flip-flop is frozen, and the Q output will remain in the same state until the clock input again goes high.

The rules of operation of the D latch are as follows: (1) when the clock input is high and the D input is high, the Q output is high; (2) when the clock is high and the D input is low, the Q output is low; and (3) whatever condition exists at the D input (high or low) at the instant the clock goes low will be transferred to the Q output and will be retained until the clock goes high again.

Figure 6.16 helps to illustrate the operation of a D latch. At T1, the D input goes high, the clock input (C) is low, and the Q output remains low. At T2, the clock goes high while the D input remains

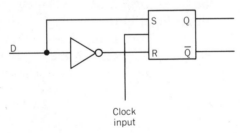

Clock
input

Figure 6.15   D latch

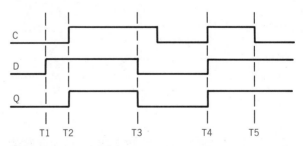

Figure 6.16   Timing diagram for a D latch

high. Since both the clock and the D input are now high, Q also goes high. At T3, the D input drops low while the clock remains high. The Q output follows the D input and goes low. At T4, the clock and the D input go high at the same time. The Q output follows the D input and goes high. At T5, the clock drops low. D is high at the instant the clock drops low, and so Q remains high.

Figure 6.17 is another timing diagram that further illustrates the operation of a D latch. At T1, the clock is high and the D input goes high. Q follows D and goes high. At T2, the D input drops low while the clock remains high, and the Q output goes low. At T3, the D input again goes high while the clock remains high. Q follows D and goes high. At T4, the clock drops low. The D input is high at the instant the clock drops low, and so Q remains high. At T5, the D input drops low. The clock is low at T5, and so the change in D has no effect on Q, and Q remains high. Between T5 and T6, the D input changes several times. Since the clock is low at this time, the output is not affected by the changes in D. At T6, the clock again goes high, and the D input is low. Since D is low, Q changes (Q goes low). At T7, the D input goes high while the clock remains high. Q follows D and goes high. At T8, the D input goes low while the clock remains high. Q goes low.

The D latch is often referred to as "transparent during high clock" since the Q output follows the D input when the clock is high. Whatever condition exists at the D input is transferred to the output when the clock is high.

POSITIVE EDGE TRIGGERED D FLIP-FLOP

Any time the clock input of a D latch is high, the latch is "transparent." The information present at the D (data) input is trans-

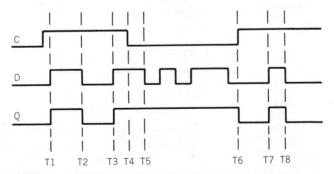

Figure 6.17 Timing diagram further illustrating the operation of a D latch

mitted to the Q output. The **edge triggered D flip-flop** is more selective than the D latch: the D input can be transferred to the Q output only at a specific point in time.

Figure 6.18 shows an edge triggered D flip-flop. The clock input is fed through a resistor-capacitor network. When the clock first goes high, current flows through the capacitor. As the capacitor charges, the charging current decreases. The values of the capacitor and the resistor are chosen so that the capacitor charges very rapidly. This circuit is known as a **differentiator**.

Figure 6.19 shows the response of the resistor-capacitor network to a clock input. When the clock goes high, the output of the resistor-capacitor network immediately goes high (positive). The capacitor very quickly charges, and the output of the resistor-capacitor network drops to zero even though the clock is still high. When the clock drops low, the output of the resistor-capacitor network goes negative and then quickly returns to zero.

Figure 6.20 is a schematic diagram of a D latch with a resistor-capacitor network connected to the clock input. Since the D input is transferred to the Q output only when the clock input is high, and

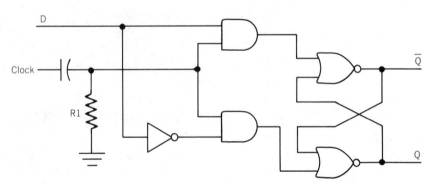

Figure 6.18   Positive edge triggered D flip-flop

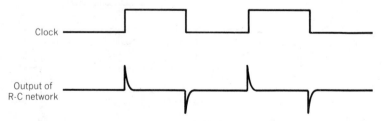

Figure 6.19   Response of a resistor-capacitor network to a clock input

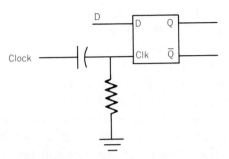

Figure 6.20   D latch with resistor-capacitor network connected to clock input to form a positive edge triggered D flip-flop

since the clock input is high for only a fraction of a second while the capacitor is charging, the D input is transferred only when the clock first goes from low to high. Once the initial spike output from the resistor-capacitor network has passed, data can no longer be transferred from the D input to the Q output even if the clock remains high.

When the clock drops low, the output of the resistor-capacitor network is a negative spike. The clock input of the flip-flop is only activated when the input is positive. The negative spike created when the clock goes from high to low is ignored, and the D input is not transferred to the Q output.

Since the data transfer occurs only at the instant the clock is going from low to high (positive), the flip-flop is said to be "positive edge triggered."

Figure 6.21 shows the timing diagram for a **positive edge triggered D flip-flop.** At T1, the D input goes high. The clock and the Q output are both low. At T2, the clock goes from low to high.

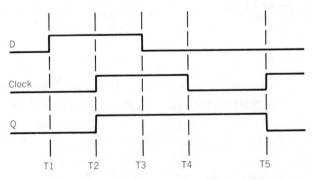

Figure 6.21   Timing diagram for a positive edge triggered D flip-flop

As the clock goes from low to high, the high condition of the D input is transferred to the Q output. At T3, the D input drops low. The clock is still high. The Q output remains high because the clock is not in transition from low to high. (If this were a D latch and the clock were high, the latch would be transparent. The data would be transferred from the input to the output. It is very important that you recognize this difference. The positive edge triggered D flip-flop transfers data only when the clock is in transition from low to high.) At T4, the D input remains low and the clock drops low. The Q output remains high. At T5, the clock goes from low to high. The D input is still low. The Q output switches low.

The logic symbol for a positive edge triggered D flip-flop is shown in Figure 6.22. The small triangle at the clock input indicates that the flip-flop is positive edge triggered.

NEGATIVE EDGE TRIGGERED D FLIP-FLOP

In addition to positive edge triggered D flip-flops there are also **negative edge triggered D flip-flops.** The negative edge triggered D flip-flop ignores the low-to-high clock transition. The negative edge triggered D flip-flop transfers data from the D input to the Q output only at the instant the clock is going from high to low.

Figure 6.23 shows one possible way of constructing a negative edge triggered D flip-flop. The difference between positive edge triggered and negative edge triggered D flip-flops is the inverter on the clock input of the negative edge triggered D flip-flop.

Figure 6.24 shows the logic symbol for a negative edge triggered D flip-flop. The triangle on the clock input tells us that the flip-flop is edge triggered. The bubble tells us that it is a negative edge triggered flip-flop.

FLIP-FLOP DESIGN

Although it is possible to build an edge triggered D flip-flop using a resistor-capacitor network, there are better designs that are

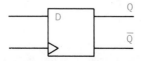

Figure 6.22 Logic symbol for a positive edge triggered D flip-flop

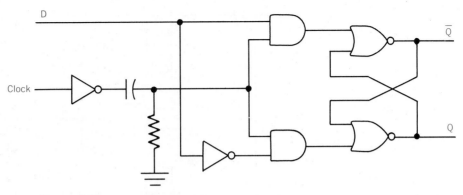

Figure 6.23 Negative edge triggered D flip-flop

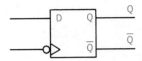

Figure 6.24 Logic symbol for a negative edge triggered D flip-flop

easier to implement when making IC chips. Large capacitors are difficult to make as parts of integrated circuits.

Figure 6.25 shows the design of a positive edge triggered D flip-flop that has been implemented without capacitors or resistors. The operating characteristics are identical to those described for the positive edge triggered D flip-flop with the resistor-capacitor network.

PRESET AND CLEAR INPUTS

In addition to the inputs described up to this point, many flip-flops also have inputs labeled PRESET and CLEAR. Figure 6.26 is a schematic drawing of a positive edge triggered D flip-flop with PRESET and CLEAR inputs.

The PRESET and CLEAR inputs are important for ensuring a proper starting point. When a flip-flop is initially turned on, the Q output may be high or low. Many computers require initialization after the power is turned on. When the computer is initialized, the flip-flops are reset or set to ensure that they are at the proper state. This is done through a special program that brings high either the PRESET inputs or the CLEAR inputs on all the flip-flops.

There are also times when a flip-flop must be set or reset in spite of the state of the clock. The PRESET and CLEAR func-

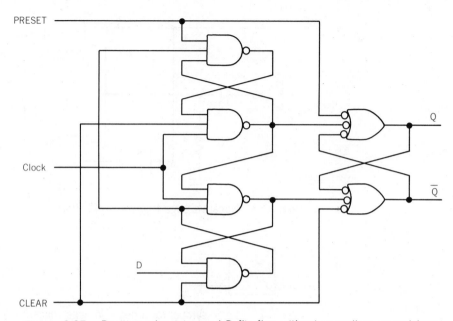

Figure 6.25   Positive edge triggered D flip-flop without capacitors or resistors

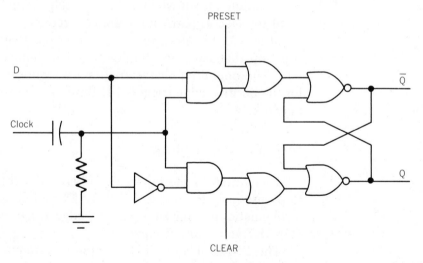

Figure 6.26   Positive edge triggered D flip-flop with PRESET and CLEAR inputs

tions are implemented by simply adding OR gates to the SET and RESET inputs.

When the PRESET input is brought high, the Q output goes high in spite of the state of the clock. When the CLEAR input is brought high, the Q output resets to 0 in spite of the state of the clock. The PRESET and CLEAR inputs cannot be brought high at the same time: this situation would attempt to set and reset the flip-flop at the same time, and this is a disallowed state.

The PRESET and CLEAR inputs are called **asynchronous** inputs because they operate independent of the clock input. The D input is called a **synchronous** input because it is dependent on the clock input.

Figure 6.27 shows the logic symbol for a positive edge triggered D flip-flop with PRESET and CLEAR inputs.

INVERTED PRESET AND CLEAR INPUTS

Many flip-flops use inverted inputs for PRESET and CLEAR. A schematic drawing of such a circuit is presented in Figure 6.28. When the PRESET input is low and the CLEAR input is high, the Q output goes high. When the clear input is low and the PRESET input is high, the Q output resets to 0. The disallowed state is when both PRESET and CLEAR inputs are low.

The logic symbol for a negative edge triggered D flip-flop with inverted PRESET and CLEAR inputs is shown in Figure 6.29. The bubbles on the PRESET and CLEAR inputs tell us that they are active low rather than active high.

J-K FLIP-FLOP

One of the most common flip-flops is the **J-K flip-flop,** which is a modification of the basic R-S flip-flop. Figure 6.30 shows one possible way of building a J-K flip-flop.

When the basic R-S flip-flop was modified to become a clocked R-S flip-flop, a pair of two-input AND gates was added. The J-K

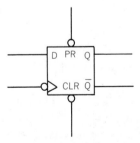

Figure 6.27   Logic symbol for a positive edge triggered D flip-flop with PRESET and CLEAR inputs

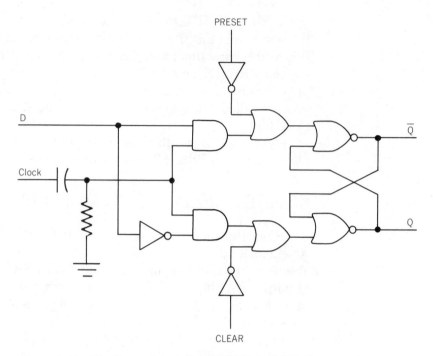

Figure 6.28   Flip-flop with inverted PRESET and CLEAR inputs

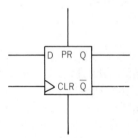

Figure 6.29   Logic symbol for a negative edge triggered D flip-flop with inverted PRESET and CLEAR inputs

flip-flop has a pair of three-input AND gates. One input of one of the AND gates is connected to one input of the other AND gate to form the clock input.

The J-K flip-flop is edge triggered. This circuit uses an R-C network for positive edge triggering.

The J and K inputs are control inputs. If the J input remains low the Q output cannot be set high, and if the K input remains low the Q output cannot be reset.

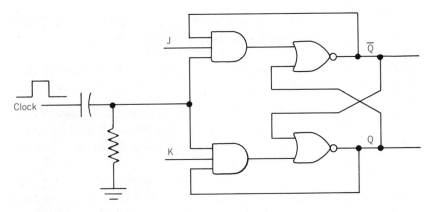

Figure 6.30 J-K flip-flop

Toggling

The Q output of the J-K flip-flop is wired to the AND gate which is driving the RESET input, and the $\overline{Q}$ output is wired to the AND gate driving the SET input.

If both the J and the K inputs are high the flip-flop is controlled by the clock input. When the Q output is low and the $\overline{Q}$ output is high, the AND gate connected to the SET input is enabled. When the clock input goes from low to high, the SET input brings the Q output high and the $\overline{Q}$ output low. With the Q output high, the AND gate connected to the RESET input is enabled. The next rising clock pulse drives the Q output low and the $\overline{Q}$ output high.

The outputs switch from low to high or from high to low with each rising clock pulse. This switching of the outputs from high to low or from low to high with each clock pulse is called **toggling**.

Figure 6.31 is a timing diagram that illustrates the concept of toggling. On the first clock pulse the Q output goes high, and on the second clock pulse the Q output goes low.

Notice that it takes two clock pulses to create one output pulse. The first clock pulse brings the Q output high, and the second clock pulse brings the Q output low. Also notice that it takes four clock pulses to generate two output pulses. The flip-flop is dividing by

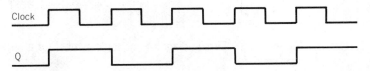

Figure 6.31 Timing diagram illustrating toggling

two. The ability of the J-K flip-flop to divide by two is important in building counting circuits, which are described later in this chapter.

Positive Edge Triggered and Negative Edge Triggered J-K Flip-Flops

The J-K flip-flop may be either positive or negative edge triggered. The timing diagrams and logic symbols for positive and negative edge triggered J-K flip-flops are shown in Figures 6.32 and 6.33, respectively.

The most common J-K flip-flop is triggered on the negative or falling clock edge. Figure 6.33 shows a **negative edge triggered J-K flip-flop.** The bubble at the clock input tells us that this is a negative edge triggered flip-flop.

The **positive edge triggered J-K flip-flop** is shown in Figure 6.32. The symbol is identical to that for the negative edge triggered J-K flip-flop except that there is no bubble on the clock input.

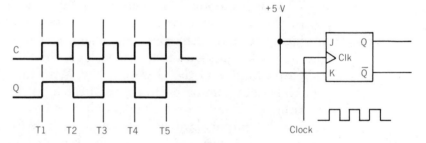

Figure 6.32   Timing diagram and logic symbol for a positive edge triggered J-K flip-flop

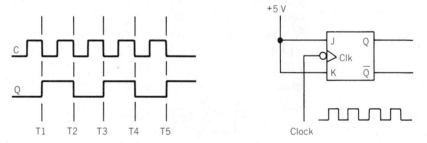

Figure 6.33   Timing diagram and logic symbol for a negative edge triggered J-K flip-flop

PRESET and CLEAR Inputs for J-K Flip-Flops

Many J-K flip-flops have PRESET and CLEAR inputs. Figure 6.34 shows one possible method of building a positive edge triggered J-K flip-flop with PRESET and CLEAR inputs. The PRESET and CLEAR inputs in this design are negated (active low or inverted).

The truth table for this flip-flop is presented in Figure 6.35. The PRESET and CLEAR inputs are independent of the clock input and the J and K inputs. When the PRESET input is low and the CLEAR input is high, the Q output goes high. The conditions of the clock and of the J and K inputs have no effect. When the CLEAR input is low and the PRESET input is high, the Q output goes low. The disallowed state is for both PRESET and CLEAR inputs to be low at the same time, because under this condition the PRESET input

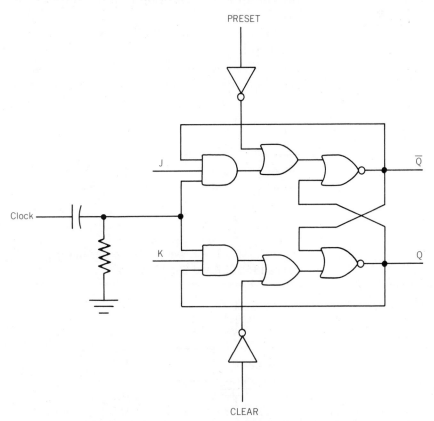

Figure 6.34 Positive edge triggered J-K flip-flop with PRESET and CLEAR inputs

$\overline{PRESET}$	$\overline{CLEAR}$	J	K	Clk	Q	$\overline{Q}$
0	1	X	X	X	1	0
1	0	X	X	X	0	1
1	1	1	0	⌐	1	0
1	1	0	1	⌐	0	1
1	1	1	1	⌐	Toggle	

X = don't care
⌐ = rising clock pulse
    (positive edge triggered)

Figure 6.35   Truth table for a positive edge triggered J-K flip-flop with PRESET and CLEAR inputs

would be attempting to drive the Q output high while the CLEAR input would be attempting to drive the Q output low, and this would cause the flip-flop to be unstable. When the PRESET and CLEAR inputs are high they have no effect on the operation of the flip-flop. The flip-flop is under the control of the clock and the J and K inputs.

Figure 6.36 shows the logic symbol for a positive edge triggered J-K flip-flop with negated PRESET and CLEAR inputs, and Figure 6.37 shows the logic symbol for a negative edge triggered J-K flip-flop with negated PRESET and CLEAR inputs.

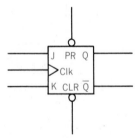

Figure 6.36   Logic symbol for a positive edge triggered J-K flip-flop with negated PRESET and CLEAR inputs

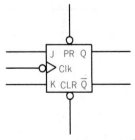

Figure 6.37   Logic symbol for a negative edge triggered J-K flip-flop with negated PRESET and CLEAR inputs

Racing

Examine Figure 6.34. As we said earlier, the Q output is connected through an AND gate to the RESET input, and the $\overline{Q}$ output is connected through an AND gate to the SET input. Assume for a moment that the J and K inputs are both high. When the clock line goes high the flip-flop toggles. If the Q output was low it now becomes high.

With the Q output high and the K input high, if the clock should still be high then all three inputs of the AND gate connected to the RESET input of the flip-flop will be high. The flip-flop resets and Q is switched low. When Q is low, $\overline{Q}$ is high. With the $\overline{Q}$ output high and the J input high, if the clock input should still be high then the $\overline{Q}$ output will switch low and the Q output will switch high. This brings us back to the condition wherein Q is high, K is high, and the clock is high, and the flip-flop switches outputs again. The flip-flop is switching back and forth at a very high rate of speed: this condition is called **racing**.

The race condition does not actually occur in a properly designed J-K flip-flop, because the resistor-capacitor network connected to the clock brings the clock input of the J-K flip-flop high for only a split second, and it takes a certain amount of time for the outputs of the flip-flop to change from high to low or from low to high. The time that it takes for the outputs to switch after the clock pulse arrives is called **propagation delay.**

If it takes more time for the outputs to switch than the length of time that the clock input is actually high, the race condition cannot occur. For example, assume that the Q output is low and the $\overline{Q}$ output is high. Both the J and K inputs are high. The flip-flop will toggle on the next clock pulse. The rising clock pulse triggers the flip-flop so that Q will switch high. Before the flip-flop actually switches, the spike created by the resistor-capacitor network will already have returned to 0. The output Q switches high, but since the clock input is no longer high the Q output cannot reset the flip-flop until the next clock pulse.

J-K MASTER/SLAVE FLIP-FLOP

Although the problem of the race condition within the flip-flop itself has been solved, race conditions can occur in circuits when flip-flops are connected in series. For example, if the Q output is connected to the clock input of the next flip-flop, a race condition may occur.

Suppose that the flip-flops are positive edge triggered. On the first low-to-high transition of the clock, the Q output goes from low to high. If the Q output is connected to the clock input of the second flip-flop, the second flip-flop will also toggle immediately, thus producing a race condition. The **J-K master/slave flip-flop** is a common solution to this problem.

Figure 6.38 shows one way to build a J-K master/slave flip-flop. There are two flip-flops that make up a master/slave flip-flop: the first is called the master flip-flop and the second is called the slave flip-flop. The master flip-flop is triggered on the positive going edge of the clock pulse, and the slave flip-flop is triggered on the negative going edge of the clock pulse.

Figure 6.39 shows the operation of the J-K master/slave flip-flop with the J input high and the K input low. On the rising edge of the clock pulse (T1), the Q output of the master flip-flop goes high and the $\overline{Q}$ output of the master flip-flop goes low. Notice (in Figure 6.38) that the Q output of the master is connected to the J input of the slave, and that the $\overline{Q}$ output of the master is connected to the K input of the slave. When the clock pulse falls (T2), the Q output of the slave goes high and the $\overline{Q}$ output of the slave goes low. At T3, the J input of the master goes low and the K input of the master goes high. At T4, the rising clock pulse switches the Q output of the master low and the $\overline{Q}$ output of the master high. Notice that the outputs of the slave did not change state at T4. When the clock pulse falls at T5, the outputs of the slave switch: the Q output goes low and the $\overline{Q}$ output goes high. The master accepts data on the rising clock edge, and the slave copies the outputs of the master on the falling clock edge.

When both the J and the K inputs of the master are high, the master/slave toggles (Figure 6.40). At T1, the Q output of the slave is low and the $\overline{Q}$ output of the slave is high. At T2, the clock goes

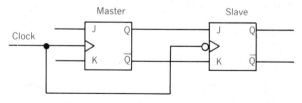

Figure 6.38 J-K master/slave flip-flop

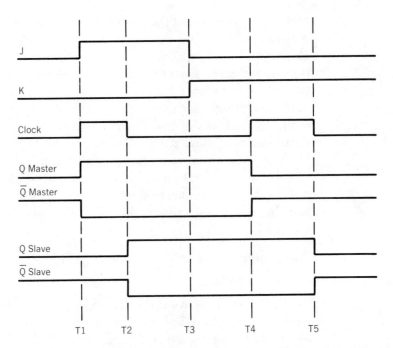

Figure 6.39 Timing diagram for a J-K master/slave flip-flop with either J or K high (but not both)

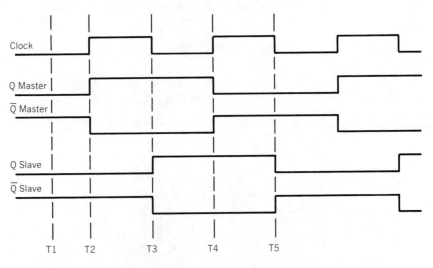

Figure 6.40 Timing diagram for a J-K master/slave flip-flop with both J and K high

high. The Q output of the master goes high and the $\overline{Q}$ output of the master goes low. The outputs of the slave are unchanged. At T3, the clock goes low. The outputs of the master are transferred to the outputs of the slave. The Q output of the slave goes high and the $\overline{Q}$ output of the slave goes low. At T4, the clock again goes high. The Q output of the master goes low and the $\overline{Q}$ output of the master goes high. The outputs of the slave remain unchanged. At T5, the clock drops low. The slave copies the outputs of the master: the Q output of the slave goes low and the $\overline{Q}$ output of the slave goes high.

The outputs continue to change with each clock pulse. The outputs of the master change state on the rising edge of the clock pulse, and the slave copies the outputs of the master on the falling edge of the clock pulse.

J-K Master/Slave Flip-Flop with PRESET and CLEAR Inputs

Figure 6.41 shows the logic symbol for a J-K master/slave flip-flop with PRESET and CLEAR inputs. Notice that the PRESET and CLEAR inputs are "bubbled" (active low). The disallowed state for the PRESET and CLEAR inputs is for both to be low at the same time. When PRESET and CLEAR are both high they have no effect on the operation of the flip-flop.

Notice that the clock input is also bubbled. The bubble on the clock input tells us that the output changes when the clock goes low.

BINARY SYSTEM

Before continuing the study of electronic circuits used in robots and other forms of automation it is important that we review the **binary system.**

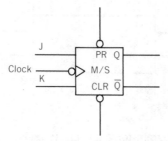

Figure 6.41   Logic symbol for a J-K master/slave flip-flop with negated PRESET and CLEAR inputs

All of our discussions of logic gates (Chapter 5) and flip-flops (present chapter) have been based on inputs and outputs being either on or off. The binary system uses only two digits in its counting system: 0 (off) and 1 (on). The number system we are more familiar with is the decimal system, which uses ten digits (0 through 9). Notice that 0 is an actual number in both the binary system and the decimal system.

Since the decimal system is based on ten digits, it is often referred to as a "base ten" system. The binary system is based on only two digits and is often called a "base two" system.

The decimal system is based on powers of 10 (Figure 6.42). The first (right-hand) column is 10 raised to the 0 power (10^0). Since any number raised to the 0 power is 1, the first column in the decimal system is the ones column. The second column is the tens column (10^1), the third column is the hundreds column (10^2), and the fourth column is the thousands column (10^3). All column "weights" are determined in this manner.

The binary system is based on powers of 2 (Figure 6.43). The first (right-hand) column is the ones column (2^0), the second column is the twos column (2^1), the third column is the fours column (2^2), and the fourth column is the eights column (2^3). All column weights are determined in this manner. (Notice that the weight of each column in the binary system is twice the weight of the preceding

10^3	10^2	10^1	10^0
Thousands	Hundreds	Tens	Ones

Figure 6.42 Column weights in the decimal system (powers of 10)

	Binary			
	2^3	2^2	2^1	2^0
Decimal	Eights	Fours	Twos	Ones
0				0
1				1
2			1	0
3			1	1
4		1	0	0
5		1	0	1
6		1	1	0
7		1	1	1
8	1	0	0	0
9	1	0	0	1

Figure 6.43 Column weights in the binary system (powers of 2) and binary equivalents for the decimal numbers 0 through 9

column, rather than ten times the weight of the preceding column as in the decimal system.)

Figure 6.44 illustrates the principle of counting in the decimal system. When we count in the decimal system we begin with the number 0 and count up to 9 in the ones column. The next number resets the 9 in the ones column to 0, and a 1 is carried to the tens column. The number in the tens column tells us how many "sets" of ten have been counted, and the number in the ones column is the remainder.

For example, if there is a 1 in the tens column and a 0 in the ones column, we have counted one set of ten and have zero units remaining. The next number is 11. There is a 1 in the tens column and a 1 in the ones column. We have counted one set of ten and have one unit remaining. The next number is 12. There is a 1 in the tens column and a 2 in the ones column. We have counted one set of ten and have two units remaining.

Now consider the number 19. There is a 1 in the tens column and a 9 in the ones column. We have counted one set of ten and

		10^1 Tens	10^0 Ones
			0
			1
			2
			3
			4
			5
			6
			7
			8
			9
		1	0
		1	1
		1	2
		1	3
		1	4
		1	5
		1	6
		1	7
		1	8
		1	9
10^2 Hundreds		2	0
		2	1
		9	8
		9	9
	1	0	0
	1	0	1

Figure 6.44 Counting in the decimal system

have nine units remaining. The next number is 20. We have counted two sets of ten and have zero units remaining.

The decimal system continues to count in this manner up to the number 99 (nine sets of ten and nine units remaining). Since there is no digit in the decimal system larger than 9, adding 1 to the ones column resets the ones column to 0 and carries a 1 to the tens column. Adding one set of ten to the nine sets of ten represented by the 9 in the tens column resets the tens column to 0 and carries a 1 to the hundreds column. The number 100 has a 1 in the hundreds column, a 0 in the tens column, and a 0 in the ones column, indicating one set of a hundred, zero sets of ten, and zero units remaining.

The same concepts are used when counting in the binary system. The difference lies in the fact that the binary system has only two digits, rather than the ten digits of the decimal system.

Figure 6.43 shows a binary chart along with the decimal equivalents. The first column is a ones column just as it is in the decimal system. Also, the first number is 0 and the second number is 1 in the binary system, just as in the decimal system. In the binary system there is no digit larger than a 1, just as there is no digit larger than a 9 in the decimal system. Since there is no digit higher than a 1 in the binary system, the next highest number resets the ones column to 0 and carries a 1 to the twos column. The binary equivalent of the decimal number 2 has a 1 in the twos column and a 0 in the ones column (10). In the binary system, this number is read as "one-zero," and not as "ten."

Binary and Decimal Notation

Since the binary equivalent of the decimal number 2 is 10, and since the number 10 is also a number in the decimal system, we need a means of indicating the system in which we are counting. We differentiate between the decimal and binary systems by using the subscripts "10" and "2" to indicate that a given number is a number in the base ten (decimal) system or in the base two (binary) system. For example, the decimal number 10 is written as "10_{10}" (meaning 10 in the base ten system), and the binary number 10 (equivalent to 2 in the decimal system) is written as "10_2" (meaning 10 in the base two system). Similarly, the decimal number 11 is written as "11_{10}" (11 in the base ten system), and the binary number 11 (equivalent to 3 in the decimal system) is written as "11_2" (11 in the base two system).

Binary-to-decimal Conversion

An easy method of converting a binary number into its decimal equivalent is to draw a binary chart with the column weights identified (Figure 6.45) and then write the binary number starting from the right (Figure 6.46). There is a 1 in the ones column, a 1 in the twos column, a 0 in the fours column, and a 1 in the eights column. Now simply add the column weights for all columns that have 1's in them: $8 + 2 + 1 = 11$. The decimal equivalent of the binary number 1011_2 is 11_{10}.

Decimal-to-binary Conversion

Converting a decimal number into its binary equivalent is a little more difficult. We will use two examples to demonstrate one method of converting a number from the decimal system to the binary system.

Suppose we wish to convert the decimal number 14 (14_{10}) to its binary equivalent. Again, draw a binary chart. The column farthest to the left must have a weight greater than the decimal number to be converted (Figure 6.47). Now find the highest-weighted column whose weight can be subtracted from the decimal number without producing a negative number, and put a 1 in that column. The highest-weighted column whose weight can be subtracted from 14_{10} is the 8's column. Put a 1 in the 8's column. The remainder is 6

2^4 16's	2^3 8's	2^2 4's	2^1 2's	2^0 1's

Figure 6.45   Binary chart

2^4 16's	2^3 8's	2^2 4's	2^1 2's	2^0 1's
	1	0	1	1

Figure 6.46   Binary equivalent of the decimal number 11

2^4 16's	2^3 8's	2^2 4's	2^1 2's	2^0 1's
	1	1	1	0

Figure 6.47   Binary equivalent of the decimal number 14

$(14 - 8 = 6)$. Now subtract the weight of the next column to the right (the 4's column) from 6 $(6 - 4 = 2)$, and put a 1 in the 4's column. Subtract the weight of the next column to the right (the 2's column) from the remainder 2 $(2 - 2 = 0)$, and put a 1 in the 2's column. The next column to the right is the 1's column. Since 1 cannot be subtracted from 0 without creating a negative number, put a 0 in the 1's column. The binary equivalent of the decimal number 14_{10} is 1110_2.

Now convert the decimal number 9 (9_{10}) to binary (Figure 6.48). The highest-weighted column whose weight can be subtracted from 9 without producing a negative number is the 8's column $(9 - 8 = 1)$. Put a 1 in the 8's column. The next column to the right is the 4's column. Since 4 cannot be subtracted from the remainder 1, write a 0 in the 4's column. The next column to the right is the 2's column. Since 2 also cannot be subtracted from 1, write a 0 in the 2's column. The next column to the right is the 1's column. Since 1 can be subtracted from 1 $(1 - 1 = 0)$, write a 1 in the 1's column. The binary equivalent of the decimal number 9_{10} is 1001_2.

BINARY CODED DECIMAL (BCD) SYSTEM

The **Binary Coded Decimal (BCD) system** is a combination of the binary system and the decimal system. The BCD system is often used in counting circuits (see next section), because electronic decoding of numbers from BCD to decimal is easier than decoding from binary to decimal.

The BCD system is similar to the binary system in that it uses only the two digits 0 and 1, but its major columns are weighted the same as the columns in the decimal system (1's, 10's, 100's, etc.). This system differs from both the binary and decimal systems in that the entries in its major columns are four-digit numbers rather than single digits.

Ten four-digit numbers are used for the major-column entries in the BCD system: 0000, 0001, 0010, 0011, 0100, 0101, 0110, 0111, 1000, and 1001. These numbers are the binary numbers 0 through 1001, which are equivalent to the decimal numbers 0 through 9. No-

2^4 16's	2^3 8's	2^2 4's	2^1 2's	2^0 1's
	1	0	0	1

Figure 6.48   Binary equivalent of the decimal number 9

tice that the highest possible number without a carry in the BCD system is 9_{10} (1001_2). The next digit resets 1001 to 0000, and 0001 is carried to the next major column.

Figure 6.49 illustrates how BCD numbers are formed by showing the BCD equivalents of the decimal numbers 9, 10, 99, and 100 ($9_{10} = 1001_{BCD}$; $10_{10} = 0001\ 0000_{BCD}$; $99_{10} = 1001\ 1001_{BCD}$; and $100_{10} = 0001\ 0000\ 0000_{BCD}$). The binary equivalents of these four numbers would be 1001_2, 1010_2, 1100011_2, and 1100100_2.

There are other counting systems used in digital electronics. Some of these systems are the hexadecimal, octal, split octal, and gray code systems. We will explain these systems whenever necessary for characterizing the operation of a particular circuit.

COUNTING CIRCUITS

There are two basic types of **counting circuits** in common use: the **ripple counter** and the **synchronous counter.** Both types of counters are based on the J-K flip-flop.

Binary Ripple Counter

A **ripple counter** is built by connecting several J-K flip-flops in series. The Q output of one flip-flop is connected to the clock input of the next flip-flop (Figure 6.50). Notice that the J and K inputs of all the flip-flops are connected to a positive 5-V power supply. This holds the J and K inputs high. With the J and K inputs high, the flip-flops will toggle.

Examine Figure 6.51. The first clock pulse applied to the first flip-flop (FF1) will bring its Q output high when the pulse goes from high to low. Notice that the J-K flip-flops selected for this circuit are negative edge triggered. The flip-flops trigger on the falling

	BCD											
Decimal	Hundreds				Tens				Ones			
	8's	4's	2's	1's	8's	4's	2's	1's	8's	4's	2's	1's
9	0	0	0	0	0	0	0	0	1	0	0	1
10	0	0	0	0	0	0	0	1	0	0	0	0
99	0	0	0	0	1	0	0	1	1	0	0	1
100	0	0	0	1	0	0	0	0	0	0	0	0

Figure 6.49 BCD equivalents of the decimal numbers 9, 10, 99, and 100

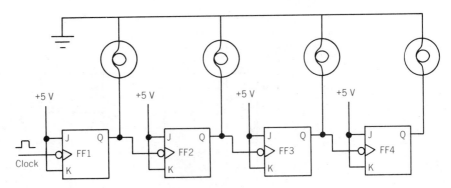

Figure 6.50 J-K ripple counter

Clock pulses	FF4	FF3	FF2	FF1
1	0	0	0	1
2	0	0	1	0
3	0	0	1	1

Figure 6.51 Truth table for a J-K ripple counter

clock pulse. When the second clock pulse makes the transition from high to low, FF1 resets — that is, the Q output goes from high to low. The transition of the Q output of FF1 from high to low sets the output of FF2 high. FF1 is low and FF2 is high (Figure 6.51). The third clock pulse will again set the Q output of FF1 high. Since the Q output of FF1 is making the transition from low to high, it has no effect on FF2. FF2 is only triggered when the Q output of FF1 is going from high to low. At this point FF1 is high and FF2 is high.

At this point, three clock pulses have been sent to the ripple counter. FF1 and FF2 are high while FF3 and FF4 are low. This corresponds to the binary counting system.

Figure 6.52 shows the timing diagram for this circuit. At T1, FF1 is high. At T2, FF1 goes low and FF2 goes high. At T3, FF1 again goes high. At T4, both FF1 and FF2 reset low and FF3 goes high. At T5, FF1 is again driven high, while FF2 remains low and FF3 remains high. You can trace the circuit as additional pulses are sent. Notice that the switching of the flip-flops corresponds precisely to the binary system.

This type of counter is called a "ripple counter" because the clock pulses "ripple" from one J-K flip-flop to the next. Additional J-K flip-flops can be connected in series with the four flip-flops shown in Figure 6.50 to create a counter as large as needed.

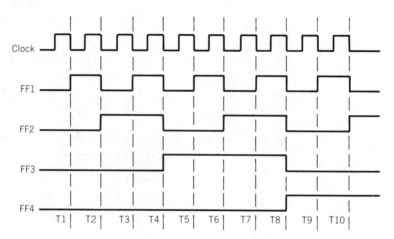

Figure 6.52 Timing diagram for a J-K ripple counter

BCD Ripple Counter

The J-K ripple counter can be modified to operate as a **BCD ripple counter** by the addition of a NAND gate (Figure 6.53). Notice that the flip-flops have been arranged from right to left to correspond with the 1's, 2's, 4's, and 8's columns of binary and BCD charts (Figure 6.43).

Note that the J-K flip-flops have CLEAR inputs. The CLEAR inputs are negated inputs (active low). The output of the NAND gate is normally high. The output of the NAND gate holds the

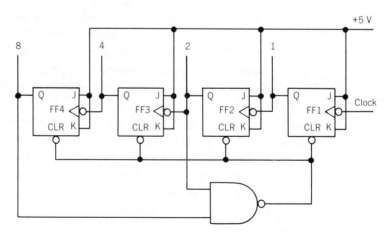

Figure 6.53 BCD ripple counter with four J-K flip-flops. Maximum count is 9_{10} before automatic resetting.

CLEAR inputs high until both inputs of the NAND gate go high. The inputs of the NAND gate are connected to the outputs of flip-flops FF2 and FF4. After ten clock pulses, both FF2 and FF4 outputs are high. With both FF2 and FF4 outputs high, the output of the NAND gate goes low, resetting all the flip-flops.

If a second set of J-K flip-flops is wired in series and the input to the second set of flip-flops is taken from the NAND gate, the second set of J-K's will receive one pulse each time the first set of J-K's resets (Figure 6.54).

The first set of J-K's will count the first ten pulses. On the tenth pulse, the first group of J-K's resets to 0 and one pulse is passed on to the second set of J-K's. The counter output shows a 1 in the tens counter and a 0 in the ones counter.

When the next pulse arrives, the ones counter again takes over. There will be a 1 in the ones counter and a 1 in the tens counter.

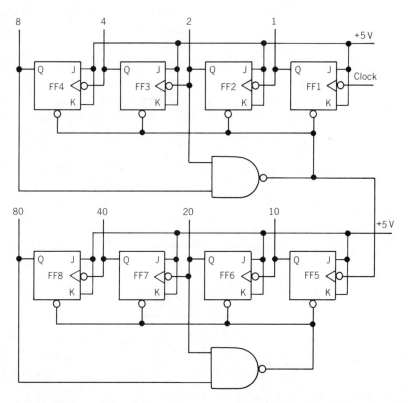

Figure 6.54 BCD ripple counter with eight J-K flip-flops. Maximum count is 99_{10} before automatic resetting.

The count is 11_{10}. After another nine pulses, the ones counter again resets and another pulse is passed on to the tens counter. The counter now shows a total of twenty pulses.

This sequence continues up to a total of 99 pulses. On the hundredth pulse, both the ones counter and the tens counter reset to 0. If another set of J-K flip-flops were connected to the tens counter, the total count possible would be 999.

Synchronous Counter

One problem with the J-K ripple counter is that it is slow. Each J-K flip-flop takes a split second for the output to go high after the clock pulse changes from high to low. The delay from the time the clock changes state to the time the output changes state is called "propagation delay." Although propagation delay is a desirable feature in the design of the J-K flip-flop (to prevent the racing condition), it is a disadvantage in a counting circuit.

When only a few flip-flops are connected in series, the propagation delay is not significant; however, if the counter is designed to count very large numbers, the ripple counter may be far too slow. The **synchronous counter,** also called the "parallel counter," overcomes this problem by switching all the flip-flops at the same time.

Figure 6.55 is a schematic diagram of a synchronous counter. The clock inputs are not wired in series as they are in the ripple counter, but rather are wired in parallel. When a clock pulse arrives, all the appropriate flip-flops shift to a new level at the same time rather than changing in sequence as they do in the ripple counter.

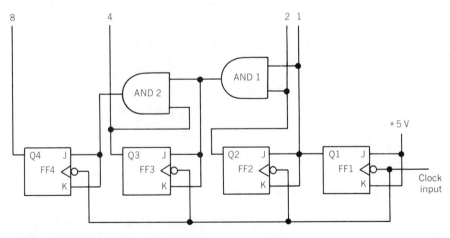

Figure 6.55 Synchronous counter

In order to achieve control of the flip-flops, AND gates have been added to the circuit. To understand the operation of the synchronous counter, examine the timing diagram shown in Figure 6.56. Notice that the J-K's in this circuit are also negative edge triggered. It is also important that you keep in mind the operation of J-K flip-flops when the J and K inputs are high and when the J and K inputs are low. When the J and K inputs are both high, the J-K flip-flop toggles. When the J and K inputs are both low, the changes in state of the clock input are ignored, and the flip-flop is disabled.

Refer again to Figures 6.55 and 6.56. At T1, J-K FF1 toggles. Output Q1 goes high. At T2, FF2 goes high and FF1 goes low. At T3, FF1 again goes high and FF2 remains high. Up to this point, the synchronous counter is acting like a ripple counter. At T4, FF1 and FF2 go low and FF3 goes high. FF1, FF2, and FF3 toggle at the same instant. The reason for this can be seen in the schematic drawing presented in Figure 6.57.

Note that both FF1 and FF2 are high at T3. With FF1 and FF2 high, both inputs of AND gate 1 are high. With the AND gate's output high, the J and K inputs of flip-flop FF3 are high. With the J and K inputs of FF3 high, FF3 switches high on the next clock pulse (T4). At the same instant FF3 is switching high, FF1 and FF2 both switch low. Both inputs to AND gate 1 are now low, its output drops low, and the J and K inputs of FF3 drop low. When the next clock pulse arrives (at T5), FF3 ignores the pulse and remains high.

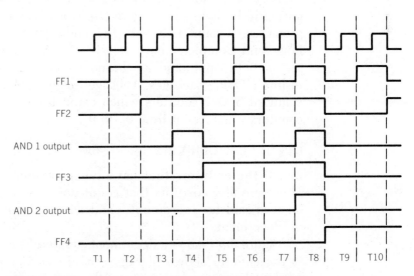

Figure 6.56 Timing diagram for a synchronous counter

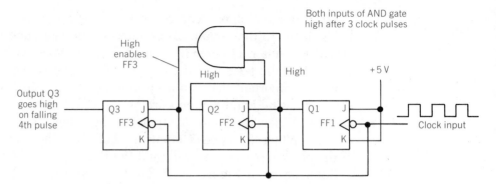

Figure 6.57   Diagram showing how all flip-flops in a synchronous counter switch at the same instant

At T5, T6, and T7 (Figure 6.56), flip-flops FF1 and FF2 operate as they did at T1, T2, and T3. At T7, both FF1 and FF2 are again high. The output of AND gate 1 is high. Since FF3 is also high, both inputs of AND gate 2 are high. With both inputs of AND gate 2 high, the J and K inputs of FF4 are high. At the next clock pulse (T8), FF4 switches high and all other flip-flops switch low.

As you can see, the pulses do not have to "ripple" from one flip-flop to the next. At T4, three flip-flops switch at the same time, and at T8, four flip-flops switch at the same time. This instant change of all flip-flops eliminates the propagation delay found in ripple counters.

Fortunately for designers and servicemen alike, counters are not normally built using discrete flip-flops. IC manufacturers make prepackaged counters in both TTL and CMOS (Transistor Transistor Logic and Complementary Metal-Oxide Semiconductor: see Chapter 5). Both ripple and synchronous counters are available in dual in-line ("DIP") packages. Packaged counters can be purchased as standard binary counters which count from 0_{10} to 15_{10} or as BCD counters which count from 0_{10} to 9_{10}.

Decrement Counters

All the counters that have been presented so far count up. There are also counters that count down. When a counter counts up it is said to "increment," and when it counts down it is said to "decrement."

Figure 6.58 shows a **decrement counter.** This counter is identical to the counter shown in Figure 6.50 except that the $\overline{Q}$ output is used to drive the clock input of the next flip-flop and the PRESET inputs of all flip-flops are wired in parallel.

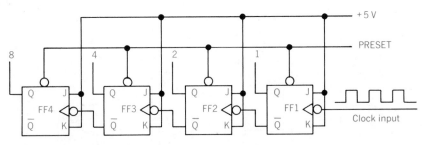

Figure 6.58 Decrement counter

Before the down count is started, the PRESET input is brought low, which presets all the flip-flops high. The total of the Q outputs of all the flip-flops is 15_{10}. On the first clock pulse, flip-flop FF1 resets. The Q output goes low and the $\overline{Q}$ output goes high. The total of the Q outputs is now 14_{10}.

On the next clock pulse, the Q output of FF1 goes high and the $\overline{Q}$ output goes low. When $\overline{Q}$ goes from high to low, flip-flop FF2 toggles. The Q output of FF2 goes low and the $\overline{Q}$ output goes high. The total output is now 13_{10}. The counter continues the counting sequence, just as the ripple counter did, until the output is 0.

Many IC manufacturers are building chips that both increment and decrement. These counters are often called **up-down counters.** The direction of counting is controlled by special inputs on the chip.

Preset Counter

The **preset counter** can be preset to any number within its counting range. Once preset to a certain number, the preset counter decrements one count on each clock pulse. The preset counter is often used to shut off automated equipment after a preset number of machine cycles. When the counter decrements to 0, it shuts off the equipment.

The schematic of a preset counter is shown in Figure 6.59. The method of presetting the counter to a predetermined count is often referred to as the JAM technique.

To preset the counter, the binary number is entered at A, B, C, and D. At the same time that the PRESET inputs are brought high the JAM input is brought high. For example, if we wish to preset the counter at 5_{10} the JAM input is brought high. This brings one input of each of the NAND gates connected to the PRESET inputs high. Now line A and line C are brought high. The outputs of the PRESET NAND gates A and C drop low, and the Q outputs of FF1 and FF3 are set high.

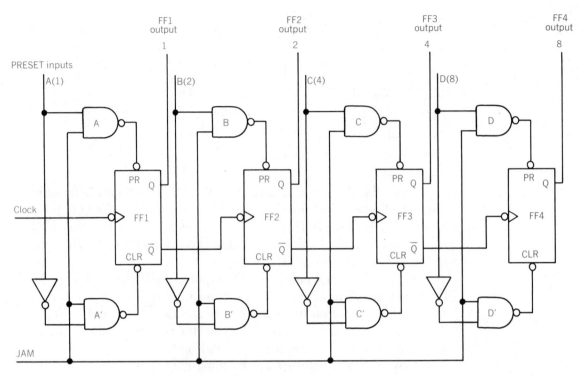

Figure 6.59 Preset counter

At the same time, the outputs of the NOT gates connected to lines B and D are high since their inputs are low. The other inputs of NAND gates A′, B′, C′, and D′ are high since they are also connected to the JAM input. This clears FF2 and FF4, and their Q outputs are low.

The counter is now preset at 5_{10}. The counter will decrement on each falling clock pulse.

If we wish to build a counter that can be preset and that will increment rather than decrement, we simply connect the Q output, rather than the $\overline{Q}$ output, to the clock input of the following flip-flop.

BINARY ADDITION

To perform the functions required of it, a robot control must be able to perform arithmetic operations as well as the counting operations described previously. This section, and the following section on binary subtraction, are not intended to be a comprehensive study of binary arithmetic, but rather a review for those who have previously studied binary arithmetic or an introduction for those who have not.

Binary addition follows the same rules as those of decimal addition. The only difference that must be kept in mind is that 1 is the highest digit in the binary system. Thus, in binary addition, $0 + 0 = 0$, $1 + 0 = 1$, and $1 + 1 = 0$ with 1 to carry.

For example, add $1001 + 0011$. First, add the digits in the far right-hand column, just as would be done in decimal addition. The digit in the far right-hand column of each number is called the **Least Significant Bit,** which is normally abbreviated **LSB**. The digit in the far left-hand column of each number is called the **Most Significant Bit,** which is normally abbreviated **MSB**. The result of adding the LSB's is 0 with 1 to carry:

$$
\begin{array}{r}
1 \\
1001 \\
+0011 \\
\hline
0
\end{array}
$$

Now, in the second column from the right, add the carried 1 plus 0 plus 1. The result is 0 with 1 to carry:

$$
\begin{array}{r}
11 \\
1001 \\
+0011 \\
\hline
00
\end{array}
$$

Next, in the third column, add the carried 1 plus 0 plus 0. The result is 1:

$$
\begin{array}{r}
11 \\
1001 \\
+0011 \\
\hline
100
\end{array}
$$

Finally, add the MSB's (1 plus 0). The result is 1:

$$
\begin{array}{r}
11 \\
1001 \\
+0011 \\
\hline
1100
\end{array}
\qquad
\left.
\begin{array}{r}
9 \\
+3 \\
\hline
12
\end{array}
\right\} \text{ Check}
$$

The final result is 1100. As you can see, addition of binary numbers proceeds just as does addition of decimal numbers.

You should be careful when adding three or more 1's. As we have already indicated, the result of $1 + 1$ is 0 with 1 to carry. When another 1 is added to the 0, the result is 1, and 1 is still carried. Thus, $1 + 1 + 1 = 1$ with 1 to carry. For example, add $111 + 011$. The second step in this problem will illustrate the adding of $1 + 1 + 1$:

$$
\begin{array}{r}
1 \\
111 \\
+011 \\
\hline
0
\end{array}
$$

$$
\begin{array}{r}
11 \\
111 \\
+011 \\
\hline
10
\end{array}
$$

$$
\begin{array}{r}
111 \\
111 \\
+011 \\
\hline
010
\end{array}
$$

$$
\begin{array}{r}
111 \\
111 \\
+011 \\
\hline
1010
\end{array}
\qquad
\left.
\begin{array}{r}
7 \\
+3 \\
\hline
10
\end{array}
\right\} \text{Check}
$$

The final result is 1010.

Half-adder

The logic circuit that can perform binary addition is the XOR gate. Recall the truth table for the XOR gate (Figure 6.60). The truth table shows that $0 + 0 = 0$, $1 + 0 = 1$, $0 + 1 = 1$, and $1 + 1 = 0$.

The XOR gate obeys all the rules of binary addition except for the carrying operation. To build a circuit that can also carry, an AND gate is connected in parallel with the XOR gate (Figure 6.61). When both inputs are high $(1 + 1)$ at the same time, the output of the XOR gate is 0 and the output of the AND gate is 1. The output of the AND gate is the carry bit, which is abbreviated "CO." This circuit is called a **half-adder,** because it has a carry output but lacks

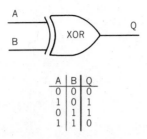

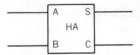

Figure 6.60 Logic symbol and truth table for an XOR gate

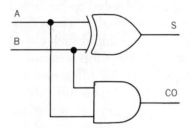

Figure 6.61 Half-adder

the circuitry to accept a carry input. The logic symbol for a half-adder is shown in Figure 6.62.

Full-adder

The **full-adder** differs from the half-adder in that it has both a carry output and a carry input (abbreviated "CI"). The full-adder is built by connecting the outputs of two half-adders to an OR gate (Figure 6.63). The logic symbol for a full-adder is shown in

Figure 6.62 Logic symbol for a half-adder

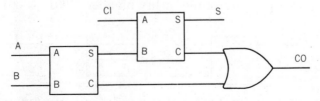

Figure 6.63 Full-adder

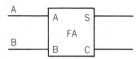

Figure 6.64   Logic symbol for a full-adder

BINARY SUBTRACTION

Binary subtraction is a more difficult process than binary addition, and the logic circuits needed to perform it are more complex. However, IC manufacturers produce chips that perform subtraction, and thus it is not necessary to build such circuits. The logic symbols for a **half-subtractor** and a **full-subtractor** are shown in Figure 6.65.

The half-subtractor finds the difference between two inputs and also has the ability to borrow. The full-subtractor can find the difference, can borrow, and also can account for borrowing from other subtractors.

Using subtractor circuits is not the common method of subtracting in computers. Subtraction is more often done by "complementing" followed by addition.

Subtracting by Complementing

Subtracting by use of half-subtractors and full-subtractors is not normally done in computers. Instead, the subtraction process is performed using half-adders and full-adders.

The operation $7 - 3 = 4$ could also be expressed as $7 + (-3) = 4$. From this expression it is obvious that subtraction can be thought of as an addition process.

To subtract one binary number from another, the number being subtracted must be **complemented**.

1's Complement

When a binary number is converted to its **1's complement**, all the 0's are changed to 1's and all the 1's are changed to 0's. The 1's complement of the binary number 110011 is 001100. Notice that the

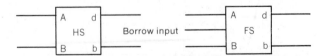

Figure 6.65   Logic symbols for a half-subtractor (left) and a full-subtractor (right)

number is being inverted. Remember that a NOT gate inverts numbers. If the input to a NOT gate is 1 its output is 0, and if its input is 0 its output is 1. Also recall that the Boolean equation for the NOT gate is $A = \overline{A}$. If the number to be complemented is called A, the complemented form of that number is called $\overline{A}$.

Now we will perform a binary subtraction using the 1's complement method. As an example we will solve the problem $10_{10} - 5_{10}$. The binary equivalent of the decimal number 10 is 1010_2, and the binary equivalent of the decimal number 5 is 0101_2. The first step in the process is to find the 1's complement of the binary number 0101_2:

$$
\begin{array}{ll}
\text{Binary number:} & 0101 \\
\text{1's complement:} & 1010 \\
\end{array}
$$

Add the 1's complement (1010) to the binary number 1010_2:

$$
\begin{array}{ll}
1010 & (10_{10}) \\
\underline{+1010} & \text{(1's complement of } 5_{10}) \\
10100 &
\end{array}
$$

The result is 10100. When this binary number is converted to decimal, the result is 20_{10}. Clearly, this is not the correct answer. The problem is that, whenever a smaller number is subtracted from a larger number using the 1's complement method, a 1 always appears in the column on the left as a result of a carry. This 1 is not actually the most significant bit (MSB). This 1 must now be added to the least significant bit (LSB) of the result. This process is called an **End Around Carry**, abbreviated **EAC**. Returning to our example we find:

$$
\begin{array}{ll}
1010 & (10_{10}) \\
\underline{+1010} & \text{(1's complement of } 5_{10}) \\
(1)0100 & \\
\underline{+1} & \text{(EAC)} \\
0101 &
\end{array}
$$

The result of the EAC is 0101. When this number is converted to decimal, the result is 5, which is the correct answer.

Another example will help demonstrate the EAC. Solve the problem $14 - 6$ in binary using the 1's complement method. The binary equivalent of the decimal number 14 is 1110_2, the binary

equivalent of the decimal number 6 is 0110_2, and the 1's complement of 0110_2 is 1001. Thus:

$$
\begin{array}{ll}
1110 & (14_{10}) \\
+1001 & (\text{1's complement of } 6_{10}) \\
\hline
(1)0111 & \\
\quad +1 & (\text{EAC}) \\
\hline
1000 &
\end{array}
$$

The binary number 1000_2 is the equivalent of the decimal number
8_{10}, and thus is the correct answer $(14 - 6 = 8)$.

The 1's complement method provides the correct answer when
the EAC is used and the result (remainder) is positive. A problem
arises when we attempt to subtract a larger number from a smaller
number. For example, solve the problem $7 - 10$ in binary using the
1's complement method. $7_{10} = 0111_2$; $10_{10} = 1010_2$; 1's complement
of $1010_2 = 0101$:

$$
\begin{array}{ll}
0111 & (7_{10}) \\
+0101 & (\text{1's complement of } 10_{10}) \\
\hline
1100 &
\end{array}
$$

The result is 1100. The decimal equivalent of this binary number is
12, which clearly is not the correct answer. When the 1's complement of a number is added to another number and there is no EAC,
the result is the 1's complement of the answer, and the answer is
negative. In this case a 1 was not carried as a result of adding the
MSB's, and so there was no EAC. Thus, to find the correct answer,
we take the 1's complement of 1100_2, convert to decimal, and add a
minus sign. The 1's complement of 1100_2 is 0011, the decimal
equivalent of 0011 is 3, and therefore the correct answer is -3.

Review of the 1's Complement Method.   In order to perform
binary subtraction using the 1's complement method:

1. Take the 1's complement of the number being subtracted.

2. Add the 1's complement to the other number.

3. If there is an EAC, perform the EAC addition. The result is the
   correct answer, and the answer is positive.

4. If there is no EAC, the result is the 1's complement of the correct answer. Complement the result and add a minus sign (the
answer is negative).

2's Complement

The **2's complement** method of subtracting binary numbers
eliminates the EAC. To find the 2's complement of a number, take
the 1's complement and add 1. For example, find the 2's complement of 1101_2:

$$
\begin{array}{ll}
0010 & \text{(1's complement of 1101)} \\
+\quad 1 & \text{(Add 1)} \\
\hline
0011 & \text{(2's complement)}
\end{array}
$$

Find the 2's complement of 1010_2:

$$
\begin{array}{ll}
0101 & \text{(1's complement of 1010)} \\
+\quad 1 & \text{(Add 1)} \\
\hline
0110 & \text{(2's complement)}
\end{array}
$$

Subtract 3 from 9 using the 2's complement method. $9_{10} = 1001_2$;
$3_{10} = 0011_2$; 1's complement of $0011 = 1100$; 2's complement of
$0011 = 1100 + 1 = 1101$:

$$
\begin{array}{ll}
1001 & (9_{10}) \\
+1101 & \text{(2's complement of } 3_{10}) \\
\hline
\text{Sign bit} \longrightarrow (1)0110 &
\end{array}
$$

The 1 at the far left, which is called a "sign bit," is not used as an
EAC. This digit is significant, however, because when a 1 is carried
from the result of adding the MSB's it tells us that the answer is positive. Thus, considering all the other bits, 0110_2 (or 6_{10}) is the result.

Subtract 6 from 4 using the 2's complement method. $4_{10} = 0100_2$;
$6_{10} = 0110_2$; 1's complement of $0110 = 1001$; 2's complement of
$0110 = 1001 + 1 = 1010$:

$$
\begin{array}{ll}
0100 & (4_{10}) \\
+1010 & \text{(2's complement of } 6_{10}) \\
\hline
\text{Sign bit} \longrightarrow (0)1110 &
\end{array}
$$

The result is (0)1110. The decimal equivalent of this binary number is 14, which obviously is not the correct answer. The problem here is that our answer must be negative since there was no carry from the result of adding the MSB's. When the answer is negative we must complement the result, just as in the 1's complement method. Because we are working in the 2's complement method, we must take the 2's complement of the result in order to arrive at the correct answer. The 2's complement of 1110 is 0001 + 1 = 0010, and the decimal equivalent of 0010_2 is 2. The correct answer is −2.

Review of the 2's Complement Method.   In order to perform binary subtraction using the 2's complement method:

1. Take the 1's complement of the number being subtracted and add 1. This produces the 2's complement.

2. Add the 2's complement to the other number.

3. If the MSB (sign bit) is 1, the result of the subtraction is positive and the sign bit is dropped.

4. If the MSB (sign bit) is 0, the result of the subtraction is negative and is the 2's complement of the correct answer. Take the 2's complement to arrive at the correct answer.

2's Complement Adder/Subtractor

Figure 6.66 shows a **2's complement adder/subtractor.** This circuit uses four full-adders and four XOR gates. The inputs to the full-adders and XOR gates are labeled with X's and Y's, respectively.

When the SUB input is low, data present at the Y inputs passes through the XOR gates without being complemented. The data present at the Y inputs is added to the data at the X inputs. The results of these additions appear at the Q outputs.

When the SUB input is high, data present at the Y inputs is inverted as it passes through the XOR gates. For example, if input Y1 is high the output of the corresponding XOR gate will be low, and if input Y1 is low the output of the XOR gate will be high. The XOR gate is obeying the rules we learned earlier. The output of the XOR gate is the 1's complement of the condition at the corresponding Y input.

Notice that the SUB input is also connected to the carry input of the first full-adder. This adds 1 to the first full-adder and forms the 2's complement.

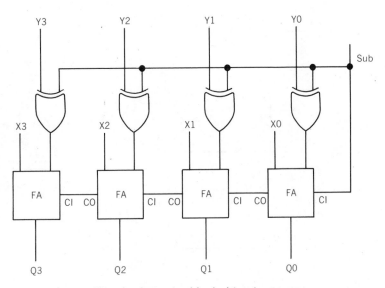

Figure 6.66 2's complement adder/subtractor

The output at Q will be the correct result if we are subtracting a smaller number from a larger one. If, however, we are subtracting a larger number from a smaller number, the output at Q will be the 2's complement of the answer. Another circuit can be connected to convert the result from the 2's complement to the correct binary answer.

DECODERS

Although the binary system provides a convenient means of counting, adding, and subtracting, the outputs of the counters are difficult to read. We are more accustomed to reading numbers in the decimal system than in the binary system.

Decoding binary numbers to decimal numbers can be done with a group of AND gates. The circuit shown in Figure 6.67 will convert the binary number 0000 to the decimal number 0. If the input to all NOT gates is 0, their outputs will be high, bringing all legs of the AND gate high. The output of the AND gate will be high. By connecting a light bulb to the output of the AND gate we can easily determine when the binary output of the counter or adder/subtractor is 0.

Conversion of the binary number 0001 to the decimal number 1 can be done using the circuit shown in Figure 6.68. In this circuit

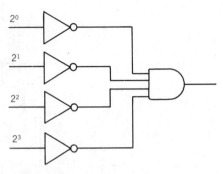

Figure 6.67   Decoding circuit that can convert the binary number 0000 to the decimal number 0

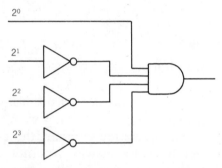

Figure 6.68   Decoding circuit that can convert the binary number 0001 to the decimal number 1

the first input leg is not inverted. All other input legs are inverted. When the first input leg is high and all other input legs are low, the output of the AND gate will be high.

To decode the binary number 0010 to the decimal number 2 we would connect inverters on the first, third, and fourth inputs of the AND gate. When all inputs are low except the second line, all inputs to the AND gate will be high. The output of the AND gate will be high.

Figure 6.69 shows a decoding circuit that can decode any binary number up to 1001_2 into the correct decimal equivalent. For example, if the input to the circuit shown is the binary number 0011, inputs A and B will be high and inputs C and D will be low. With inputs A and B high, the top two legs of AND gate #3 will be high. With inputs C and D low, the outputs of their inverters will be high. These inverters are connected to the bottom two legs of AND gate #3. All the inputs to AND gate #3 are high, and its output goes high. This turns on the #3 light.

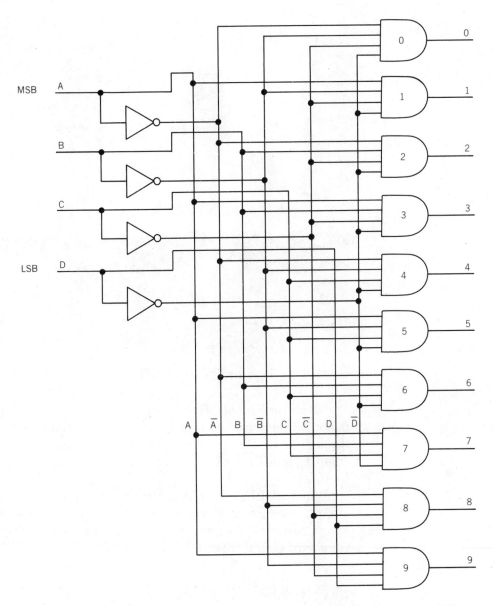

Figure 6.69 Decoding circuit that can convert any binary number up to 1001 to its decimal equivalent

LIGHT EMITTING DIODES

Bulbs can be connected to the output of the decoder or, as an alternative, **Light Emitting Diodes (LED's)** can be used (Figure 6.70).

Figure 6.70   Light emitting diodes (LED's)

LED's are much like the diodes discussed in Chapter 5. The major difference is that an LED lights up when current is passed through it. LED's draw very little current, which makes them more compatible with the limited amount of current that can be handled by IC's.

When an LED is used as the output for a decoder or other IC circuit, a resistor should be included to protect the circuit from excess current draw since the LED is a low-resistance path when it is forward biased (Figure 6.71).

SEVEN-SEGMENT DISPLAYS

A special display called a **seven-segment display** (Figure 6.72) is often used as the output device on robots and other automated equipment. You are probably most familiar with seven-segment displays used in calculators. Calculators that have small red displays are a type of LED seven-segment display.

Figure 6.73 shows the way in which LED's are positioned to form a seven-segment display. By lighting various LED's, all digits from 0 to 9 can be formed, as well as some letters of the alphabet and some special symbols.

Notice that the segments in the seven-segment display are labeled A through G. The designation of segments has become stan-

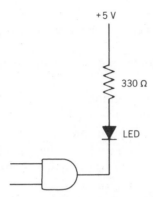

Figure 6.71 LED with protective resistor

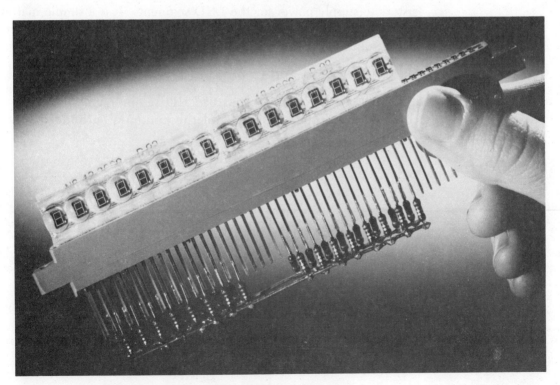

Figure 6.72 Seven-segment LED displays

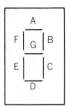

Figure 6.73   Positioning of LED's to form a seven-segment display

dard, and the letters identifying the seven segments are always in the same arrangement.

If voltage is applied to segments A, B, and C at the same time, the display will show the number 7. If segments A, B, C, D, and G are lighted, the display will form the number 3.

Special decoders are needed to decode binary numbers and bring the appropriate lines high to form the numbers. These decoders have become common and are available from many IC chip manufacturers. Many of these chips can supply sufficient power to illuminate the display without the need to install "driver" transistors to supply power for the LED segments. It is a simple matter of connecting the decoder to the counting circuit and connecting the output of the decoder to the seven-segment display.

Common Anode and Common Cathode LED Displays

There are two basic types of LED seven-segment displays: the **common anode display** and the **common cathode display** (Figure 6.74).

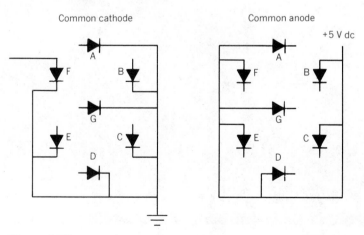

Figure 6.74   Common cathode (left) and common anode (right) displays

In the common cathode seven-segment display, the cathodes of all seven LED's are connected together. In the common anode display, all the anodes are connected together. Both displays look identical, and it is almost impossible to tell them apart. The only way to be certain is to check the manufacturer's number.

Common cathode and common anode displays are not interchangeable. When servicing equipment be certain to replace the display with the proper type. The wrong type of display will fail to light because the LED's will be reverse biased.

GAS DISCHARGE DISPLAY

There is one major drawback to LED displays. The light output of an LED display is quite low. In direct sunlight or in a brightly lighted room the display can be difficult if not impossible to read.

The **gas discharge display** is much brighter than the LED display. It produces a bright orange display. The gas discharge display is bright enough to be seen in most situations. This improved readability does not come free. Operation of the gas discharge display requires from 150 to 200 V. This voltage is far beyond the capacity of a decoder chip. The gas discharge display requires driver transistors, which add to the complexity and cost of the circuit.

NUMITRON DISPLAY

The **Numitron display,** which is made by RCA, is also a seven segment display. The Numitron display uses small filaments similar to those in a bulb. The filaments operate on 3.5 to 5 V and draw only slightly more current than an LED. There are IC decoder chips that supply adequate current to light the Numitron without the need for driver transistors.

FLUORESCENT DISPLAYS

There are also seven-segment **fluorescent displays.** Fluorescent displays glow with a blue-green light. These displays require 20 to 40 V. IC chips cannot handle this voltage, and driver transistors are required.

5 × 7 MATRIX

The **5 × 7 matrix** is made up of 35 individual dots that can be lighted (Figure 6.75). The advantage of the 5 × 7 matrix is that all letters of the alphabet, as well as numbers and special symbols, can be formed by this display. By bringing various rows and columns high and low, all of the individual dots can be lighted. Special decoders are needed to drive the 5 × 7 matrix.

LIQUID CRYSTAL DISPLAY

Another type of display is the **Liquid Crystal Display (LCD)**. The LCD does not emit light as do the other displays but rather reflects light. For this reason the LCD is very visible in the high-light situations which often exist in manufacturing plants. In addition, operation of the LCD requires very little power. For these reasons, the LCD has become very popular for use in electronic watches and calculators.

Liquid crystals are liquids which under certain temperature conditions have the properties of crystals. Liquid crystals have elongated molecules. These molecules can be thought of as rods. Liquid crystals are placed between transparent glass and black glass. When the crystals are aligned parallel to the glass plates, light striking the crystals is reflected and the display appears white or light gray. When an electric field is applied, the crystals realign. Light passes through the liquid crystal and strikes the black glass plate, and the display appears black.

By arranging liquid crystal cells in the shape of a seven segment display, all digits, as well as some letters and special symbols, can be

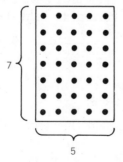

Figure 6.75

formed, just as they are in LED seven segment displays. The liquid crystal display can also be made in the form of a dot matrix. This arrangement allows for the formation of the entire alphabet as well as numbers and a large variety of symbols.

CLOCK CIRCUITS

There is one final circuit to study in this chapter before looking at a simple robot circuit. The **clock circuit** is used to create the regular string of pulses needed to synchronize the timing of the many circuits in a control.

The basic circuit for a clock is shown in Figure 6.76. This circuit should look familiar, because it is almost identical to the R-S flip-flop studied earlier in this chapter. The major difference between this circuit and the R-S flip-flop is the timing capacitors in the clock circuit. The capacitors charge and discharge, switching the transistors on and off. The switching of the transistors on and off creates the regular string of pulses.

There is one major problem with this circuit: it is very sensitive to temperature changes. A slight change in temperature will change the clock rate, and therefore this circuit is not used very often.

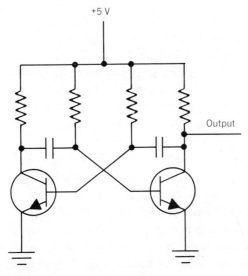

Figure 6.76   Basic clock circuit

Crystal Controlled Clock

In the **crystal controlled clock,** a quartz crystal is used to control the clock frequency. The quartz crystal is sliced into very thin sheets. When a voltage is applied across the crystal, the crystal vibrates. The frequency of the clock is determined by the thickness of the crystal.

The vibrations of the quartz crystal are amplified and used to control the output of the clock circuit. The frequency of the crystal controlled clock is high, often several million pulses per second. The output of the clock can be connected to a ripple counter, which operates as a frequency divider. By choosing the appropriate output from the counter, almost any frequency can be generated.

The 555 IC Chip

Another common method of constructing a clock circuit is the use of the **555 IC chip.** The 555 chip is more stable than the circuit shown in Figure 6.76 but not as stable as the crystal controlled clock.

The schematic diagram of a 555 chip is shown in Figure 6.77. The major components are two voltage comparators and an R-S flip-flop. As you know, the flip-flop is stable. When the SET input is high the output is high, and when the RESET input is high the output is low. If the R and S inputs are brought high alternately, the output of the flip-flop will be a regular string of pulses. The voltage comparators serve the function of switching the R and S inputs of the R-S flip-flop.

There are two basic types of voltage comparators. One type of comparator switches high when the input voltages are equal. The other type of comparator switches high when the voltages are unequal. In the 555 chip, the outputs of the comparators go high when the input voltages are equal.

The discharge (pin 7) of the 555 chip is fed back to the comparators through a resistor-capacitor network (Figure 6.78), and a regular string of pulses is generated. The frequency of the clock circuit is changed by adjusting potentiometer R2.

The 555 chip is extremely versatile. Changes can be made in its outboard components and wiring that allow it to be used for many other functions such as timers and "debouncing switches."

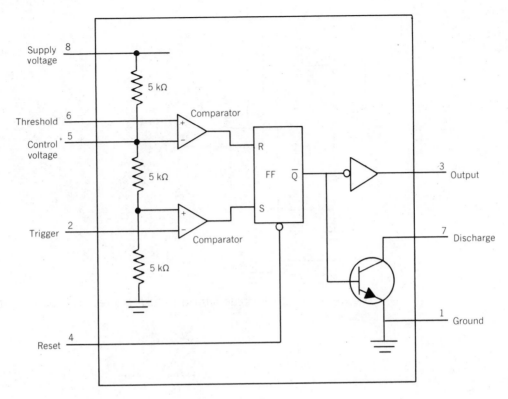

Figure 6.77   555 IC chip

PICK-AND-PLACE ROBOT

Figure 6.79 shows a three-axis robot. The robot has a waist for turning, a shoulder for elevation, and arm extension for reaching. The gripper at the end of the arm is pneumatic while the other major moves are powered by hydraulics. A hydraulic motor is used for waist rotation, and hydraulic cylinders are used for shoulder elevation and reach. In this section we will examine the hydraulic, pneumatic, and electronic circuits which make up this robot as well as interfacing of the robot with the soap box filling machine.

Hydraulic and Pneumatic Schematics

Figures 6.80 and 6.81 show the hydraulic and pneumatic circuits for the robot.

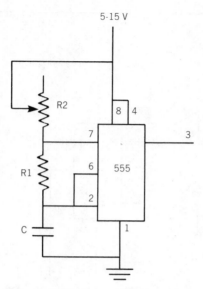

Figure 6.78 Clock circuit formed by adding a resistor-capacitor network to a 555 IC chip

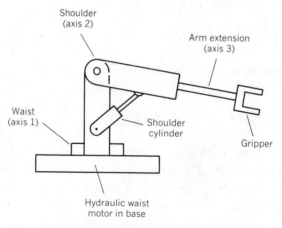

Figure 6.79 Three-axis pick-and-place robot

The hydraulic pump is a positive displacement pump. It feeds a filter, a check valve, an accumulator, and a relief valve. From this point the system branches into three separate circuits: one for waist rotation, another for shoulder elevation, and a third for arm extension.

Each circuit contains a closed center, closed port solenoid control valve. The control valves are spring-centered. In addition to

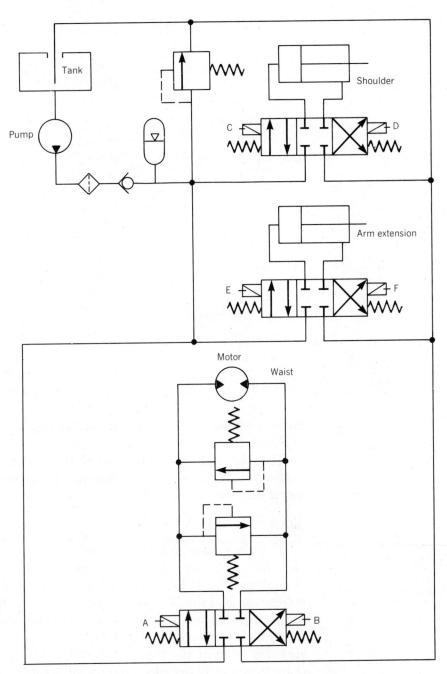

Figure 6.80 Hydraulic circuit for the robot shown in Figure 6.79

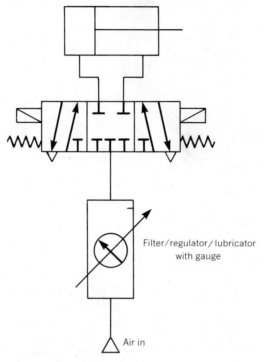

Figure 6.81   Pneumatic circuit for the robot shown in Figure 6.79

these components, the waist circuit has two additional relief valves to absorb shock loads.

The pneumatic circuit (Figure 6.81) has a filter/regulator, a solenoid control valve, and a cylinder. The solenoid control valve is spring-centered.

Block Diagram

The block diagram for the robot control is shown in Figure 6.82. The clock frequency is 1 pulse per second (1 hertz or 1 Hz). The clock drives the D latch. When the enable input is high, the clock pulses that feed the D input are passed to the output. The output of the D flip-flop feeds a ripple counter. The outputs of the ripple counter are connected to a 1 of 16 decoder. A 1 of 16 decoder operates the same as the decimal decoder described earlier in this chapter. A 1 of 16 decoder brings a different line high for each possible binary input. The decoder feeds the program panel, which contains the program steps for the robot.

The program panel feeds a 1 of 10 decoder. The 1 of 10 decoder decodes the program instructions and drives the hydraulic and

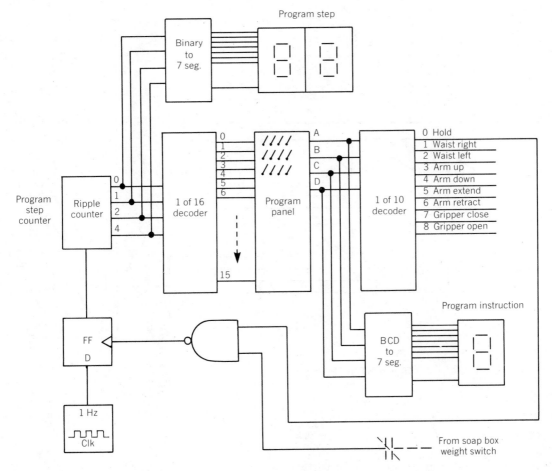

Figure 6.82 Block diagram for robot control

pneumatic solenoids. The control also contains an instruction code display. This display tells the operator what action the robot is taking. Finally, there is a display that tells the operator which step in the program the robot is executing.

Program Panel

The program panel is shown in Figure 6.83. The panel is made up of toggle switches. Setting the toggle switches establishes the moves that the robot will execute. The program instruction codes are shown in Figure 6.84. If the first instruction desired is for the robot to extend its arm, the switches in row 1 are set so that A is on, B is off, C is on, and D is off. The robot will now execute an arm extension as the first step in the program.

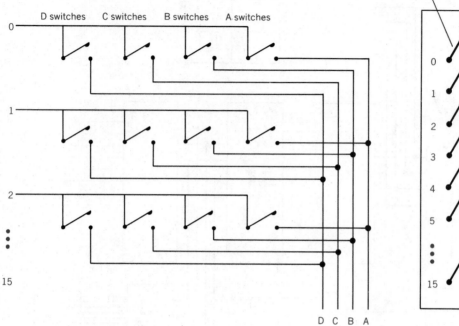

Figure 6.83   Program panel

```
                    D C B A
                    \ | | /
0 — Hold            0000
1 — Waist  right    0001
2 — Waist  left     0010
3 — Arm  up         0011
4 — Arm  down       0100
5 — Arm  extend     0101
6 — Arm  retract    0110
7 — Gripper  close  0111
8 — Gripper  open   1000
```

Figure 6.84   Program instruction codes

If the second step in the program is for the gripper to open, the second row of switches is set so that A, B, and C are off and D is on. This procedure of setting switches is done for each step desired in the program.

Notice that the code being used is a standard binary number. A "hold" instruction is 0000, and the decimal equivalent is 0. An "arm extension" instruction is 0101, and the decimal equivalent is 5. The decimal equivalent of the instruction code is shown on the seven-segment LED display labeled "program instruction."

Stepping Through a Program

To step through the program entered at the program panel, the clock drives the D latch. The binary output of the ripple counter is decoded by the 1 of 16 decoder. The output lines of the 1 of 16 decoder come high one at a time as the ripple counter counts the clock pulses. The output lines of the 1 of 16 decoder are wired to the program panel. Line 1 of the decoder is connected to row 1 of the program panel, line 2 of the decoder is connected to row 2 of the program panel, and so on.

As the 1 of 16 decoder decodes the output from the ripple counter, each subsequent row on the program panel becomes active. The output of the program panel is connected to a 1 of 10 decoder. The output lines of the 1 of 10 decoder are connected to the various hydraulic and pneumatic solenoids that control the movements of the robot. Line 1 of the decoder is wired to solenoid A. This solenoid opens the control valve in the waist circuit, and the robot swings to the right. Line 2 of the decoder is wired to solenoid B, which drives the robot to the left. Each solenoid is wired to the appropriate output of the decoder so that the robot responds properly to each program instruction.

Line 0 of the decoder is not wired to a solenoid but rather to the NAND gate. When instruction 0_{10} (binary 0000) is executed, one input to the NAND gate is driven high. If the other input to the NAND gate is high, the output of the NAND gate drops low. When the output of the NAND gate is low, the clock input of the D latch is low. The latch will no longer pass pulses to the ripple counter, and the robot "holds" its position.

The second input of the NAND gate from the "soap box weight switch" is called a "robot input." The inputs of a robot are used to transmit to the robot information about its environment. We will cover inputs in detail in Chapter 10. When the box is empty, the robot must wait ("hold"). After the box is full the weight switch opens. Power is no longer applied to the NAND gate. The output of the NAND gate goes high, and clock pulses pass through the D latch and the ripple counter. The robot continues through the program steps.

Figure 6.85 shows the layout of the soap box filler and the robot. The starting position (home position) of the robot is arm left and down. The arm is retracted and the gripper is open.

The program steps for loading and unloading the soap box filler are as follows:

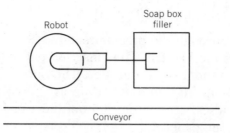

Figure 6.85 Soap box filler and robot

Step	Instruction code	Switch setting
0	Arm up (3)	0011
1	Arm extended (5)	0101
2	Gripper closed (7)	0111
3	Arm retracted (6)	0110
4	Swing left (2)	0010
5	Arm extended (5)	0101
6	Gripper open (8)	1000
7	Hold (0)	0000
8	Gripper closed (7)	0111
9	Arm retracted (6)	0110
10	Swing right (1)	0001
11	Arm extended (5)	0101
12	Gripper open (8)	1000
13	Arm retracted (6)	0110
14	Arm down (4)	0100

With the switches set, the robot will step through the program. On execution of program step 0, the robot's arm is raised through activation of solenoid A (see Figure 6.80). With solenoid A open, hydraulic fluid flows into the base of the cylinder and the arm raises. One second later the clock advances the ripple counter and the robot executes the next step in the program. Solenoid A is turned off and solenoid F is turned on. Fluid flows into the base of the arm extension cylinder. The robot reaches. One second later the clock advances the counter and the next step in the program is executed. The gripper closes around the empty box. The robot executes program steps 3, 4, 5, and 6, locating the box under the soap dispenser. Step 7 "holds" the robot while the soap box is filled.

When the soap box is full the weight switch opens, shutting off the soap dispenser. The weight switch is also wired to the NAND gate in the robot. When the weight switch opens, the output of the NAND gate goes high, which allows the robot to continue through the rest of the program steps. The robot completes the program by returning to its home position.

QUESTIONS FOR CHAPTER 6

1. The simplest type of flip-flop is a _____ flip-flop.

2. Write the truth table for the flip-flop above.

3. What is the value of a clocked flip-flop?

4. What is the difference between a D latch and a D flip-flop?

5. Draw the symbol for a negative edge triggered D flip-flop with inverted PRESET and CLEAR inputs.

6. Explain the operation of a J-K flip-flop with the J input high and the K input low, with K high and J low, and with J and K high at the same time.

7. What is the advantage of the master/slave flip-flop?

8. Convert the following numbers to their decimal equivalents:

$$1111_2 = \text{_____}$$

$$1000_2 = \text{_____}$$

$$1001_2 = \text{_____}$$

$$1101_2 = \text{_____}$$

9. Convert the following numbers to their BCD equivalents:

$$100_{10} = \text{_____}$$

$$576_{10} = \text{_____}$$

$$21_{10} = \text{_____}$$

$$452_{10} = \text{_____}$$

10. Draw the circuit for a ripple counter.

11. What is the advantage of a synchronous counter?

12. Convert the following numbers to their binary equivalents, and then find the 2's complement of each:

$12_{10} =$ _____   2's complement = _____

$9_{10} =$ _____   2's complement = _____

$5_{10} =$ _____   2's complement = _____

$3_{10} =$ _____   2's complement = _____

13. For the following subtraction problems, convert the numbers to binary and then subtract using the 1's and 2's complements:

$$\begin{array}{r} 25 \\ -22 \\ \hline \end{array}$$

$$\begin{array}{r} 9 \\ -3 \\ \hline \end{array}$$

$$\begin{array}{r} 7 \\ -10 \\ \hline \end{array}$$

14. Explain the advantages and disadvantages of the seven-segment LED.

CHAPTER SEVEN

Operational Amplifiers, DAC's, and ADC's

The circuits examined up to this point have been digital circuits. Digital circuits operate at saturation or cut-off, meaning that they are fully on or fully off. Circuits of another class, called **analog circuits,** operate between saturation and cut-off.

The **Operational Amplifier (Op Amp)** is one of the most popular analog circuits. It was developed in the 1940's to perform mathematical computations and was the basis for the development of the first computer. The first computers were analog rather than digital.

Op Amps were originally built using vacuum tubes, resistors, and capacitors. In the 1950's, transistors were substituted for the vacuum tubes, and engineers started applying the Op Amp to industrial control applications.

In 1965, Robert Widlar developed the first **integrated circuit (IC) Op Amp** while doing research for Fairchild Semiconductor. Widlar's design was marketed as the μA709. The μA709 was so well designed that it is still a popular choice today. The development of Op Amps has continued, and today there are more than 2000 designs on the market.

The reason for the popularity of the Op Amp is its versatility. By connecting just a few components to the IC chip it can be made to perform addition, subtraction, multiplication, division, logarithms, and integral and differential calculus, as well as to amplify small signals to levels that can be used for control of other devices. The most popular of all designs, which is a modification of Widlar's original design, is called the μA741 (Figure 7.1). The μA741 requires a few less components connected to the external leads of the chip and can produce the virtually ideal operating characteristics which Widlar was seeking.

Figure 7.1   The μA741 IC Op Amp

IDEAL OP AMP CHARACTERISTICS

Before studying the actual operating characteristics and applications of an Op Amp, we will consider the characteristics of the ideal Op Amp.

Infinite Gain

The ideal Op Amp would be able to accept very small input signals and amplify them thousands of times. Even if the signal were almost nonexistent, the amplifier would be able to amplify the signal to whatever output level was needed. The ideal Op Amp would have infinite gain.

Infinite Input Impedance

The ideal Op Amp would have very high input impedance (resistance). Input impedance is important for Op Amps because a low input impedance reduces the input signal. If the input signal is very small and the input impedance is not high with respect to the source impedance, the input signal will be reduced to the point where it is lost. The higher the input impedance the smaller the effect on the input signal. The ideal Op Amp would have infinite input impedance.

Zero Distortion

It is critically important that the output signals of an Op Amp are not distorted. When a very small signal enters the Op Amp, the amplifier may modify the signal so that the output is no longer a faithful reproduction of the input. If this happens, the signal is said to be distorted. Figure 7.2 shows an input signal and a distorted output signal.

Many amplifiers are able to amplify some signals without distortion but not all signals. The ideal Op Amp would amplify constant-voltage signals as well as signals that oscillate at infinity.

Zero Output Impedance

The ideal Op Amp would have zero output impedance. A high output impedance reduces the effective output of an amplifier. The closer the output impedance is to zero the better the output will be transferred to the next circuit or load.

IC OP AMPS

The ideal characteristics of an amplifier cannot be attained, but the IC Op Amp comes closer to the ideal than any other design. It is not necessary to analyze the design of Op Amps or study their schematics in order to put the Op Amp to work in control circuits. It is,

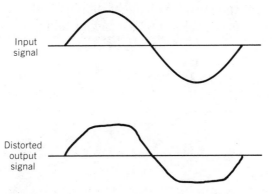

Input signal

Distorted output signal

Figure 7.2 Input signal and distorted output signal

however, important to be able to predict the output of an Op Amp based on its operating characteristics and on how its characteristics are modified by connection of external components.

OP AMP SCHEMATIC SYMBOL

The schematic symbol for an Op Amp is shown in Figure 7.3. Notice that there are five leads attached to the amplifier. There are two input leads, one output lead, and two power supply leads.

Power Supply Inputs

The power supply leads deliver power to the Op Amp. The Op Amp is made up of many transistors and resistors. The transistors and resistors must be powered in order to operate.

Normally the power needed by an Op Amp is ±15 V. It is common practice to omit the power supply leads in schematic drawings. You should remember, however, that in an actual circuit the Op Amp must be supplied with power.

Inverting Input

The input leads of the Op Amp are labeled minus (−) and plus (+). The minus input is an "inverting input." A positive voltage applied to the inverting input exits the Op Amp's output as a negative voltage.

If the input signal is a sine wave and is connected to the inverting input, the output signal will be a sine wave that is 180° out of phase with the input (Figure 7.4). As the input signal's voltage rises positive, the output voltage of the Op Amp goes negative. After the input voltage peaks at 90°, it begins falling toward 0 and the output

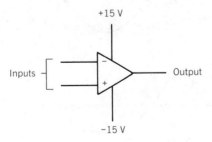

Figure 7.3   Schematic symbol for an Op Amp

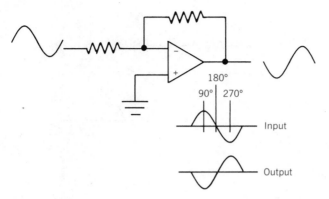

Figure 7.4 When input signal is connected to inverting input terminal, output is 180° out of phase with input

voltage rises from its maximum negative value toward 0. When the input voltage is 0 the output voltage is 0. As the input voltage goes negative the output goes positive. At 270°, the input voltage reaches the most negative voltage and the output is at the highest positive voltage. Finally, the input voltage rises from its most negative point and the output voltage falls from its most positive point toward 0.

Noninverting Input

The input labeled + (plus) is the noninverting input. A signal applied to the noninverting input of an Op Amp is reproduced at the output. If the input signal is positive the output signal is positive, and if the input signal is negative the output signal is negative. If the signal applied to the noninverting input is a sine wave, the output of the Op Amp will be in phase with the input. As the input voltage goes positive the output voltage goes positive, and as the input voltage goes negative the output voltage goes negative (Figure 7.5).

DIFFERENTIAL INPUTS

The inverting and noninverting inputs are called **differential inputs.** The Op Amp amplifies only the difference between the inverting input and the noninverting input. If the inverting input has a 2-V positive voltage applied to it and the noninverting input has a 5-V positive voltage applied to it, the Op Amp will amplify only the difference, or +3 V (Figure 7.6).

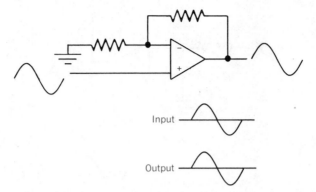

Input

Output

Figure 7.5 When input signal is connected to noninverting input terminal, output is in phase with input

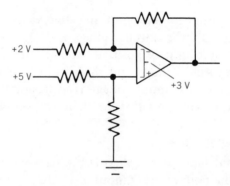

+2 V

+5 V

+3 V

Figure 7.6 Output resulting from differential inputs

Common Mode Rejection

The characteristic of amplifying only the difference between the inverting and noninverting inputs is called **common mode rejection.** The Op Amp rejects all signals that are "common" to both inputs (signals that are identical).

If the signals applied to the inverting input and the noninverting input are identical, the output will be 0. Figure 7.7 shows an Op Amp with 3 V applied to both the inverting and noninverting inputs. The output of the Op Amp is 0.

Figure 7.8 shows an Op Amp with 3-V dc signals applied to both the inverting and noninverting inputs. In addition to the 3-V dc inputs there is a sine wave applied to the inverting input terminal. The Op Amp rejects the 3-V dc inputs applied to both the inverting and noninverting inputs and amplifies only the sine wave.

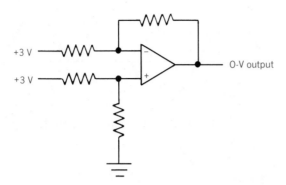

Figure 7.7 Zero output due to common mode rejection

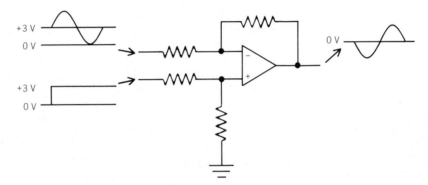

Figure 7.8 Example of common mode rejection where 3 V dc is subtracted from the sine wave input

Electrical Noise

Common mode rejection is an important feature when an Op Amp is installed in a robot or other automated system control, because of the high levels of **electrical noise** that often are present in industrial plants. Electrical noise is defined as unwanted voltage fluctuations. Electrical noise can be generated by external sources or within the control itself. The very fast switching of logic circuits can generate noise.

Noise can be shown as small ripples on a sine wave (Figure 7.9). On the left is a pure sine wave and on the right is the same sine wave with noise added.

Figure 7.10 shows a sine wave signal with noise being fed to the inverting input of an Op Amp. Just the noise is being fed to the non-inverting input. The output shows a pure sine wave. All the noise is being rejected by the Op Amp because of common mode rejection.

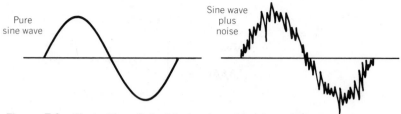

Figure 7.9   Illustration of electrical noise added to a sine wave

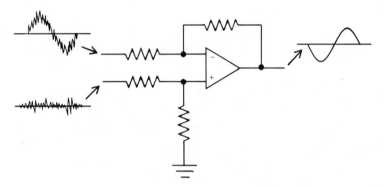

Figure 7.10   Noise-free output due to common mode rejection

Maximum Output Voltage

The supply voltage requirement of most Op Amps is ± 15 V. It is reasonable to assume that the output voltage of the Op Amp cannot exceed the supply voltage, and this is indeed the case. In fact, the output voltage is always somewhat less than the supply voltage due to internal losses in the Op Amp. If the supply voltage is ± 15 V, the **maximum output voltage** is normally between ± 13 and ± 14 V. The Op Amp's output is at a maximum when the amplifier is "saturated." If the input voltage is high enough to drive the Op Amp into saturation, any further increase in the input voltage will not increase the output voltage.

Op Amp Gain

The amount that the input signal is amplified is called **gain**. The formula used to compute the gain of an Op Amp is:

$$A_v = \frac{V_{out}}{V_{in}}$$

where A_v is the gain of the amplifier, V_{out} is the output voltage, and V_{in} is the input voltage. Op Amp gain ranges from 10,000 to 200,000 times. Since the gain of an Op Amp is so high, a very small input voltage will drive the Op Amp into saturation.

Suppose that an Op Amp has a gain of 100,000 and that the maximum output voltage is 14 V at saturation. The maximum input voltage can be calculated by rearranging the gain formula:

$$V_{in} = \frac{V_{out}}{A_v}$$

$$V_{in} = \frac{14}{100,000}$$

$$V_{in} = 0.00014 \text{ V}$$

$$= 0.14 \text{ mV}$$

$$= 140 \ \mu\text{V}$$

VIRTUAL GROUND

Figure 7.11 shows an Op Amp with the noninverting input terminal grounded. Since the noninverting input terminal is grounded, the input voltage on the noninverting terminal is 0. An input of 0.00014 V (0.14 mV) is applied to the inverting terminal, and the output of the Op Amp is −14 V (saturated). Since the maximum input voltage is so small, the inverting terminal is considered to be "virtually grounded." The inverting terminal is said to be at **virtual ground.** The input terminal "appears" to be grounded with respect to voltage. If an oscilloscope is connected to the inverting

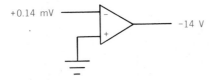

+0.14 mV   −14 V

Since input voltage is almost 0 V,
it is called "virtual ground" and
can be considered a "short to voltage."

Figure 7.11 Op Amp with noninverting input grounded and inverting input at "virtual ground"

terminal, no signal will appear on the scope. "Virtual ground" applies only to input voltage, not to current. In fact, the input "appears" to be open or have infinite resistance with respect to current. Engineers often refer to this unusual characteristic of an Op Amp as a "short to voltage" and an "open to current."

INPUT IMPEDANCE

The ideal Op Amp would have infinite **input impedance**. Actual Op Amps do not approach this ideal but do have very high input impedances. The actual input impedances that are available vary from 10,000 to 1,000,000 Ω (10 kΩ to 1 MΩ). Since the input impedance is so high, almost no current flows into the input terminals (open to current).

Returning to our previous example of an Op Amp with a gain of 100,000, and assuming an input impedance of 100,000 Ω (100 kΩ), we can compute the total current flowing into the input terminal. Recall that we computed the input voltage to be 0.00014 V. Applying Ohm's law (I = V/R), we find that the total input current is:

$$I = \frac{0.00014}{100,000}$$

$$I = 0.0000000014 \text{ A}$$

$$= 14 \times 10^{-10} \text{ A}$$

$$= 0.14 \text{ nA}$$

$$= 140 \text{ pA}$$

The total input current flowing into the input is 140 pA (Figure 7.12).

OUTPUT IMPEDANCE

The ideal Op Amp would have zero **output impedance**. Again, actual Op Amps do not approach the ideal but do have very low output impedances. The output impedances of commercial IC Op Amps vary from 50 to 200 Ω. If external components are added to an Op Amp, its output impedance will approach 0.

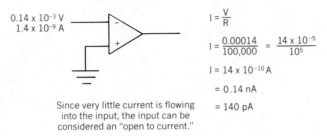

Input impedance = 100 kΩ (100,000 Ω)

0.14 x 10^{-3} V
1.4 x 10^{-9} A

$I = \dfrac{V}{R}$

$I = \dfrac{0.00014}{100,000} = \dfrac{14 \times 10^{-5}}{10^5}$

$I = 14 \times 10^{-10}$ A

$= 0.14$ nA

$= 140$ pA

Since very little current is flowing into the input, the input can be considered an "open to current."

Figure 7.12 Computation of maximum input current to an Op Amp

OPEN LOOP AND CLOSED LOOP GAIN

Op Amps have a maximum gain of 200,000 before addition of external feedback resistors. When there are no feedback components the Op Amp is said to be operating in the **open loop** mode. Op Amp gain can be controlled by addition of input and feedback resistors. When a feedback resistor is added, the Op Amp is said to be operating in the **closed loop** mode.

Figure 7.13 shows an Op Amp with the input signal feeding the inverting input. The gain of the Op Amp is -10. The gain is written as a negative number because an input signal applied to the inverting input is inverted by the Op Amp. For example, with a positive input signal of 1 V the output will be -10 V. With the gain

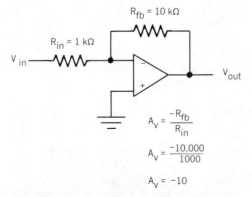

$R_{fb} = 10$ kΩ

$R_{in} = 1$ kΩ

V_{in}

V_{out}

$A_v = \dfrac{-R_{fb}}{R_{in}}$

$A_v = \dfrac{-10,000}{1000}$

$A_v = -10$

Figure 7.13 Computation of gain in a closed loop Op Amp from known values of input and feedback resistors

reduced to -10, the maximum input voltage of the Op Amp is 1.4 V before the Op Amp is saturated. Notice that the only components added to control the gain of the Op Amp are an input resistor and a feedback resistor.

The formula for determining the gain of an inverting Op Amp given the values of the input resistor and the feedback resistor is:

$$A_v = \frac{-R_{fb}}{R_{in}}$$

where A_v is the voltage gain of the Op Amp, R_{fb} is the value of the feedback resistor, and R_{in} is the value of the input resistor.

Figure 7.14 shows an Op Amp with a gain of -5. The input resistor has a value of 2000 Ω (2 kΩ), and the feedback resistor has a value of 10,000 Ω (10 kΩ). Thus:

$$A_v = \frac{-10,000}{2000} = -5$$

Mathematical Proof of the Inverting Op Amp Gain Formula

Figure 7.15 shows a closed loop Op Amp. The feedback resistor is labeled R_{fb} and the input resistor is labeled R_{in}. For this example we will assume that a positive voltage is being applied to the input resistor. The noninverting input is tied to ground.

Recall that the inverting input is at "virtual ground" with respect to voltage (Figure 7.16). With the inverting terminal at virtual ground, any voltage applied to the input resistor R_{in} will be dropped across the input resistor. This can be expressed mathematically as

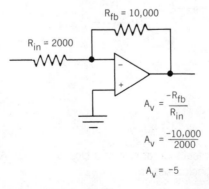

Figure 7.14   Another example of the computation of gain

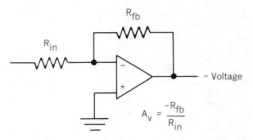

Figure 7.15 Closed loop Op Amp with positive voltage applied to inverting input through input resistor

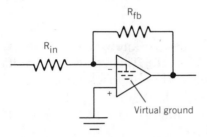

Figure 7.16 Closed loop Op Amp, showing inverting input at "virtual ground"

$V_{in} = V_{Rin}$, where V_{in} is the input voltage and V_{Rin} is the voltage dropped across the input resistor.

The total current flowing through the input resistor can be computed using Ohm's law $(I = V/R)$. For example, if the value of the input resistor is 1000 Ω (1 kΩ) and the input voltage is 5 V, the input current is:

$$I = \frac{5}{1000}$$

$$I = 0.005 \text{ A (or 5 mA)}$$

The voltage applied to the input resistor is positive. Since the input resistor is connected to the inverting input of the Op Amp (Figure 7.16), the output of the Op Amp is a negative voltage. The negative output voltage of the Op Amp is "fed back" through the feedback resistor (Figure 7.17). Since the inverting terminal is at "virtual ground" the total output voltage is dropped across the feedback resistor R_{fb}. This can be expressed mathematically as $V_{out} = V_{Rfb}$, where V_{out} is the output voltage of the Op Amp and V_{Rfb} is the voltage dropped across the feedback resistor.

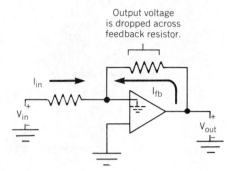

Figure 7.17   Closed loop Op Amp, showing output voltage dropped across feedback resistor

From Ohm's law we know that voltage is equal to current times resistance ($V = IR$). Then, the output voltage equals the feedback current times the feedback resistance. Expressed mathematically this would be $V_{out} = (I_{fb})(R_{fb})$, where V_{out} is the output voltage, I_{fb} is the feedback current, and R_{fb} is the resistance of the feedback resistor.

Current Analysis

Recall that the input impedance (resistance) of an Op Amp is very high. Since the input impedance is so high, virtually no current flows into the Op Amp. Recall that in Figure 7.12 we found that the input current to the Op Amp is only slightly more than one billionth of an ampere. Since virtually no current is flowing into the Op Amp, we can analyze the circuit by assuming that the current flowing through the input resistor must also be flowing through the feedback resistor. The input and feedback resistors therefore appear to be in series (Figure 7.18).

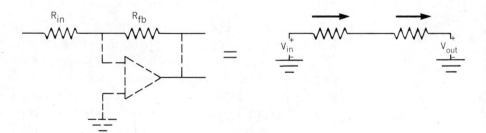

Figure 7.18   Input and feedback resistors appear to be in series when current flow is being analyzed

In a series resistor circuit all the current flowing through one resistor is flowing through the other resistor (Kirchoff's current law). The current flowing through the input resistor R_{in} is equal to the current flowing through the feedback resistor R_{fb}. This is expressed mathematically as $I_{in} = -I_{fb}$, where I_{in} is the input current and I_{fb} is the current flowing through the feedback resistor. Combining this equation with Ohm's law gives:

$$I_{in} = \frac{V_{in}}{R_{in}} = -I_{fb} = \frac{-V_{out}}{R_{fb}}$$

This can be rearranged as:

$$\frac{V_{out}}{V_{in}} = \frac{-R_{fb}}{R_{in}}$$

Recall that gain is defined as $A_v = V_{out}/V_{in}$. This equation can be substituted into the previous equation:

$$A_v = \frac{-R_{fb}}{R_{in}}$$

DIRECT CURRENT OUTPUT OFFSET VOLTAGE

The ideal Op Amp would have an output voltage of 0 when the input voltage is 0. In reality, a small **output offset voltage** appears at the output of the Op Amp when the input voltage is 0. There are four reasons for this: (1) input bias current, (2) input offset current, (3) input offset voltage, and (4) drift.

Input Bias Current

As we have said, the input impedance of an Op Amp is extremely high but is not infinite. In many Op Amps, the current flowing into the input terminal is 140 pA. This small current flowing into the input terminal, which is converted into a small output voltage, is called the **input bias current.**

Input Offset Current

In Op Amps the input bias currents flowing into the inverting and noninverting terminals are not equal. The difference between the two input currents is called the **input offset current.** This difference causes a small output voltage at the output terminal.

The input offset current can be compensated for by adding a resistor to the noninverting terminal and grounding it (Figure 7.19). The value of the compensating resistor can be found by using the following formula:

$$R_c = \frac{R_{in} \times R_{fb}}{R_{in} + R_{fb}}$$

where R_c, R_{in}, and R_{fb} are the values of the compensating, input, and feedback resistors, respectively. Notice that this is the formula used to compute the resistance of any two resistors in parallel. Since the objective of the compensating resistor is to bring the output voltage of the Op Amp to 0 V, we will consider the output to be at 0 V or "virtual ground." If the output is at virtual ground, the feedback resistor and the input resistor will appear to be in parallel. Computing this resistance gives us the input impedance at the inverting terminal. If the same impedance is placed between the noninverting terminal and ground, both inputs will have the identical impedance and the inputs will be in balance, thus eliminating the input offset current.

Input Offset Voltage

In the internal circuitry of the ideal Op Amp, the components would be perfectly balanced. Unfortunately, this is not the case in reality, and the slight differences that exist cause an **input offset voltage.** The current compensating resistor described in the previ

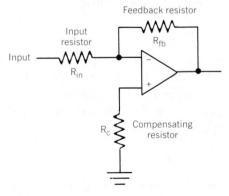

Figure 7.19 Compensation for input offset current by addition of a resistor

ous section will not correct this problem. Compensation for input offset voltage requires additional components (see "Nulling," below).

Drift

The values of the components in an Op Amp can change due to age or temperature. Slight changes in component values can cause input offset voltage. The quality of the Op Amp IC and the environment can cause component values to **drift**. Drift can be eliminated by **nulling** the Op Amp.

Nulling

To correct for input offset voltage, many Op Amps are provided with external connections. A potentiometer is connected across the terminals that are called "offset null" or "balancing" terminals. Figure 7.20 shows how to connect the potentiometer. Notice that the resistor is connected across the balancing or null terminals and that the wiper arm is connected to the negative power supply.

Figure 7.21 shows a method of nulling an Op Amp when balancing terminals are not provided. A potentiometer is connected to the noninverting terminal through an input resistor. Notice that resistor R_c is still used in the circuit to compensate for the input offset current.

When a circuit board is to be replaced in a robot or other automated system and when the board contains an Op Amp, the service manual should be consulted to determine if the board requires nulling. If the board does require offset nulling, be certain to

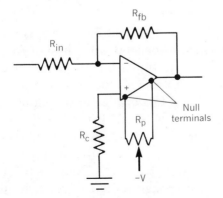

Figure 7.20 Compensation for input offset voltage by "nulling" when balancing terminals are provided

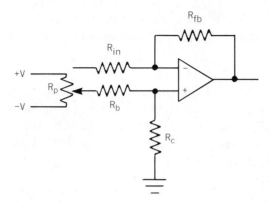

Figure 7.21 Compensation for input offset voltage by "nulling" when balancing terminals are not provided

follow the manufacturer's recommendations to ensure that nulling is done properly. If you fail to null the Op Amp, the robot may operate erratically.

SLEW RATE

Another important operating characteristic of Op Amps is the **slew rate.** The slew rate is the rate of change of the output after an input voltage is applied. The slew rate can be stated mathematically as the change in output voltage divided by the change in time:

$$SR = \frac{\Delta V_{out}}{\Delta t}$$

This slow ramping up of the output voltage is caused by external or internal capacitance associated with the Op Amp. The capacitance may be caused by the transistors or by other components. Since capacitors take time to charge, the output of the Op Amp cannot respond instantaneously to a change in the input voltage.

The Greek letter "delta" (Δ) is used in the formula to indicate a change. The formula is read, "slew rate equals delta V over delta t." The slew rate is measured in volts per microsecond (V/μs). For example, the slew rate of some Op Amps is 0.5 V/μs.

The output will change by 0.5 V for each microsecond that an input voltage is applied. Figure 7.22 shows the output wave form that results when a square wave is applied to the input of an Op Amp. The output of the Op Amp is not a perfect square wave because it takes 8 μs for the output to reach 4 V. If the input pulse duration is less than 8 μs, the output does not have time to reach the full output voltage before the input voltage begins to fall (Figure 7.23).

Stability

In the ideal world, the Op Amp would be perfectly stable — that is, the Op Amp would never go into a self-induced oscillation. Unfortunately, this is not the case. Op Amps may go into oscillation due to internal capacitance or improper compensation. Although nothing can be done with regard to internal capacitance, oscillation normally can be eliminated through proper compensation.

OP AMP ADDER

Op Amps can perform the mathematical operation of addition. When an Op Amp is used for adding it is called an **Op Amp adder** or a "summing amp." Figure 7.24 shows a two-input summing amp. The input resistors are wired to the inverting input. The point at

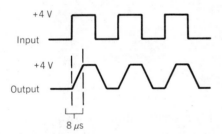

Figure 7.22   Output wave form that results from square wave input due to slew rate

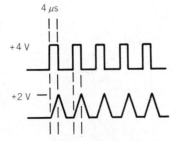

Figure 7.23   Imperfect square wave output that results from insufficient duration of input pulse

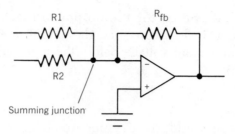

Figure 7.24   Two-input Op Amp adder ("summing amp")

which the resistors are tied together is called the "summing junc-tion." Any voltage applied to the inputs will be added together at the summing junction.

Figure 7.25 shows a summing amp. If a voltage of 2 V is applied to resistor R1, the entire 2 V will be dropped across resistor R1 since the inverting input is at virtual ground. The value of resis-tor R1 is 1000 Ω (1 kΩ).

The current flowing through resistor R1 can be found by Ohm's law:

$$I = \frac{V}{R}$$

$$I = \frac{2}{1000}$$

$$I = 0.002 \text{ A (or 2 mA)}$$

The value of the second input resistor (R2) is also 1000 Ω (1 kΩ). It has 3 V applied to it, and this voltage is dropped across the re-sistor since it is also at virtual ground. The current flowing through R2 is:

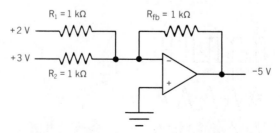

Figure 7.25   Summing amp with input resistors of equal value

$$I = \frac{V}{R}$$

$$I = \frac{3}{1000}$$

$$I = 0.003 \text{ A (or 3 mA)}$$

The total current flowing through the input resistors can be found using Kirchoff's current law, which states that the total current leaving a node is equal to the sum of the currents entering the node. This formula can be written $I_t = I_{R1} + I_{R2}$, where I_t is the total current, I_{R1} is the current through resistor R1, and I_{R2} is the current through resistor R2. Substituting into the formula we find:

$$I_t = I_{R1} + I_{R2}$$

$$I_t = 0.002 \text{ A} + 0.003 \text{ A}$$

$$I_t = 0.005 \text{ A (or 5 mA)}$$

The total current flowing through the input resistors is 5 mA. Recall that in an Op Amp the input resistors and the feedback resistor appear to be in series when current is being analyzed and therefore, in general, the sum of the input currents is equal to the feedback current. This is expressed mathematically as:

$$I_{fb} = I_{in1} + I_{in2} + \ldots$$

where I_{fb} is the total feedback current, and I_{in1}, I_{in2}, etc., are the input currents.

In an inverting Op Amp the feedback current equals negative output voltage divided by feedback resistance. Stated mathematically:

$$I_{fb} = \frac{-V_{out}}{R_{fb}}$$

By substitution:

$$I_{in1} + I_{in2} = \frac{-V_{out}}{R_{fb}}$$

From Ohm's law we know that $I = V/R$, and thus, by substitution:

$$\frac{V_{in1}}{R_{in1}} + \frac{V_{in2}}{R_{in2}} = \frac{-V_{out}}{R_{fb}}$$

By rearranging this formula we find that:

$$V_{out} = -R_{fb}\left(\frac{V_{in1}}{R_{in1}} + \frac{V_{in2}}{R_{in2}}\right)$$

By using this formula, given the feedback resistance, the input voltages, and the values of the input resistors, we can find the output voltage of any summing amp. Using the values given in Figure 7.25 (where $V_{in1} = 2$, $V_{in2} = 3$, $R_{fb} = 1000$, $R_{in1} = 1000$, and $R_{in2} = 1000$), we have:

$$V_{out} = -1000\left(\frac{2}{1000} + \frac{3}{1000}\right)$$

$$V_{out} = -1000\left(\frac{5}{1000}\right)$$

$$V_{out} = -5 \text{ V}$$

From this example we can conclude that if (and only if) all of the input resistors, and the feedback resistor, are of the same value, the output voltage will be the sum of the input voltages.

Figure 7.26 shows a summing amp with input resistors of different values. Our general formula still applies:

$$V_{out} = -R_{fb}\left(\frac{V_{in1}}{R_{in1}} + \frac{V_{in2}}{R_{in2}}\right)$$

$$V_{out} = -1000\left(\frac{2}{1000} + \frac{2}{2000}\right)$$

$$V_{out} = -1000(2 \text{ mA} + 1 \text{ mA})$$

$$V_{out} = -1000(3 \text{ mA})$$

$$V_{out} = -3 \text{ V}$$

This example shows that the output of a summing amp is not the simple sum of the input voltages when the values of the resistors are not identical. Using input resistors of different values is called "weighting." The input voltage at V_{in1} carries the full "weight" and is recognized by the Op Amp as 2 V. Input V_{in2} is recognized as having

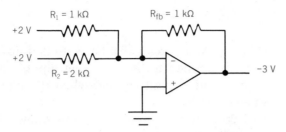

Figure 7.26 Summing amp with input resistors of different values

only half the "importance" of V_{in1} and therefore has an effective input of 1 V. The total output is −3 V. Weighting is used in circuits that convert binary numbers into analog voltages. We will study these circuits later in this chapter.

There is one final resistor combination for a summing Op Amp. The feedback resistor can have a different value than the value of the input resistors. Increasing the resistance of the feedback resistor increases the gain of the amplifier. The Op Amp not only will sum the input voltages but also will multiply them. Figure 7.27 shows a summing amp with 1000-Ω input resistors and a 2000-Ω feedback resistor. This Op Amp will sum the input voltages and multiply the summed voltage by a factor of 2. If both inputs have 1 V applied to them, the output voltage will be −4 V:

$$V_{out} = -R_{fb}\left(\frac{V_{in1}}{R_{in1}} + \frac{V_{in2}}{R_{in2}}\right)$$

$$V_{out} = -2000(0.001 + 0.001)$$

$$V_{out} = -2000(0.002)$$

$$V_{out} = -4 \text{ V}$$

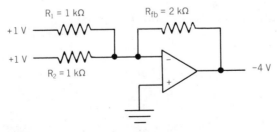

Figure 7.27 Summing amp with 1000-Ω input resistors and a 2000-Ω feedback resistor

NONINVERTING OP AMP

In all the Op Amp circuits we have studied up to this point, the input voltage is applied to the inverting inputs: when the input is negative the output is positive, and when the input is positive the output is negative.

Figure 7.28 shows a **noninverting Op Amp.** The output of the noninverting Op Amp is in phase with the input: if the input is positive the output is positive, and if the input is negative the output is positive.

Gain Formula for Noninverting Op Amps

The formula for computing the gain of a noninverting Op Amp is similar but not identical to that for the inverting Op Amp. Recall that the gain formula for the inverting Op Amp is:

$$A_v = \frac{-R_{fb}}{R_{in}}$$

where A_v is the gain, R_{fb} is the value of the feedback resistor, and R_{in} is the value of the input resistor. The formula for the gain of a noninverting Op Amp is:

$$A_v = \frac{R_{fb}}{R_A} + 1$$

where A_v is the gain, R_{fb} is the value of the feedback resistor, and R_A is the value of the resistor connected between the inverting ter-

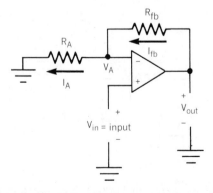

Figure 7.28   Noninverting Op Amp

minal and ground. Notice that the only differences between the two formulas are that in the formula for the noninverting Op Amp there is no minus sign and the resistor ratio is increased by 1.

Mathematical Proof of the Noninverting Op Amp Gain Formula

This formula can be derived quite easily by recalling some of the principles of the inverting Op Amp. Recall that the voltage difference between the noninverting terminal and the inverting terminal is always 0 V. The noninverting Op Amp uses the noninverting terminal as its input. When voltage is applied to the noninverting terminal, the inverting terminal is at "virtually the same potential." If the input voltage is 2 V the inverting terminal will be at approximately 2 V (Figure 7.29).

Resistor R_A is grounded. The input voltage (V_{in}) at the noninverting terminal appears at the inverting terminal and is dropped across resistor R_A.

The current can be computed using Ohm's law: $I_A = V_{in}/R_A$.

The voltage drop across the feedback resistor can be calculated using Ohm's law: $V_{fb} = I_{fb} R_{fb}$, where V_{fb} is the voltage drop across the feedback resistor, I_{fb} is the feedback current, and R_{fb} is the resistance of the feedback resistor. The feedback current is equal to the current in R_A and therefore $V_{fb} = I_A R_{fb}$.

Combining the equations $V_{fb} = I_A R_{fb}$ and $I_A = V_{in}/R_A$ yields:

$$V_{fb} = \left(\frac{R_{fb}}{R_A}\right) V_{in}$$

The output voltage is equal to the voltage drop across the feed-

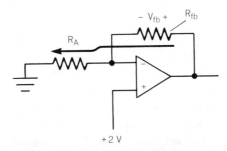

Figure 7.29 Noninverting Op Amp, showing inverting terminal at virtually the same potential as noninverting terminal

back resistor plus the input voltage. Stated mathematically:

$$V_{out} = V_{fb} + V_{in}$$

By combination and substitution:

$$V_{out} = \left(\frac{R_{fb}}{R_A}\right)V_{in} + V_{in}$$

By factoring:

$$V_{out} = \left(\frac{R_{fb}}{R_A} + 1\right)V_{in}$$

By rearrangement:

$$\frac{V_{out}}{V_{in}} = \frac{R_{fb}}{R_A} + 1$$

Recall that gain is defined as:

$$A_v = \frac{V_{out}}{V_{in}}$$

By substitution:

$$A_v = \frac{R_{fb}}{R_A} + 1$$

COMPARISON OF INVERTING AND NONINVERTING OP AMPS

There are several important differences between inverting and noninverting Op Amps:

(1) The output of the noninverting Op Amp is in phase with the input: as the input voltage rises, the output voltage rises; and as the input voltage falls, the output voltage falls. The output of the inverting Op Amp is 180° out of phase with the input: as the input voltage goes positive, the output voltage goes negative; and as the input voltage goes negative, the output voltage goes positive.

(2) The gain of a noninverting Op Amp is never less than 1, whereas the gain of the inverting Op Amp can be less than 1.

(3) The input impedance of the noninverting Op Amp is the input impedance of the Op Amp itself. Since the input of the Op Amp is very high, the input impedance is very high. The input impedance of the inverting Op Amp, on the other hand, is a function of the value of the input resistor. The lower the resistance of the input resistor, the lower the input impedance of the Op Amp.

DIFFERENCING OP AMP

The Op Amp can be used not only to sum input voltages but also to find the difference between two input voltages. Figure 7.30 is the schematic diagram for a **differencing Op Amp** or "difference amp."

Note that in the difference amp neither the inverting input nor the noninverting input is grounded. When signals are applied to both inputs at the same time, only the difference between the two inputs is amplified. If the input signals are identical, the output voltage is 0. Recall that this principle was described earlier as "common mode rejection."

If all the resistors are of the same value, the formula for computing the output voltage is:

$$V_{out} = V_{in2} - V_{in1}$$

If the values of the input resistors are equal, and if the values of re-

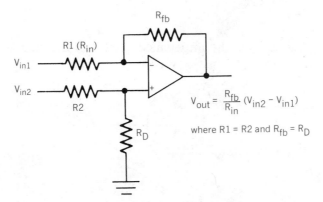

$$V_{out} = \frac{R_{fb}}{R_{in}}(V_{in2} - V_{in1})$$

where R1 = R2 and $R_{fb} = R_D$

Figure 7.30   Schematic diagram for a differencing Op Amp ("difference amp")

sistors R_{fb} and R_D are equal to each other but different from those of the input resistors, the above formula for computing the output voltage will not apply. The more general formula is:

$$V_{out} = \left(\frac{R_{fb}}{R_{in}}\right)(V_{in2} - V_{in1})$$

where $R_1 = R_2$ and $R_{fb} = R_D$ (Figure 7.30).

OP AMP INTEGRATOR

Another important application of the Op Amp is the advanced mathematical operation called integration. Although a full description of integration is beyond the scope of this text, it is possible to understand the output wave forms given an input, and to apply the **Op Amp integrator** to robot applications.

Figure 7.31 is a schematic diagram of an Op Amp integrator. The circuit resembles that of a normal inverting Op Amp. The only difference is that a capacitor has been substituted for the feedback resistor.

When a dc voltage is applied to the input of the Op Amp there is initially a high feedback current through the capacitor. The capacitor is presenting a path of very low impedance to current. The gain of the Op Amp is 0, and the output voltage is 0. This can be seen by substituting values into the gain formula for an inverting Op Amp:

$$A_v = \frac{-R_{fb}}{R_{in}}$$

Assume that the impedance to current flow is 1 Ω and that the value of the input resistor is 1000 Ω. Then:

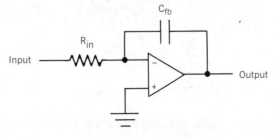

Figure 7.31   Schematic diagram of an Op Amp integrator

$$A_v = \frac{-1}{1000}$$

$$A_v = -0.001$$

As the capacitor begins to charge, the current charging the capacitor goes down. The capacitor is a high-impedance path to current flow. The output of the Op Amp increases. If the impedance of the capacitor is now equivalent to 1,000,000 Ω (1 MΩ), then:

$$A_v = \frac{-1,000,000}{1000}$$

$$A_v = -1000$$

The gain of the Op Amp is -1000, and the output of the Op Amp has reached saturation.

When a dc voltage is applied to the input of an integrator, the output of the integrator is proportional to the input voltage and to the length of time the voltage is applied. If the input voltage is held constant for a long enough time, the output of the amplifier will reach saturation. The formula for the output voltage of an Op Amp integrator is:

$$V_{out} = -\left(\frac{t}{RC_{fb}}\right)V_{in}$$

where R is input resistance, C_{fb} is capacitance, V_{in} is input voltage, and t is time. This formula is applicable only if the input voltage is constant.

Figure 7.32 shows a graph of input voltage compared with output voltage. At T1, a negative dc voltage is applied to the input of the integrator. The output voltage begins to climb. At T2, the Op Amp saturates. The output voltage is at its maximum. Notice that the input voltage is negative and that the output voltage rises from 0 V to the maximum positive voltage.

Figure 7.33 shows the output of an integrator wired to a dc motor. When a negative voltage is applied to the input of the integrator, the output voltage of the integrator begins to increase. As the output voltage increases, the speed of the motor increases. The speed of the motor continues to increase until the Op Amp saturates. The acceleration of the motor is controlled by the output of the integrator.

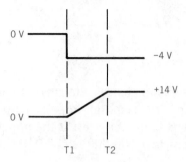

Figure 7.32 Input voltage vs. output voltage for an Op Amp integrator

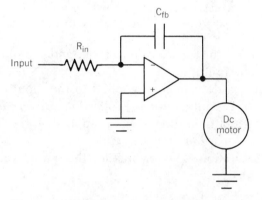

Figure 7.33 Op Amp integrator with output wired to a dc motor

Figure 7.34 shows the output of the integrator when the input is a square wave. When the input voltage goes positive, the output of the integrator begins to go negative. When the input to the integrator goes negative, the output of the integrator begins to climb. When the input to an integrator is a square wave, the output is a "sawtooth."

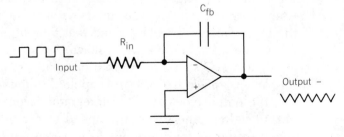

Figure 7.34 Sawtooth output of an Op Amp integrator resulting from a square wave input

OP AMP DIFFERENTIATOR

Differentiation is another high-level mathematical operation which can be performed by an Op Amp. Figure 7.35 is the schematic diagram of an **Op Amp differentiator.** Notice that a capacitor is connected to the input of the Op Amp rather than being connected in the feedback loop, which was the case for the integrator.

The output of a differentiator is proportional to the rate of change of the input voltage. Differentiation should not be confused with difference. Difference is simply subtraction, whereas differentiation is a function of the rate of change of the input voltage.

If the input voltage to a differentiator is a dc voltage and remains constant, the output of the differentiator is 0 (Figure 7.36). If the input voltage changes slowly but at a constant rate, the output of the differentiator will be a small constant dc voltage. If, on the other hand, the input voltage is changing rapidly but at a constant rate, the output of the differentiator will be a higher constant voltage (Figure 7.37).

Figure 7.38 shows a sawtooth input to a differentiator. The output of the differentiator is a square wave. Now compare the output of the differentiator in Figure 7.38 with that of the integrator in Figure 7.34. When the input to an integrator is a square wave, the out-

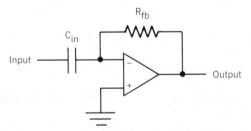

Figure 7.35 Schematic diagram of an Op Amp differentiator

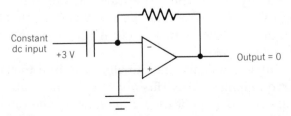

Figure 7.36 Op Amp differentiator with zero output resulting from a constant dc input

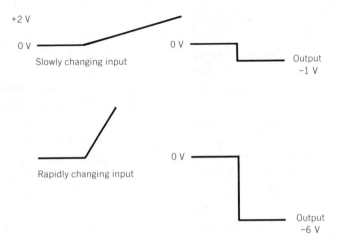

Figure 7.37   Outputs of an Op Amp differentiator resulting from slowly and rapidly changing input voltages

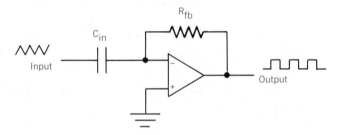

Figure 7.38   Square wave output of an Op Amp differentiator resulting from a sawtooth input

put is a sawtooth. When the input of a differentiator is a sawtooth, the output is a square wave.

Figure 7.39 shows the result of using a square wave as the input signal to a differentiator. A square wave is changing its voltage level as rapidly as possible. The differentiator responds by outputting a high voltage. As soon as the input voltage stabilizes, the output of the differentiator returns to 0. A square wave input creates a series of voltage spikes.

Integrators and differentiators are used to control the outputs of servo amplifiers used in robots and other automated controls. We will cover the application of integrators and differentiators in Chapter 9.

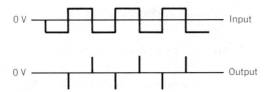

Figure 7.39   Result of a square wave input to an Op Amp differentiator

DIGITAL-TO-ANALOG CONVERTERS

The outputs of computers are binary numbers. If a computer is to be used to control motors or servo valves, its binary output must be converted to an analog voltage. The circuit used to make this conversion is the **digital-to-analog converter,** normally abbreviated **D/A** or **DAC**.

Binary Weighted Resistor Ladder

The major component used to convert digital information into analog information is the Op Amp. By selection of resistors of the proper values, the DAC can be made to accurately convert a binary number into an analog voltage.

The resistors used in a DAC are arranged into a "ladder." This arrangement of resistors is called a **binary weighted resistor ladder** (Figure 7.40). The circuit shown in Figure 7.40 can accept any bi-

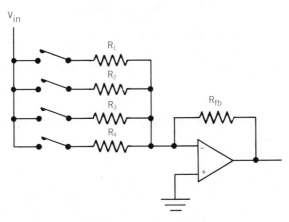

Figure 7.40   Binary weighted resistor ladder

nary number between 0000 and 1111 and convert it into a unique voltage.

The resistor values in this circuit will be selected so that the binary number 0000 will generate an output of 0 V and the binary number 1111 will generate the highest voltage. For example, if the circuit is set such that the binary number 1111 (15_{10}) is equal to -7.5 V and the number 0000 is equal to 0 V, then 0001 (1_{10}) will equal -0.5 V ($\frac{1}{15}$ of -7.5 V), 0010 will equal -1.0 V, and 0011 will equal -1.5 V. As the binary number increases by 1, the output voltage of the DAC increases by -0.5 V.

Choosing Resistor Values.   Figure 7.41 shows the values of the resistors as R, 1/2R, 1/4R, and 1/8R. The values of the resistors are the reciprocals of the column weights in the binary system (Figure 7.42). The least significant bit (LSB) in the binary system has a decimal value of 1, and the value of the resistor is 1/1 or R. The next most significant bit in the binary system has a decimal value of 2, and the value of the resistor is 1/2R. The next bit in the binary system has a value of 4 in the decimal system. The value of

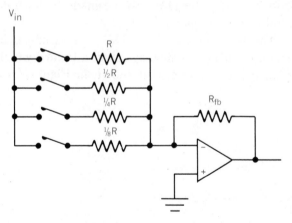

Figure 7.41   Binary weighted resistor ladder showing relative resistor values

MSB			LSB
$2^3 = 8$	$2^2 = 4$	$2^1 = 2$	$2^0 = 1$
$\frac{1}{8}$R	$\frac{1}{4}$R	$\frac{1}{2}$R	$\frac{1}{1}$R

Figure 7.42   Chart illustrating that the values of resistors in a binary weighted resistor ladder are the reciprocals of the column weights in the binary system

the resistor is 1/4R. The most significant bit (MSB) is 8 in the decimal system. The value of the resistor is 1/8R.

Figure 7.43 shows a DAC with input resistor values of 8000 Ω (R), 4000 Ω (1/2R), 2000 Ω (1/4R), and 1000 Ω (1/8R).

To convert the binary number 0001 (1_{10}) to an analog output voltage, a voltage is applied to resistor R (Figure 7.44). For this example, the input voltage will be 4 V. Recall that the formula for computing the output voltage of a summing Op Amp is:

$$V_{out} = \left(\frac{-R_{fb}}{R_{in}}\right)V_{in}$$

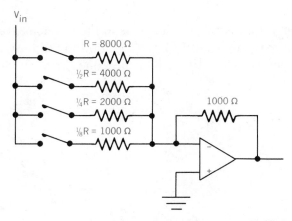

Figure 7.43 DAC with input resistor values of 8000, 4000, 2000, and 1000 Ω

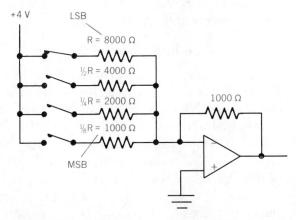

Figure 7.44 Circuit shown in Figure 7.43 with a voltage applied to resistor R

Thus, by substitution:

$$V_{out} = \left(\frac{-1000}{8000}\right)4$$

$$V_{out} = -0.5 \text{ V}$$

If the binary output is 0010 (2_{10}), resistor 1/2R is connected to the 4-V supply and all other resistors are open. The output of the DAC is:

$$V_{out} = \left(\frac{-1000}{4000}\right)4$$

$$V_{out} = -1 \text{ V}$$

The output is twice as large as in the previous example. Now, if the binary number 0011 (3_{10}) is inputted to the DAC, both the first and second resistors are high at the same time. The output voltage is:

$$V_{out} = \left(\frac{-R_{fb}}{R_{in}}\right)V_{in} + \left(\frac{-R_{fb}}{\frac{1}{2}R_{in}}\right)V_{in}$$

$$V_{out} = \left(\frac{-1000}{8000}\right)4 + \left(\frac{-1000}{4000}\right)4$$

$$V_{out} = -0.5 + -1.0$$

$$V_{out} = -1.5 \text{ V}$$

The output of the DAC is three times larger than when only the first resistor was brought high. The DAC is generating output voltages that are proportional to the binary input.

R-2R Resistor Ladder

There is one major problem associated with the binary weighted resistor ladder described above. If the circuit is expanded for conversion of larger binary numbers, the values of the resistors will become very high. High-value resistors are difficult to fabricate if the DAC is an IC chip. This problem is overcome by the **R-2R resistor ladder.** Figure 7.45 shows the design of the R-2R resistor ladder. (Note that when the R-2R DAC is implemented as an IC chip the switches are electronic. They are switched by applying a voltage to the data input lines of the DAC.) There are only two values for the resistors: R and 2R. If each of the R resistors has a value of 1000 Ω (1 kΩ), then each of the 2R resistors has a value of 2000 Ω (2 kΩ).

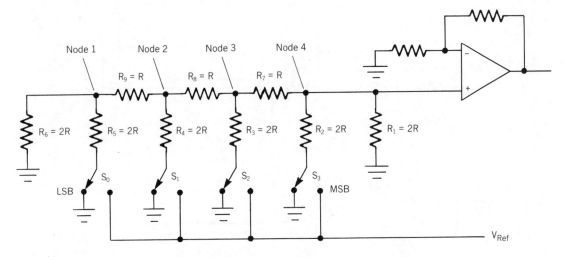

Figure 7.45 R-2R resistor ladder

Note that the resistor ladder is connected to the noninverting terminal of an Op Amp. If the R-2R network were connected to the inverting terminal of the Op Amp, it would not operate properly. Resistor R_1 would be connected between the inverting terminal and ground. Since the inverting terminal is at "virtual ground," we could say that resistor R_1 was connected "between ground and ground." Resistor R_1 would have no effect on the circuit, and the circuit would not operate as a binary resistor ladder.

Examine Figure 7.45 again. All of the switches are set at their grounded positions. If we examine the circuit we will find that there is an equivalent resistance of 2R to the left of, to the right of, and below each node. For example, examine node 1. Resistor R_6, which is to the left of node 1, is a 2R resistor. Resistor R_5, which is connected between node 1 and ground, is also a 2R resistor. However, when we look to the right of node 1 we see a resistor network. The total resistance of this network is not immediately apparent. To determine the resistance of this network, begin with resistors R_1 and R_2. These resistors are both 2R resistors and are in parallel. Figure 7.46 shows the equivalent circuit. With resistors R_1 and R_2 in parallel, the equivalent resistance of the R_1-R_2 network is R. We will call this network (and its equivalent resistance) R_A. R_A is in series with resistor R_7 (Figure 7.47). The resistance of R_7 is R. The equivalent resistance of the R_A-R_7 network (equivalent resistance of R_A plus resistance of R_7) is 2R. We will call this network (and its equivalent resistance) R_B. R_B is in parallel with resistor R_3 (Figure 7.48).

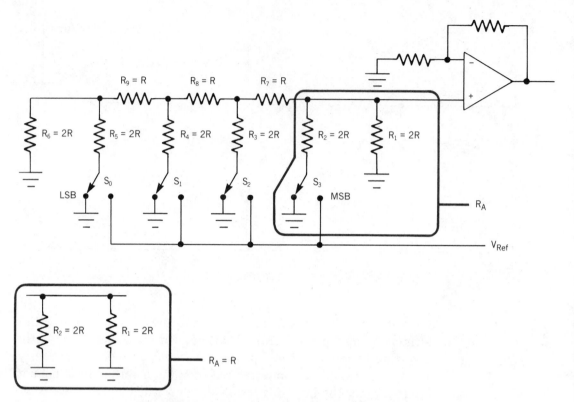

Figure 7.46   Equivalent resistance of $R_1 \| R_2 = R$

The equivalent resistance of the R_B-R_3 network is R. We will call this network (and its equivalent resistance) R_C. R_C is in series with resistor R_8. The equivalent resistance of the R_C-R_8 network is 2R (Figure 7.49). We will call this network (and its equivalent resistance) R_D. R_D is in parallel with resistor R_4. The equivalent resistance of the R_D-R_4 network is R. We will call this network (and its equivalent resistance) R_E (Figure 7.50). R_E is in series with resistor R_9. The equivalent resistance of the R_E-R_9 network is 2R. You can now see that there is a resistance of 2R to the left of, to the right of, and below node 1 (Figure 7.51).

If we perform a similar analysis for any other node in the resistor network, we will find that there is a resistance of 2R to the left of, to the right of, and below that node. For example, look at node 3 in Figure 7.45. Resistor R_3, which is connected between node 3 and ground, is a 2R resistor. The network to the left of node 3 also has a

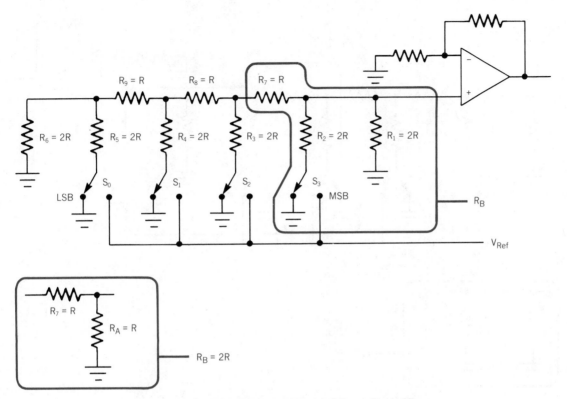

Figure 7.47 Equivalent resistance of $R_A + R_7 = 2R$

resistance of 2R. To prove this, begin with resistors R_6 and R_5. Each of these resistors has a value of 2R. Since they are in parallel, the combined resistance of R_6 and R_5 is R. The R_6-R_5 network is in series with resistor R_9, which also has a value of R. The combined resistance of the R_6-R_5-R_9 network is 2R. By continuing the evaluation in this manner, you will find that the total resistance of the resistor network to the left of node 3 is 2R. Now evaluate the network to the right of node 3. Begin the analysis with resistors R_1 and R_2. Since R_1 and R_2 are in parallel, their equivalent resistance is R. The R_1-R_2 network is in series with resistor R_7, which has a value of R. Thus the total resistance to the right of node 3 is also 2R. It is left to the student to examine the other nodes and prove that the total resistance to the left of, to the right of, and below each node is 2R.

Figure 7.52 shows switch S_1 switched from its ground position to the reference voltage V_{Ref}. With switch S_1 "on," the reference

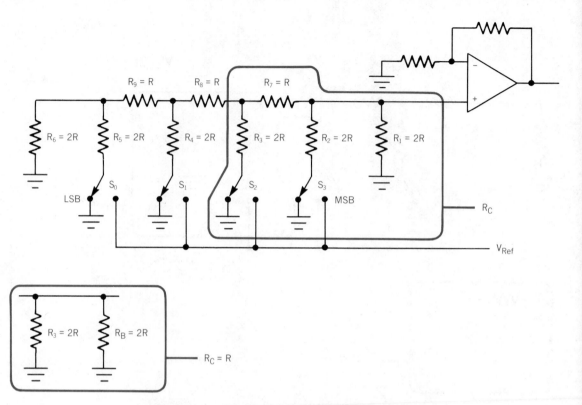

Figure 7.48   Equivalent resistance of $R_B \| R_3 = R$

voltage sees a total resistance of 3R. Recall that there is a resistance of 2R to the left of, to the right of, and below any node. Since the networks to the left and right of node 2 are both grounded, their resistances are in parallel (Figure 7.53). The combined resistance of the left and right networks is R. From this you can see that we have built a voltage divider network. The voltage at the node will be $V_{Ref}/3$. If the reference voltage is 12 V, then the voltage at the node will be 4 V. When a switch is turned on, the voltage of the node associated with that switch will be 4 V.

In Figure 7.54 the MSB switch (S_3) has been turned on. The voltage at node 4 is 4 V, and therefore the voltage at the input of the Op Amp is 4 V.

In Figure 7.55 switch S_2 has been turned on while all other switches remain off. The voltage at node 3 is 4 V. Looking to the

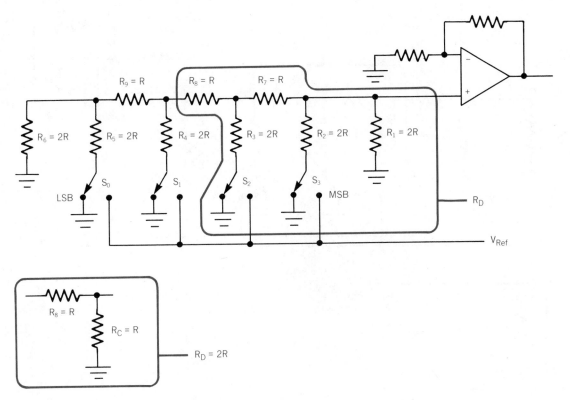

Figure 7.49 Equivalent resistance of $R_C + R_8 = 2R$

right of node 3 we find a voltage divider network made up of resistors R_7, R_2, and R_1. The equivalent circuit is shown in Figure 7.56. The voltage at the input of the Op Amp is 2 V, which is ½ the voltage produced by the MSB.

If switch S_1 is turned on the voltage at node 2 will again be 4 V, but looking to the right we find another voltage divider network. If we analyze this network we will find that the voltage at the Op Amp is 1 V, or ¼ that of the MSB.

Finally, if switch S_0 is turned on while all other switches remain off, the voltage at node 1 will be 4 V, but looking to the right we find another voltage divider network. The voltage at the Op Amp will be 0.5 V, or ⅛ that of the MSB.

Since the weights are 1 for the MSB followed by ½, ¼, and ⅛, we know that the R-2R network is operating as a DAC.

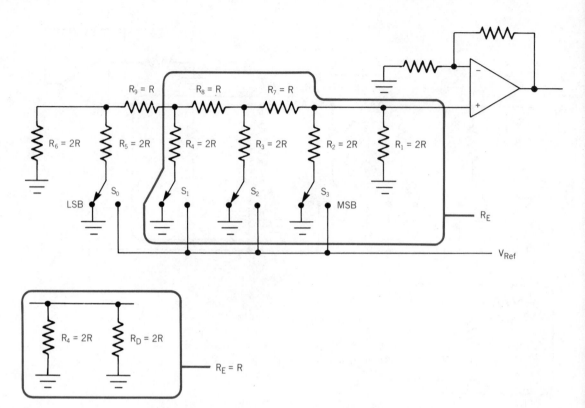

Figure 7.50   Equivalent resistance of $R_D \| R_4 = R$

IC DAC's

It is no longer necessary for engineers to design and build their own DAC's, because IC manufacturers now do the job for them. The resistors, the switches, and the Op Amp are all fabricated on an IC chip. The switches in a DAC are electronic. When a small voltage is applied to the input, the switch turns on. For example, the output of a binary counter can be connected to the "switch" inputs of an IC DAC. If the output of the counter is 0001_2, the LSB switch will turn on. If the reference voltage is 8 V, the output of the DAC will be 0.5 V. If the output of the counter is 1000_2, the MSB switch will turn on. If the reference voltage is 8 V, the output of the DAC will be 4 V. Finally, if the output of the counter is 1111_2, then all the switches will be turned on and the output of the DAC will be 7.5 V (15×0.5).

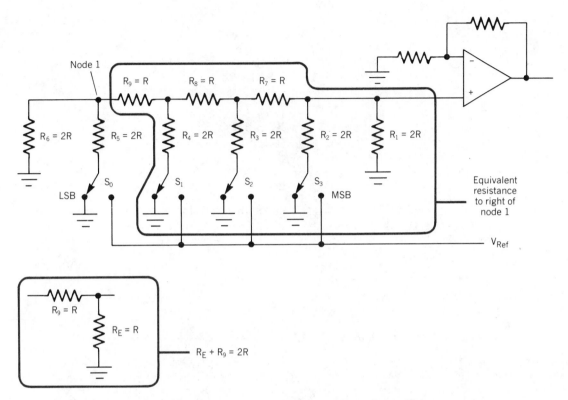

Figure 7.51 Equivalent resistance of $R_E + R_9 = 2R$

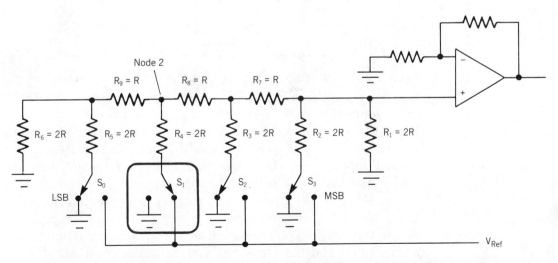

Figure 7.52 R-2R resistor ladder with S_1 switched to reference voltage

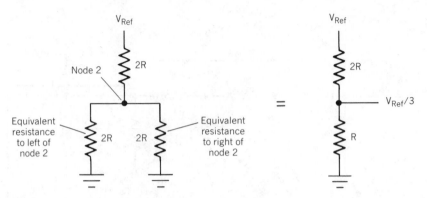

Figure 7.53   Equivalent circuit for R-2R resistor ladder with S_1 switched to reference voltage

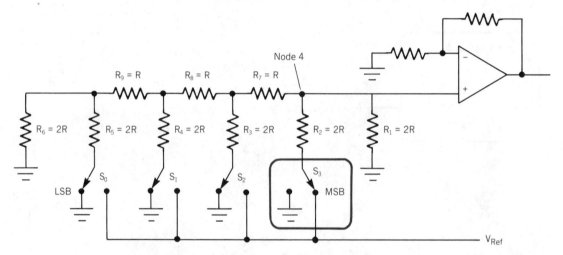

Figure 7.54   R-2R resistor ladder with S_3 (MSB) switched to reference voltage

   In addition to the resistors, the switches, and the Op Amp, many IC DAC's also include a temporary register to store the binary data that is transmitted as well as a "strobe input." Data present at the "switch" inputs of the DAC is not transferred into the DAC until the strobe input receives a pulse. When the strobe input goes high, data is transferred to the register. The register in its simplest form is a group of R-S flip-flops. There is one flip-flop for each input. The data in the register is converted to an analog voltage through the resistor ladder and the Op Amp. The output of the DAC remains constant until new data is present at the input of the DAC and the strobe line is again pulsed.

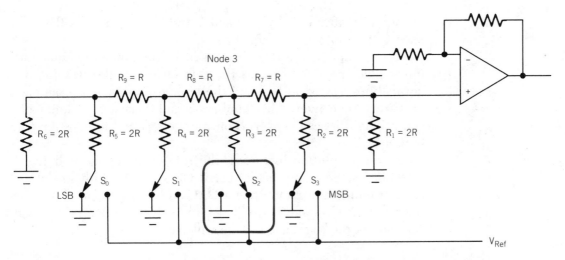

Figure 7.55 R-2R resistor ladder with S_2 switched to reference voltage

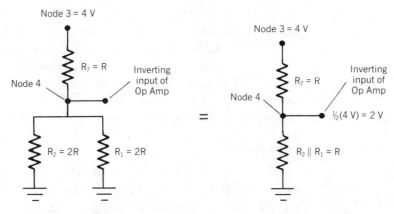

Figure 7.56 Voltage divider network with 4 V applied to node 3

Resolution

The DAC's that have been studied up to this point have been capable of converting binary numbers with 4 bits. The largest possible 4-bit binary number is 1111 (15_{10}). The DAC divides the reference voltage into 15 equal parts.

Many of the DAC's that are available as IC's can decode binary numbers with more than 4 bits. Many of the IC DAC's can convert 8-, 12-, and 16-bit numbers. An 8-bit DAC divides the reference voltage into 255 equal parts, a 12-bit DAC divides the reference voltage into 4095 equal parts, and a 16-bit DAC divides the reference voltage into 65,535 equal parts.

The **resolution** of a DAC is defined as the number of equal parts into which the DAC divides the reference voltage. Resolution is normally expressed in terms of the number of bits. For example, a DAC that can convert 4-bit numbers is said to have a resolution of 4, and a DAC that can convert 8-bit numbers is said to have a resolution of 8.

As the resolution increases, the reference voltage is being divided into smaller and smaller parts. Often a DAC with a resolution of 12 is no more accurate than a DAC with a resolution of 8. When the reference voltage is divided into such small parts, any slight variation of the reference voltage will cause an error.

ANALOG-TO-DIGITAL CONVERTERS

Digital computers accept only digital information for processing. Often the sensors that are connected to a robot produce a voltage or a current rather than a number. The computer which is part of the robot controller cannot accept the information in this form. The analog information must be converted to digital information. The **analog-to-digital converter** performs this function. The standard abbreviation for an analog-to-digital converter is **A/D** or **ADC**.

Comparator

The simplest of all ADC's is the voltage **comparator**. A comparator is simply an Op Amp that is operating without a feedback resistor. The term "comparator" appears in Figure 1.23 (Chapter 1), which is a diagram of the feedback system of a servo controlled robot. The comparator in Figure 1.23 is actually a differencing Op Amp, and not an Op Amp operating in the open loop mode. It is customary to use the term "comparator" in discussions of servo robots. The comparators referred to in the following sections of this chapter are open loop Op Amps. In Chapter 9, servo systems are described in detail, and the term "comparator" is again used to refer to a differencing Op Amp.

Recall that many Op Amps have gains of 10,000 to 20,000 when they are operating in the open loop mode. Figure 7.57 shows an Op Amp with the inverting input grounded. When a positive voltage of less than one millivolt (1 mV) is applied to the noninverting input,

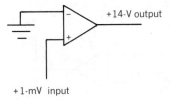

Figure 7.57 Op Amp operating without a feedback resistor and with the inverting input grounded (this Op Amp is called a "comparator")

the Op Amp goes into saturation. The output is approximately 14 V. If the input voltage should go negative by less than 1 mV, the Op Amp will again go into saturation and the output will be approximately −14 V. The output of the Op Amp is 0 only when the input is exactly 0.

If a variable resistor (potentiometer) is connected to the inverting input, the Op Amp can compare the input voltage with reference voltages other than 0 (Figure 7.58). If the potentiometer is set to a reference voltage of 6 V, any voltage applied to the noninverting input that is less than 6 V will drive the Op Amp into negative saturation (−14 V). When the input voltage is exactly 6 V, the output of the Op Amp will be 0 V. When the input voltage exceeds 6 V by less than 1 mV, the Op Amp will go into positive saturation. The output of the Op Amp will be approximately 14 V.

Flash ADC

If several voltage comparators are connected in parallel and each is adjusted to go positive at a different voltage, an analog voltage applied to the network will be converted into a binary number (Figure 7.59). In this circuit a single reference voltage is fed to the comparators through a voltage divider network.

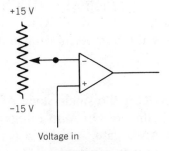

Figure 7.58 Op Amp comparator with variable reference voltage

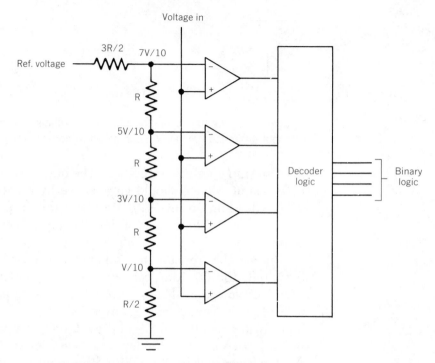

Figure 7.59   Flash ADC

When an unknown voltage is applied to the input, one or more comparators will go high. If the input voltage is small, only the first comparator will go high. As the voltage rises the second comparator will go high. If the input voltage continues to rise the third comparator will also go high.

The outputs of the comparators are connected to a decoder network. The output of the decoder network is a binary number. The binary number is a digital representation of the analog input voltage. This type of ADC is called a **flash ADC.** The flash ADC was given its name because it is the fastest of all ADC's. The accuracy of the flash ADC is limited by the number of comparators in the circuit.

Single-slope ADC

A more common ADC is the **single-slope ADC** (Figure 7.60). A single-slope ADC contains an Op Amp integrator, a comparator, gate control logic, an AND gate, a clock, a counter, and often an output device such as a seven-segment LED.

Figure 7.61 shows a timing diagram for a single-slope ADC. At T1, the control logic brings one leg of the AND gate high and opens

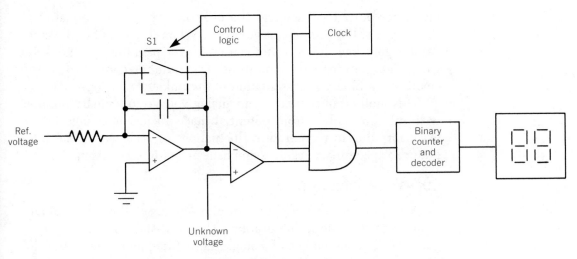

Figure 7.60 Single-slope ADC

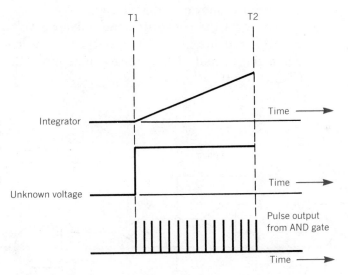

Figure 7.61 Timing diagram for a single-slope ADC

switch S1. The output voltage of the integrator begins to climb. At the same time, an unknown voltage is applied to one input of the comparator. When the output voltage of the integrator and the unknown voltage are not equal, the output of the comparator is high. The clock pulses that are connected to the third input of the AND gate pass through the AND gate and are counted.

The binary counter continues to count pulses until the output of the integrator and the unknown voltage are equal. At the instant the integrator output voltage and the unknown voltage are equal (T2), the output of the comparator goes low. When the output of the comparator goes low, clock pulses can no longer pass through the AND gate. The counter stops counting. The number contained in the counter is a binary representation of the unknown analog voltage.

This method of converting an analog voltage to a binary number is simple but suffers from potential inaccuracies. A change in the clock rate due to temperature fluctuations or a change in the integrator's capacitor will cause significant errors in conversion.

Dual-slope Integrator

A better method of converting an analog voltage to a binary number is dual-slope integration. The **dual-slope integrator** uses many of the components of a single-slope integrator. The major difference between the single-slope integrator and the dual-slope integrator is the control logic. The schematic diagram of a dual-slope integrator is shown in Figure 7.62.

To begin the conversion the unknown voltage is applied to the input of the integrator. The output of the comparator goes high when the output of the integrator is more than several millivolts (Figure 7.63).

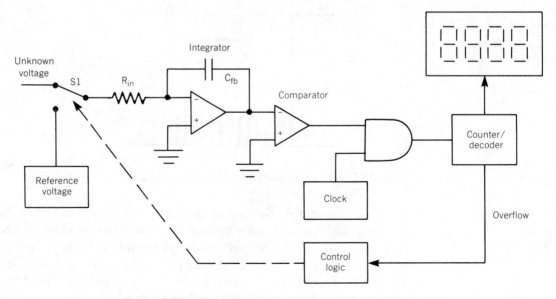

Figure 7.62 Dual-slope integrator

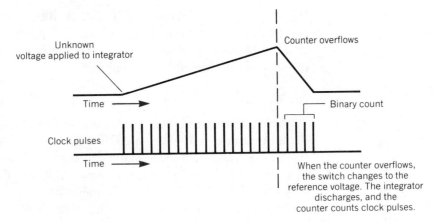

Figure 7.63 Timing diagram for a dual-slope integrator

When the output of the comparator goes high, clock pulses pass through the AND gate to the binary counter. The binary counter counts pulses until so many pulses have been counted that the counter overflows. When the counter overflows, it resets to 0 and sends a signal to the control logic. The control logic switches the input from the unknown voltage to a reference voltage. The polarity of the reference voltage is opposite that of the unknown voltage. The reference voltage discharges the capacitor at a controlled rate. During the discharge cycle, the counter is counting the clock pulses. When the output of the integrator is 0, the comparator goes low, bringing one leg of the AND gate low. Clock pulses can no longer pass through the AND gate, and the counter stops counting. The number stored in the counter is the binary representation of the analog voltage.

The major disadvantage of the dual-slope integrator is its slow conversion rate. The integrator must charge and discharge to complete the conversion cycle.

The major advantage of the dual-slope integrator over the single-slope integrator is its accuracy of conversion. The single-slope integrator depends on the value of the feedback capacitor remaining constant. On the other hand, the dual-slope integrator is independent of changes in the value of the feedback capacitor. The rate at which the capacitor is charged is dependent on the input voltage and not on the value of the capacitor. When the counter overflows, it resets; the input is switched to a reference voltage whose polarity is opposite that of the unknown voltage, and the capacitor discharges at a constant rate. During the time that the capacitor is discharging, the counter is accumulating pulses. When the capacitor has discharged to 0 V, the counter stops and the number in the counter is proportional to the unknown voltage.

Servo ADC

A faster and more accurate means of converting analog signals to digital numbers is the **servo ADC.** The servo ADC is also known as the "ramp ADC." The schematic diagram of the servo ADC is shown in Figure 7.64. Notice that the servo ADC uses a digital to analog converter (DAC) as one of its components.

To begin the conversion, an unknown voltage is applied to the noninverting input of a voltage comparator. A start pulse is applied to the "beginning of conversion" (BOC) input line of the control logic. The output of the comparator is high. The AND gate passes clock pulses to the binary counter. The output of the binary counter is also wired to the DAC. On each clock pulse, the counter increments one and the DAC's output voltage rises (Figure 7.65). When the output voltage of the DAC is greater than the unknown voltage, the comparator drops low. The AND gate can no longer pass clock pulses, and the conversion is complete.

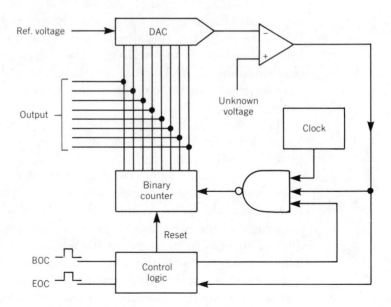

Figure 7.64 Servo ADC

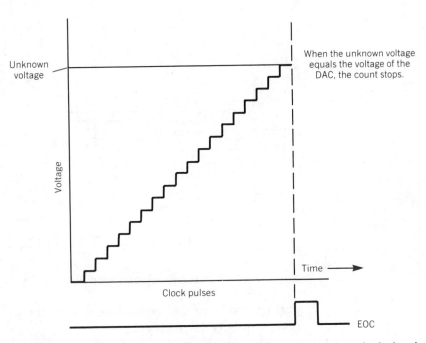

Figure 7.65 Servo ADC, showing increases in DAC output with each clock pulse

Note that the control logic also has an output labeled "EOC." EOC stands for "end of conversion." The EOC line goes high when the output of the DAC matches the unknown voltage. The EOC output is needed to inform other circuits connected to the ADC that the conversion is complete. While the DAC is stepping toward the unknown voltage, the output of the binary counter could be interpreted as the correct conversion. The output of the counter is not correct until the EOC line goes high.

Successive Approximation ADC

The **successive approximation ADC** is the most popular ADC on the market today. Its conversion time is shorter than that of the servo ADC, but its control circuitry is more complex. Figure 7.66 is a schematic diagram of a successive approximation ADC.

The operation of the successive approximation ADC is much like the game "Twenty Questions." In this game, by asking a series of questions and receiving only "yes" or "no" answers, a player attempts to determine what object another player has in mind. Similarly, the successive approximation ADC tests its guess by out-

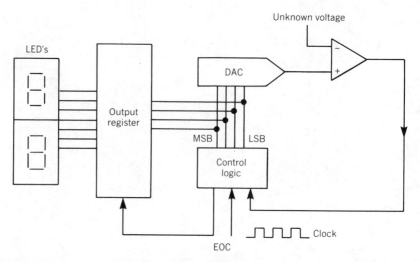

Figure 7.66 Successive approximation ADC

putting a voltage and comparing it with the unknown voltage. The process proceeds as follows.

The control logic brings the MSB high (1) on the first clock pulse while all other bits remain low (0). This number is stored in the control logic and is also outputted to the DAC. The DAC converts the number to a voltage, which is then compared with the unknown voltage. If the output voltage of the DAC is greater than the unknown voltage, the comparator drops low and the bit is reset to 0. If the output of the DAC is less than the unknown voltage, the MSB remains at 1.

The next most significant bit is set to 1 and the comparison is made again. If the voltage generated by the DAC is greater than the unknown voltage, the comparator again drops low and this bit is reset to 0. If the voltage generated by the DAC is less than the unknown voltage, the bit remains at 1.

This sequence of setting bits and testing continues until all bits have been tested. When all bits have been tested, the control logic outputs an EOC to the register. The number stored in the control logic is transferred to the output register and the conversion is complete.

QUESTIONS FOR CHAPTER 7

1. List the ideal operating characteristics of an Op Amp.

2. What is the gain of an inverting Op Amp with an input voltage of 5 V and an output voltage of 10 V?

3. In an inverting Op Amp, by how many degrees is the output out of phase with the input? Why?

4. What term is used to denote the noise rejection of an Op Amp?

5. What is meant by the term "virtual ground"?

6. What is "input offset current," and what can be done to correct this problem?

7. What is meant by the term "slew rate"?

8. Compute the gains of inverting Op Amps with resistors of the following values (hint: the Op Amps for parts b and c are summing Op Amps):

   (a) $R_{in} = 1\ k\Omega$ (b) $R_{in1} = 2\ k\Omega$ (c) $R_{in1} = 1\ k\Omega$
       $R_{fb} = 10\ k\Omega$ $R_{in2} = 3\ k\Omega$ $R_{in2} = 1\ k\Omega$
                         $R_{fb} = 5\ k\Omega$ $R_{fb} = 10\ k\Omega$

9. What is the gain formula for a noninverting Op Amp?

10. Compute the gains of noninverting Op Amps with resistors of the following values:

    (a) $R_{in} = 2\ k\Omega$ (b) $R_{in} = 1\ k\Omega$ (c) $R_{in} = 1\ k\Omega$
        $R_{fb} = 6\ k\Omega$ $R_{fb} = 20\ k\Omega$ $R_{fb} = 0\ \Omega$

11. Draw a differencing Op Amp with a gain of 1 and compute the values of the resistors if $R_{in1} = 2\ k\Omega$.

12. What is the output of an Op Amp integrator when the input is 10 V dc, $C = 25\ \mu F$, $R = 100\ k\Omega$, and the input voltage is applied for 1 second?

13. What is the output of an Op Amp differentiator when the input is a sawtooth?

14. Draw the schematic diagram of an R-2R DAC and explain its operation.

15. Draw the schematic diagram of a dual-slope integrator ADC and explain its operation.

16. Draw the schematic diagram of a successive approximation ADC and explain its operation.

CHAPTER EIGHT

Memories and Microprocessors

MEMORIES

Robots need to store information in the form of programs and data. Sometimes the information is stored for several years, whereas at other times it is stored for only a few milliseconds. Data that must be stored for a long time is stored on magnetic disks or magnetic tapes or in electronic circuits called ROM's. Information that is to be stored for only a short time is stored in flip-flops.

REGISTERS

A flip-flop can store only one bit of information. The bit of information can be either high or low (1 or 0). By arranging flip-flops into groups, thousands of bits of information can be stored.

When flip-flops are arranged into a group, the group of flip-flops is called a **register**. There are four modes by which information can be fed into and read out of a register: Serial In Serial Out (SISO), Serial In Parallel Out (SIPO), Parallel In Serial Out (PISO), and Parallel In Parallel Out (PIPO).

Serial In Serial Out (SISO) Register

Figure 8.1 is a schematic diagram of a **serial in serial out (SISO) register**. This type of register is often referred to as a **shift register**. Data is fed into the register one bit at a time. The first bit is loaded into the first flip-flop. When the second bit is loaded, the first bit is "shifted" from the first flip-flop to the second flip-flop, and so forth.

The SISO shift register is often built using four D flip-flops wired in series. Recall that a D flip-flop accepts data when the clock

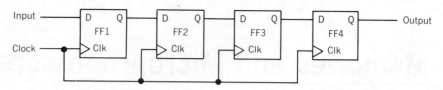

Figure 8.1 SISO shift register

changes from low to high (positive edge triggered). At all other times, the high or low (1 or 0) condition of the input has no effect on the output of the flip-flop.

Timing Diagram for a SISO Register

The timing diagram for a four-bit SISO shift register is shown in Figure 8.2. Notice that all the flip-flops are wired in series. The output of the first flip-flop is connected to the input of the second flip-flop, the output of the second flip-flop is connected to the input of the third flip-flop, and the output of the third flip-flop is connected to the input of the fourth flip-flop. Notice that all the clock inputs of the flip-flops are connected in parallel. All the flip-flops are "clocked" at the same time.

At the first clock transition (T1), the data (D) input of the first flip-flop is high (1). Since D is high during the first clock transition, the first flip-flop is set high.

At T2, the output of the first flip-flop is high (1). Since the output of the first flip-flop is connected to the input of the second flip-

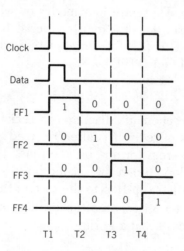

Figure 8.2 Timing diagram for a four-bit SISO shift register

flop, the input of the second flip-flop is also high. When the clock goes from low to high, the second flip-flop is set high. At the same instant, the D input of the first flip-flop goes low, and the output of the first flip-flop also goes low (0). The data has been transferred to the second flip-flop.

At T3, the output of the second flip-flop is high (1). The output of the first flip-flop is low (0). The input to the first flip-flop is 0. At T3, the clock goes from low to high. The 0 at the input of the first flip-flop is shifted into the first flip-flop, the 0 at the input of the second flip-flop is shifted into the second flip-flop, and the 1 at the input of the third flip-flop is shifted into the third flip-flop.

At T4, the input to the first flip-flop is again low. The data again shifts to the right on the low-to-high clock transition. A 0 is loaded into the first flip-flop, a 0 is shifted into the second flip-flop, a 0 is shifted into the third flip-flop, and the 1 in the third flip-flop is shifted into the fourth flip-flop. The register is now fully loaded.

The number stored in the register is 0001. To read this number out of the register, four more clock pulses must be applied. At the first clock pulse, the 1 present in the fourth flip-flop is shifted out and the 0's present in the other flip-flops shift to the right. After four clock pulses, all the data that was loaded has been shifted out.

This type of register is often referred to as a **First In First Out (FIFO) shift register.** The first bit of data loaded into the register is the first bit that is read out after the register is fully loaded.

The SISO shift register is the slowest shift register. The SISO shift register shown in Figure 8.1 requires four clock pulses to load the register and four clock pulses to read out the data stored in the register.

Many SISO shift registers are constructed with more than four flip-flops. Any number of flip-flops can be connected in series. The more flip-flops that are connected in series, the more data than can be stored, but the slower the register becomes. There are SISO shift registers built on a single IC chip with 4096 flip-flops connected in series. For such a register, 4096 clock pulses are required to fully load the register and 4096 clock pulses are required to read out the data stored in the register.

The amount of time it takes to load the register can be computed using the formula $T = (1/f) \times n$, where T is the total time, in seconds, needed for loading or readout, f is the frequency of the clock in cycles per second (Hz), and n is the number of flip-flops in the register.

If the clock is operating at 1000 Hz and the register has 4096 flip-flops, then:

$$T = \frac{1}{1000} \times 4096$$

$$T = 4.096 \text{ sec}$$

The total time required for loading of the register is 4.096 sec, and the total time required for readout of the data stored in the register is 4.096 sec.

Although the SISO shift register is slow, its lack of speed can be used as an advantage. At times it is necessary to slow the speed of data in order to synchronize the data with other data in the computer. The amount that the data is delayed can be controlled by the clock frequency, by the number of flip-flops connected in series to form the register, and by the presence or absence of the clock pulses.

Recirculating Shift Register

Another important characteristic of the SISO shift register is that data entered into the register is lost in the process of data readout. Recall the four-bit SISO register described earlier. The number loaded into the register was 0001. When the data was read out of the register the data input was low. The clock sent four pulses. The data was read out and all the flip-flops were reset to 0. If it is necessary to preserve the data stored in a SISO shift register, the output of the last flip-flop is connected to the input of the first flip-flop through an OR gate. The result is a **recirculating shift register** (Figure 8.3).

The recirculating shift register operates exactly the same as the SISO shift register except that the output is fed back to the input and reloaded into the register. When a recirculating shift register is being used it is critical that an accurate count of the clock pulses be recorded. This count is the only way of knowing where the data is located in the register. For example, if the recirculating shift register has been loaded with the number 0001, on the next clock pulse

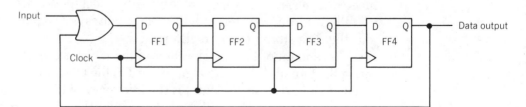

Figure 8.3   Recirculating shift register

the 1 will exit the last flip-flop and be re-entered into the first flip-flop. The number in the recirculating shift register is now 1000. On the next clock pulse the number in the register will be 0100. After a total of four clock pulses, all the data will have been read out of the register and the number stored in the register will again be 0001.

Serial In Parallel Out (SIPO) Register

The serial in serial out (SISO) shift register can easily be converted into a **serial in parallel out (SIPO) shift register.** Figure 8.4 shows a SIPO register. By connection of a wire to the output of each flip-flop, the register is given a parallel output.

To load the SIPO register, data is entered through the serial input and is shifted from one flip-flop to the next on each clock pulse. After four clock pulses the register is fully loaded. The data stored in the flip-flop can be read immediately at the parallel outputs. It is not necessary to clock the register four times, as is necessary with the SISO register, to read the data. The data stored in the register is always present at the parallel outputs. The data is not destroyed in the process of data readout as it is in the SISO register.

Parallel In Serial Out (PISO) Register

Figure 8.5 is the schematic diagram for a **parallel in serial out (PISO) register.** Notice that the D flip-flops used in this circuit have PRESET and CLEAR inputs. The PRESET and CLEAR inputs are inverted inputs. The PRESET and CLEAR inputs are active when they are low and disabled when they are high. The PRESET inputs are held high by connecting an inverter (NOT gate) to each PRESET input. The CLEAR inputs are all connected together, and one inverter is used to hold the CLEAR inputs high.

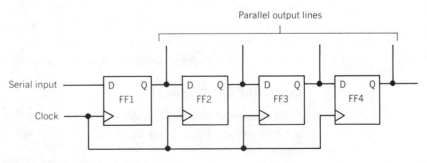

Figure 8.4   SIPO shift register

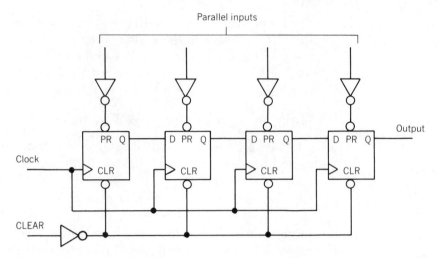

Figure 8.5   PISO register

To clear all the flip-flops to 0, the input of the inverter connected to the CLEAR input is pulsed. All the flip-flops reset to 0.

The register can be loaded by entering data through the inverters connected to the PRESET inputs. To load the number 1001 into the register, the first and fourth inverters are brought high. Both inputs can be brought high at the same time. The data is instantly entered into the register.

To read the data out, four clock pulses are sent to the clock input. The data shifts out of the register one bit at a time, just as in the SISO register.

Parallel In Parallel Out (PIPO) Register

The **parallel in parallel out (PIPO) register** (Figure 8.6) is similar to the PISO register. To convert the PISO register to PIPO, a wire is connected to the output of each D flip-flop. Data can be entered into the register by pulsing the input inverters. The instant data is loaded into the register through the PRESET inputs, it is present at the parallel outputs. This is the fastest of all registers.

Combination Register

A **combination shift register** is shown in Figure 8.7. This register can be used for serial or parallel inputs and for serial or parallel outputs.

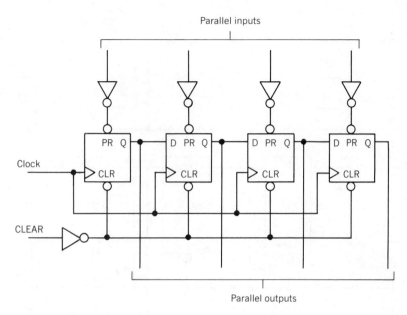

Figure 8.6 PIPO register

The combination register is built with four J-K flip-flops. These flip-flops include both PRESET and CLEAR inputs. The PRESET and CLEAR inputs are active low (inverted).

An inverter (NOT gate) is connected to the CLEAR inputs just as in the PISO register (Figures 8.5 and 8.6). Before the register is loaded, the input of the inverter is brought high. The output of the inverter drops low and all the flip-flops are cleared. After the flip-flops have been cleared, data can be entered through either the serial input or the parallel inputs.

Serial Loading.   To load the combination register using the serial input, data is placed on the serial input line. As an example, consider loading the number 0001. When the input on the serial input line is high, the J input is high. The data also feeds an inverter. The input of the inverter is high and the output is low. The K input is low. When the clock input goes from low to high, the Q output of the first flip-flop goes high. If the data at the serial input now goes low, the Q output of the first flip-flop will reset to 0 and the Q output of the second flip-flop will switch high on the next low-to-high clock transition. If the data remains low and the clock switches from low to high two more times, the register will be fully loaded. The

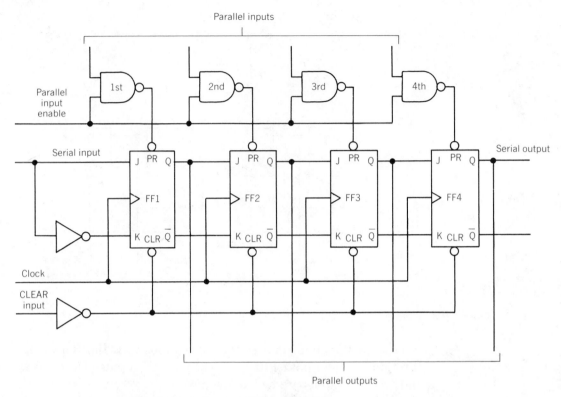

Figure 8.7 Combination shift register

number stored in the register is 0001. It takes four clock pulses to fully load the register.

Parallel Loading.   To load the combination register using the parallel inputs, data is placed on the parallel input lines. Notice that the parallel inputs are connected to one input of each NAND gate. The other inputs of the NAND gates are connected together. This input is called the **parallel input enable.** The data present at the parallel inputs is not passed to the PRESET inputs until the parallel input enable is high. If the input to the fourth NAND gate is high and all other NAND inputs are low, when the parallel input enable is high the output of the fourth NAND gate drops low. The Q output of flip-flop No. 4 goes high. All other Q outputs remain low.

The register is now loaded with the number 0001. The register contains the same data as in the previous example, but in this case the data was loaded the instant the parallel input enable was brought high. Recall that the serial input required four clock pulses to fully load the register. Loading this register with the parallel input is four times as fast as with the serial input but requires four wires.

Serial Output.   The serial output operates the same as the outputs in the previous examples. Each time the clock input goes from low to high the data is shifted to the right. Four clock pulses shift all of the data out of flip-flop No. 4.

Parallel Outputs.   The parallel outputs are active at all times. If the register is being loaded using the serial input, invalid data will appear on the parallel output lines while the register is loading. When the register is fully loaded, the data on the parallel output lines will be valid.

Applications for the Combination Register.   The combination register is often used to convert serial data into parallel data and to convert parallel data into serial data.

Many robots are connected to other robots or to a host computer. The data is normally transferred in a serial format. Although the parallel format for transferring data would be much faster, it would require many wires for connection of one robot to another robot or to the host computer, which would result in large cables and a higher probability of broken wires. The serial format requires only two wires. One wire is used to carry the data, and the other wire is a ground connection which completes the circuit.

Although data is normally transferred in a serial format, the robot's controller operates on the data in parallel format. If the robot is receiving data from another robot or from a host computer, the data enters a SIPO register. When the register is fully loaded, the data on the parallel output lines of the register is read by the robot's microprocessor and is stored in the robot's memory.

If the robot is outputting data to another robot or to a host computer, the data is normally sent out in a serial format. The robot's microprocessor loads the PISO register through the parallel inputs. The register is then clocked and the data is outputted one bit at a time through the serial output line. When all the data has been shifted out of the register, the microprocessor reloads the register through the parallel input lines and the data is again outputted through the serial output.

READ/WRITE MEMORIES

Simple registers do not contain enough data storage to be used as the main memories of robot controllers. If many registers are combined, however, a memory that is large enough to store the necessary program and data information can be constructed and used as a robot's main memory (Figure 8.8).

Each register in a memory can have data stored in it and read out from it. To write data into any register, the write enable line is brought high. To select the specific register into which data is to be written, address lines are used. Entering a binary code for the specific register allows data to be written into that register. This is called "addressing."

The registers used in memories also have read enable inputs. When the read enable line is brought high and the address lines are high, the data in a particular register is read out. Reading data out of these registers is nondestructive. The data stored in the register is not altered or destroyed by being read out.

The number of registers that can be written into or read out from is often referred to by the letter "K." If a memory has 1024_{10} registers, it is called a 1K memory. If a memory has 2048_{10} registers, it is called a 2K memory. A 64K memory has 65,536 (or 1024×64) registers.

The number of flip-flops in each register is also specified. For example, if each register has four flip-flops and there are 1024 registers, the memory is called a $1K \times 4$ memory (the "times" sign is read as "by"). If each register in the memory has eight flip-flops and the memory has 1024 registers, the memory is called a $1K \times 8$ memory.

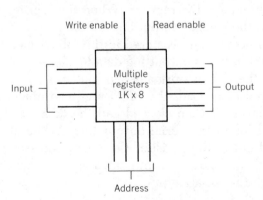

Figure 8.8 Memory formed by combining multiple registers

When data is stored and transferred eight bits at a time, each group of eight bits is normally referred to as a "byte" of data. A 1K × 8 memory is also called a 1 kilobyte (1Kb) memory. A 64K × 8 memory is called a 64Kb memory.

Random Access Memory (RAM)

A **Random Access Memory (RAM)** is a group of registers that can be individually addressed, written into, and read out. RAM's are used as the main memories of computers and robot controllers. Figure 8.9 is a schematic of a simple RAM. This RAM is an 8 × 1 memory. There are eight individual flip-flops. Each flip-flop can be individually addressed, written into, and read out. Each flip-flop can store a single bit of data.

In addition to the eight flip-flops, the memory also contains a decoder. The decoder has three address lines, a read enable line, and a write enable line. The three address lines are binary coded inputs. If address lines A0 and A1 are brought high, flip-flop 03 is addressed. If A0 and A1 are high while the write enable line is high, data present at the data input is written into flip-flop 03. If address

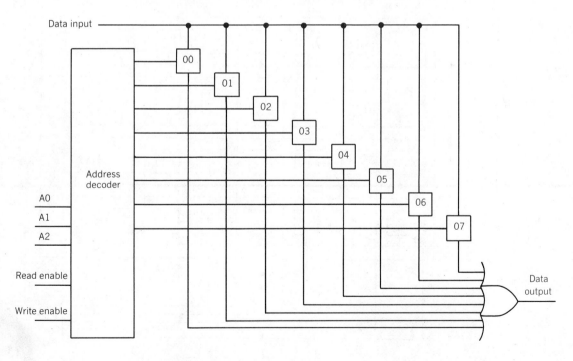

Figure 8.9 Random access memory (RAM)

lines A0 and A2 are high and the write enable line is high, data present at the data input will be written into flip-flop 05.

Data can also be read from the memory when the read enable line is high. If address lines A0 and A1 are high and the read enable line is high, data stored in flip-flop 03 will be outputted and will appear at the data output.

Many memories use registers that store more than 1 bit of data at each memory address. The memory shown in Figure 8.10 contains 32 flip-flops. The flip-flops are connected to form four eight-bit parallel in parallel out (PIPO) registers. This memory can be described as a 4 × 8 memory.

Each eight-bit register is identified as a **memory cell.** The term "memory cell" is frequently used instead of the term "register."

To write a byte of data into memory cell 1, address line A0 is brought high. The write enable line is also brought high. Whatever data is present on the data lines is written into memory cell 1. If address line A0 is high along with the read enable line, the data in memory cell 1 will be outputted to the data lines. Note that the read

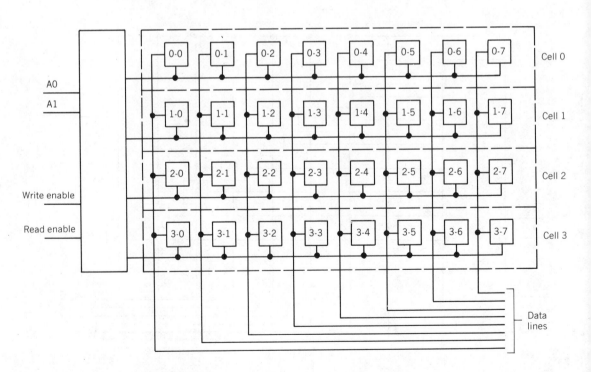

Figure 8.10   4 × 8 RAM

enable and write enable lines cannot both be high at the same time. If they were, a "race" condition would result.

Every memory cell can be individually addressed in any order, and every memory cell can have data written into it and read out from it.

One-dimensional and Two-dimensional Addressing

The type of RAM that we have been describing is called a **one-dimensional addressing** or **linear addressing** RAM. An individual address line is connected between the decoder and each memory cell.

One-dimensional addressing is adequate for small memories, but as the size of the memory increases it becomes increasingly difficult to design IC's with individual address wires for each memory cell. If one-dimensional addressing were used for a memory containing 64K memory cells, the decoder would need 65,536 individual address lines.

Two-dimensional addressing greatly reduces the number of wires coming from the decoder and is much less complex than one-dimensional addressing. Figure 8.11 illustrates the design of a memory which uses two-dimensional addressing. The memory is arranged as a matrix with four rows and four columns, for a total of 16 memory cells.

To address a memory cell, two separate addresses are entered. One address identifies the row, and the other address identifies the column, of the memory cell. If we wish to address memory cell 1-2, row address input X1 is brought high. All memory cells in row 1 have their X address inputs high. Column address input Y2 is also brought high. All memory cells in column 2 have their Y address inputs high. The only memory cell that has both the X and Y address inputs high is memory cell 1-2, and therefore this is the only memory cell which can be written into or read out from.

Two-dimensional addressing is sometimes called "X-Y addressing" or "X-Y selection."

Memory Expansion

Many robot controllers require more memory than is currently available on a single IC chip. This problem can be overcome by connecting several memory chips in parallel.

Figure 8.12 shows a memory that has been expanded by connecting all of the address lines in parallel. Each chip in this memory

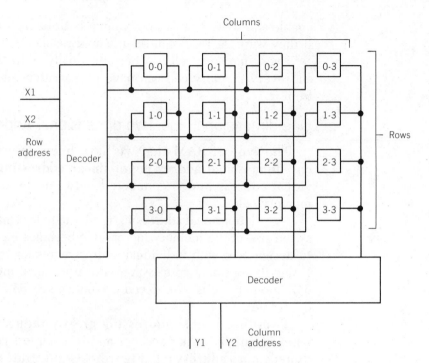

Figure 8.11   Memory with two-dimensional addressing

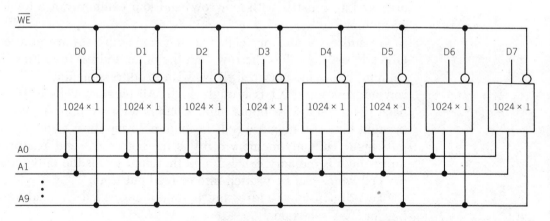

Figure 8.12   Memory expansion by parallel connection of address lines (to increase the number of bits per address)

is a 1024×1 memory chip. By connecting the address lines in parallel the memory is made to operate as a 1024×8 (1Kb) memory.

Notice that the chips use a single line for both reading and writing data. To control whether the data lines are being used for writing data into the memory cells or reading data out from the memory cells, a control line labeled "$\overline{\text{WE}}$" has been added. The letters WE stand for "Write Enable." The bar tells us that this line is active low. In other words, when the $\overline{\text{WE}}$ line is low, data present on the data lines will be written into the memory. When $\overline{\text{WE}}$ is high, data stored in the memory will be read from the memory cells and placed on the data lines (D0, D1, . . . , D7).

To expand the number of addressable memory cells, the data input and data output lines are connected in parallel (Figure 8.13). The memory address lines are also connected in parallel as well as the $\overline{\text{WE}}$ lines. In addition, the memory chips also have control lines labeled "$\overline{\text{CS}}$." The $\overline{\text{CS}}$ lines are "Chip Select" lines. The $\overline{\text{CS}}$ lines are active low. When the $\overline{\text{CS}}$ line is low, the memory chip is turned on and data can be written into and read out of the chip. When the $\overline{\text{CS}}$ line is high, the memory chip appears to be disconnected from the other memory chips. If the address lines, the data lines, and the $\overline{\text{WE}}$ line have data on them, the data is ignored by the memory chip until the $\overline{\text{CS}}$ line goes low.

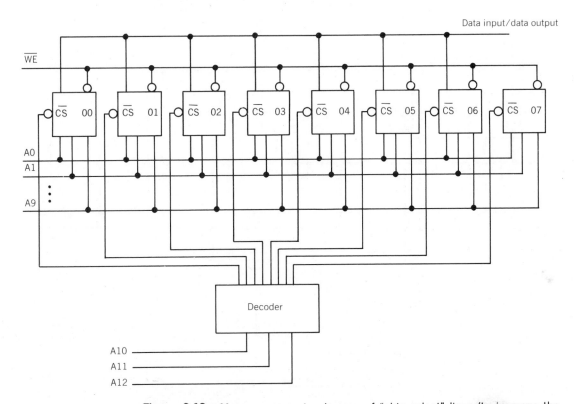

Figure 8.13 Memory expansion by use of "chip select" lines (to increase the number of addresses)

Notice that the $\overline{CS}$ lines are connected to a decoder. To select a memory cell, a thirteen-bit binary address is used. The lower 10 bits (lines A0 to A9) are used to select a memory cell, and the three most significant bits are used for the $\overline{CS}$ line. The binary address 000 0000000001 will select memory cell 1 of chip 00 (Figure 8.14).

The binary address 001 0000000001 also selects memory cell 1, but the $\overline{CS}$ line selects chip 01 (Figure 8.15). By utilizing memory cell addressing as well as chip selecting, control designers can create memories as large as necessary. For example, if a 64Kb (64K × 8) memory is needed, the designer can choose 4Kb (4K × 8) chips. If 16 chips are connected in parallel and chip selecting is used, 65,536 (64K) memory cells can be addressed.

Static and Dynamic RAM's

The memories we have been examining up to this point are called **static memories.** When data is written into the memory, the data will be retained in the memory as long as power is maintained.

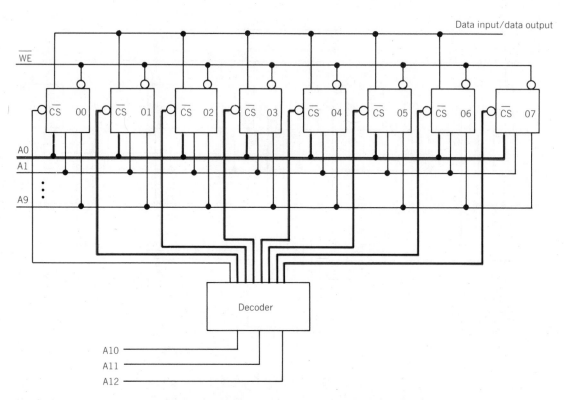

Figure 8.14   Selection of memory cell 1 of chip 00

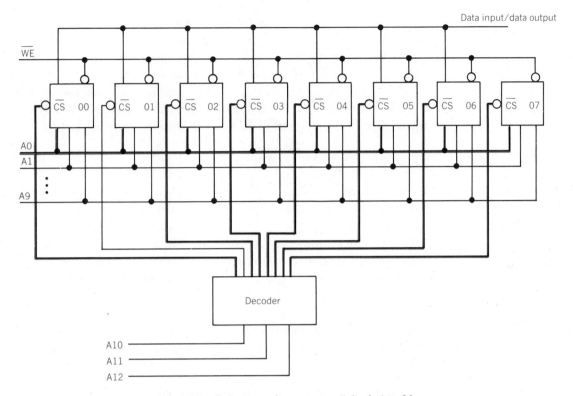

Figure 8.15 Selection of memory cell 1 of chip 01

Another type of memory is called a **dynamic memory.** When data is written into a dynamic memory, the data is very short-lived. The memory will only retain the data for 2 ms (0.002 sec). Dynamic memories are built using small capacitors rather than the flip-flops used in static memories. The small capacitor is charged to store a 1 and discharged for a 0. The small charge on the capacitor quickly drains off, and the memory loses its data. To preserve the data stored in a dynamic memory, the memory is read out at least every 2 ms and the data is rewritten into the memory. The process of reading out the contents of a dynamic memory and rewriting the data into the memory is called **refreshing.** Special circuits are used to refresh dynamic memories. Some of the circuits needed to refresh the memory may be part of the memory chip, but additional circuits, such as counters and a clock, are also needed.

Because they require so much additional circuitry, one may wonder why dynamic RAM's are ever used. Actually, there are several reasons why designers choose dynamic RAM's. Dynamic RAM's require far less power than static RAM's. In addition, the dynamic RAM memory cell is much simpler than the static RAM cell. This simplicity saves space on the chip. Many more memory cells can be packed onto a chip if it is a dynamic memory rather than a static memory. As robot controllers demand more and more memory to accomplish their more sophisticated tasks, the savings in power and space provided by dynamic RAM's can become very important.

NONVOLATILE MEMORIES

The RAM memories we have examined up to this point have all been **volatile**. A volatile memory loses its stored data when the power is removed. This is true of both static and dynamic RAM's. When power is removed from a static RAM, the flip-flops are disabled. When power is removed from a dynamic RAM, the RAM can no longer be "refreshed." In as little as 2 ms, the contents of the memory are lost.

There is also a class of memories that are **nonvolatile**. A nonvolatile memory retains the data stored in it even after the power is removed. The simplest of all nonvolatile memories is the Read Only Memory (ROM).

Applications for Nonvolatile Memories

Most robots and other automated machines use ROM's as an important part of the control. Some of the applications that ROM's are used for are as follows:

(1) *Look-Up Table*. Many controllers require mathematical tables to be stored. For example, a ROM is available that has the sine values for preselected angles from 0 to 90°.

(2) *Character Generator*. ROM's store the dot pattern necessary to form the alphanumeric characters that appear on a control CRT.

(3) *Keyboard Encoder*. A ROM is used to convert the keystrokes on a typewriter keyboard into a code that can be understood by a microprocessor.

(4) *Computer Memory*. Standard programs are often needed to coordinate the movements of the various axes of the robot. These specialized programs are often stored in a ROM and utilized each time the axes of the robot are commanded to move by a user's program.

Read Only Memory (ROM)

The **Read Only Memory (ROM)** is a nonvolatile memory — that is, it retains its data even after power is removed. Figure 8.16 shows

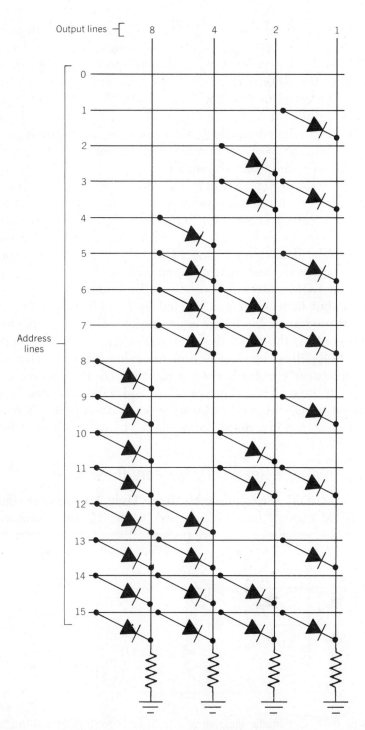

Truth table

Address line	8	4	2	1
0	0	0	0	0
1	0	0	0	1
2	0	0	1	0
3	0	0	1	1
4	0	1	0	0
5	0	1	0	1
6	0	1	1	0
7	0	1	1	1
8	1	0	0	0
9	1	0	0	1
10	1	0	1	0
11	1	0	1	1
12	1	1	0	0
13	1	1	0	1
14	1	1	1	0
15	1	1	1	1

Figure 8.16 Schematic diagram and truth table for a read only memory (ROM)

the construction of a simple ROM. The memory is constructed as a diode matrix. When an address line is brought high, the data stored at that address is outputted. If address line 4 is brought high, the output of the ROM will be 0100. (If address line 4 goes high, the diode connected to that line turns on, yielding a 0100.) If address line 14 is brought high, the output of the ROM will be 1110. This ROM is constructed as a decimal to binary encoder. The input to the address lines is a decimal number, and the output is the binary representation of the decimal number.

The ROM in Figure 8.16 is limited to a single address line. Most ROM's use multiple address lines just as do the RAM's examined earlier. Figure 8.17 shows a ROM that allows any combination of address lines to be brought high. Notice that diode D3 is not connected. The data stored in the ROM is formed through a combination of connected and nonconnected diodes.

Most ROM's are not constructed by building a discrete diode matrix, but instead are constructed by IC chip manufacturers and are not programmable by the user. If a special ROM is needed for a robot control, the robot designer writes the truth table that accurately describes the data output for every address. The IC chip manufacturer creates a special photo mask that connects some diodes in the matrix and leaves other diodes disconnected. The chip is made using the manufacturing process described in Chapter 5. The data stored in a ROM is determined by the mask and cannot be altered by the user.

Programmable Read Only Memory (PROM)

The **PROM (Programmable Read Only Memory)** is similar to the ROM except that it is shipped by the IC chip manufacturer

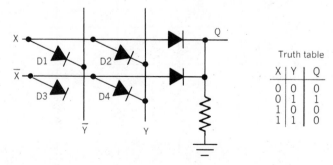

Figure 8.17 Schematic diagram and truth table for a ROM with multiple address lines

without any data stored in it. When the PROM is received, data can be "programmed" one time. After the data has been stored in the PROM it cannot be altered.

Figure 8.18 is a schematic diagram of a PROM. This PROM uses multiple-emitter transistors. The emitters are connected to fuses. To program the PROM, an address is selected by bringing the appropriate address line high and applying a high reverse bias voltage to the data line. The high voltage on the data line blows the fuse, disconnecting one of the emitters. Once the fuse is blown, the data is permanently stored. If an error is made in programming the

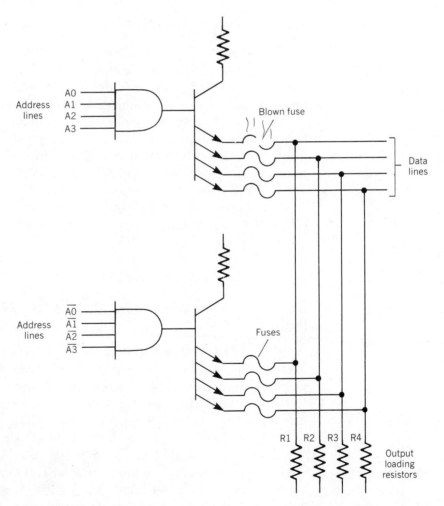

Figure 8.18 Programmable read only memory (PROM). This memory is programmed by blowing fuses.

PROM, it cannot be reprogrammed. The only solution is to throw the PROM away and start over. To help ensure that errors are not made in programming a PROM, engineers normally write a computer program to bring the appropriate address lines high and blow the fuses at those addresses. Once the correct program has been written, it can be used to program as many PROM's as are needed. Resistors R1 through R4 are output loading resistors.

PROM's are used as an alternative to ROM's by many robot manufacturers since the required quantity of specialized ROM's is normally too low to justify the cost of designing a photo mask and etching the ROM's.

Erasable Programmable Read Only Memory (EPROM)

The **EPROM (Erasable Programmable Read Only Memory)** is similar to the PROM except that it can be reprogrammed if the stored data needs to be altered due to an engineering change or because an error was made in the process of programming.

The procedure for programming an EPROM differs from manufacturer to manufacturer and for this reason will not be explained in detail here. Basically, a memory cell is addressed and a high voltage is applied for a very short period. The high voltage charges the cell. The cell will retain the charge for ten years or more.

To erase the information stored in the EPROM, a high-intensity ultraviolet light is directed at the memory chip through a quartz window for approximately 20 minutes (Figure 8.19). To protect the EPROM from being erased by natural ultraviolet light, the quartz window is covered by a piece of tape.

Electrically Erasable Programmable Read Only Memory (EEPROM)

The **EEPROM (Electrically Erasable Programmable Read Only Memory)** is similar to the EPROM. The EEPROM can be programmed by addressing a memory cell and applying a high voltage to the data lines. To erase the data, the same memory cell is again addressed and a voltage is applied (Figure 8.20).

The EEPROM should not be thought of as a substitute for the RAM even though data can be written into the EEPROM and erased electrically. Programming of the EEPROM and alteration of existing data are both very slow. The EEPROM also requires a high voltage for erasing and writing data, which raises the power consumption of the memory. This is not desirable in a computer.

Figure 8.19 Erasable programmable read only memory (EPROM)

Figure 8.20 Electrically erasable programmable read only memory (EEPROM)

MASS STORAGE DEVICES

In addition to RAM's and ROM's, most robots and other auto-mated controls have some form of mass data storage. The most common mass storage devices in use today are the magnetic tape, the floppy disk, and the hard disk.

Magnetic Tape Storage

The magnetic tape is a commonly used device for storing robot programs and data. The most common magnetic tape storage system in use today is the cassette tape.

Because magnetic tape is a nonvolatile storage medium, robot programmers often will make copies of programs on magnetic tape after the programs have been tested and verified. When a robot is programmed, its program is stored in a RAM, which is a volatile memory. If power is lost, the program is lost. If a copy of the program has been made on magnetic tape, the programmer can simply put the program tape in a tape player and reload the program into the robot's RAM memory.

Figure 8.21 is an illustration of a magnetic recording head. When voltage is applied to the winding of the recording head, a magnetic field is formed. The field is concentrated at the gap in the recording head. As the magnetic tape passes over the head, the magnetic field created by the head is transferred to the magnetic oxide on the tape.

Nonreturn to Zero (NRZ).   In the nonreturn to zero (NRZ) system, the polarity of the magnetic field of the recording head, and there-fore the polarity of the magnetic field of the tape, is determined by the direction in which the voltage is applied to the winding of the recording head. Polarity in one direction is recognized as a 1, and

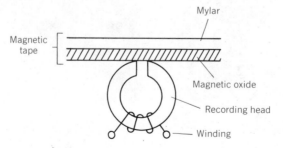

Figure 8.21   Magnetic recording head

polarity in the other direction is recognized as a 0. The advantage of the NRZ system is that both the 0 and 1 outputs are voltages. If a 0 were simply off (no voltage), noise on the tape might be interpreted as a 1, giving false data.

The tape is played back by passing it over the head. The small magnetic fields on the tape are picked up by the head and converted to a voltage. This small voltage is amplified and recorded into the RAM memory as 1's and 0's.

This system of recording and playing back the data on the tape is referred to as the "nonreturn to zero" system because 1's and 0's are recorded as magnetic polarities rather than as simple on's and off's.

Kansas City Standard (KCS).   The Kansas City Standard (KCS) uses audio signals rather than the polarity of the magnetic field to create the 1's and 0's. Four cycles at 1200 Hz is a logic 0, and eight cycles at 2400 Hz is a logic 1 (Figure 8.22). Notice that four cycles at 1200 Hz require the same length of time as eight cycles at 2400 Hz.

The KCS works well even on inexpensive cassette tape players. The major drawback of the KCS compared with the NRZ is speed. The KCS is much slower. For this reason, most industrial robots use the NRZ.

Floppy Disks

The **floppy disk** has become a very popular means of mass storage for robots. The floppy disk operates at a speed much higher than that of the cassette tape.

A floppy disk (Figure 8.23) is a circular piece of mylar coated with a high-quality magnetic oxide much the same as the oxide used to make magnetic tapes. The floppy disk gets its name from the fact that it is flexible.

Floppy disks come in three sizes: the standard floppy disk (8-in. diam), the mini-floppy disk (5.25-in. diam), and the micro-floppy disk (3.25-in. diam).

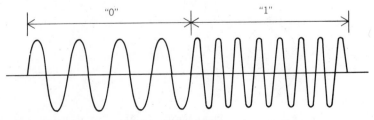

Figure 8.22   Kansas City standard (KCS)

Figure 8.23   Floppy disk (without protective jacket)

Both the standard floppy disk and the mini-floppy disk have a large hole in the center. This hole (Figure 8.24) is used to spin the disk when it is placed in the **floppy disk drive.**

Both the standard floppy disk and the mini-floppy disk also contain a small hole. This small hole, which is called an "index hole" (Figure 8.24), is used to locate the beginning of the disk. The disk drive has an LED (light emitting diode) and a phototransistor. When the floppy disk is slid into the floppy disk drive, the disk is between the LED and the phototransistor. The disk is spun by the drive at 360 rpm. When the light from the LED strikes the phototransistor

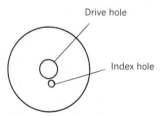

Figure 8.24   Floppy disk, showing drive hole and index hole

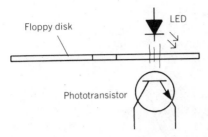

Figure 8.25 Method of identifying the first sector of a floppy disk using an LED and a phototransistor

it conducts, indicating the beginning of the disk (Figure 8.25). Every time the index hole passes the phototransistor, the transistor switches on, sending a pulse to the disk drive's controller indicating that the disk is at the beginnning.

A mini-floppy disk is divided into ten equal parts called "sectors" (Figure 8.26). The index hole is located at the beginning of the first sector. The sectors are used by the floppy disk controller and the microprocessor to locate data recorded on the disk. In addition to the ten sectors, the mini-floppy disk is divided into 40 tracks. The data is stored on these tracks (Figure 8.27).

The standard (8-in.) disk is divided into 26 sectors and 77 tracks. The standard floppy disk can store 2½ times more data than the mini-floppy disk.

Floppy disks (both mini-floppy and standard) are sealed in plastic jackets (Figure 8.28) which protect the fragile disk from dust and fingerprints. The plastic jacket that holds the disk has several openings. The jacket has a large hole to allow access to the drive hole of

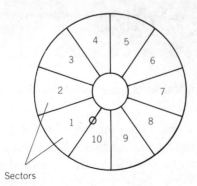

Sectors

Figure 8.26 Sectors of a mini-floppy disk

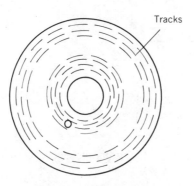

Figure 8.27   Tracks of a mini-floppy disk

Figure 8.28   Floppy disk in its protective jacket. Note holes in jacket (see text).

the floppy disk, a small opening for access to the index hole, and an oblong hole to permit the magnetic head to contact the disk. Since the head is in contact with the disk, both the head and the disk undergo wear. When the oxide on the disk wears or when the head is worn as a result of contact with the disk, both read and write errors occur.

Care should be exercised in handling floppy disks. If the magnetic oxide on the surface of the disk is touched, the oxide will be destroyed and the disk will be rendered useless. Also, if the disk should be bent, the magnetic oxide will flake off, destroying the stored data.

Another problem with the floppy disk is its sensitivity to dust and dirt. A small speck of dirt on a floppy disk can scratch the magnetic oxide on the surface of the disk and destroy the data. It is extremely important that floppy disks to be used on robots be kept in their protective shipping sleeves and be stored in sealed disk storage containers when not in use.

Hard Disk (Winchester Disk)

The **hard disk,** also known as the **Winchester disk,** records data in the same manner as the floppy disk but is rotated at a much higher rate. Hard disks are rotated at speeds from 1000 to 3600 rpm, as compared to 360 rpm for floppy disks. Because the hard disk is rotated at a much higher rate than the floppy disk, data can be written into and read out from the hard disk much faster.

When data is being written into or read out from a hard disk, the head is floating on a thin cushion of air above the disk. This cushion of air is created by the high rotational speed of the disk. Since the head does not make contact with the disk, there is virtually no wear of either the head or the disk. Electrical surges and other failures can occasionally cause the head to strike the disk. Such an occurrence, which is referred to as a "head crash," can cause damage to the disk, to the head, or to both. To preclude the possibility of data being lost as the result of a head crash, the data stored on a hard disk is frequently transferred to a backup tape.

In addition to its capability of storing more data per inch of track, the hard disk is also normally larger than the floppy disk. Even the smallest hard disk can record ten times as much data as a floppy disk, and the largest hard disks are capable of recording hundreds of times as much data as a floppy disk.

Hard disks are used by some robot manufacturers to store program data. Continuous path robots, because of the large numbers of program points that must be stored, are normally the only robots that utilize the hard disk.

MICROPROCESSOR CONTROL

Figure 8.29 is a block diagram of a robot **microprocessor control**. The main elements in this control are a microprocessor unit (MPU), a random access memory (RAM), a read only memory (ROM), five input/output (I/O) ports, a servo amplifier, a dc drive motor, a floppy disk drive, a teach pendant, a keyboard, and a cathode ray tube (CRT) display. All these components are connected together by a data bus, a control bus, and an address bus.

Data Bus

The **data bus** is a parallel group of wires. All the data, including the program instructions, is carried on this group of wires. The data

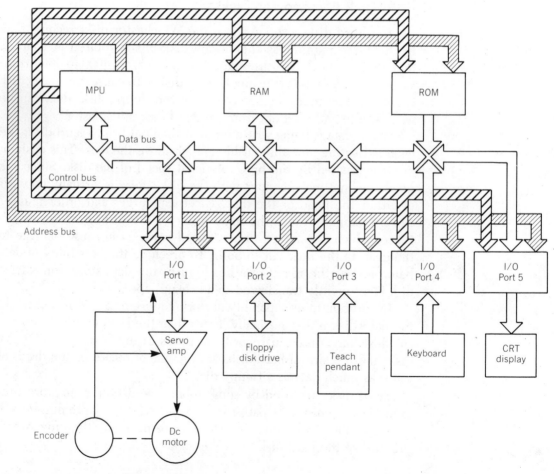

Figure 8.29   Block diagram of a robot microprocessor control

bus is referred to as a **bidirectional bus.** A bidirectional bus carries data in two directions. For example, the MPU can send data out through the data bus and can also receive data from the data bus.

The data bus may have 4, 8, 12, 16, and in some cases 32 parallel wires. The more wires in the data bus the more data that can be transmitted at one time. The size of the data bus is determined by the design of the MPU. The most common data bus structures have either 8 or 16 wires.

The size of the data bus determines, in part, the speed at which data can be processed. A processor that can transmit 16 bits (2 bytes) of data at one time will be faster than one that can transmit only 8 bits (1 byte) of data at a time.

Address Bus

The MPU sends out the "address" of data that is stored in the RAM or the ROM on the **address bus.** Notice that the address bus is also connected to the I/O ports. The I/O ports are used to interface all the "peripheral" equipment, such as the floppy disk drive and the keyboard with the MPU. Each port is assigned an "address." When the MPU outputs the address of a port, the port is turned on, and data on the data bus can be sent out through the port, or the peripheral device can send data through the port and place the data on the data bus.

The address bus is a unidirectional bus. The MPU always sends out the address desired. An address is never sent to the MPU through the address bus.

Control Bus

The **control bus** carries the control signals that control the RAM, the ROM, and the I/O ports. For example, if the MPU needs data from the RAM, the address of the data is placed on the address bus and a signal to "read" data is placed on the control bus. The data is placed on the data bus and read by the MPU.

The control bus is also connected to the I/O ports. After the I/O port is selected by the address carried on the address bus, the control bus controls whether data is being read from the port or sent to the port.

Microprocessor Operation

Refer again to Figure 8.29. A series of variations on this figure (Figures 8.30 to 8.51) will be used to explain the operation of the

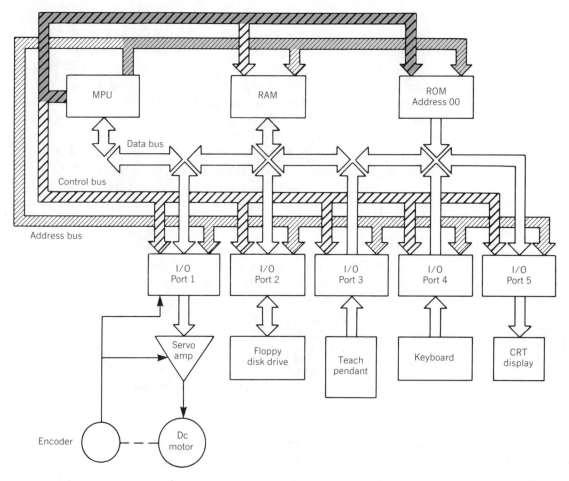

Figure 8.30   MPU outputs ROM address 00 on the address bus and the read control signal on the control bus.

robot control. We will trace the flow of instructions and data to illustrate the operation of the control. For the sake of simplicity, not all of the program control instructions have been included (the complete program contains more than 200 steps). In Figures 8.30 to 8.51, placement of a signal on the data bus, the address bus, or the control bus has been indicated by darkening that portion of the appropriate bus between the source of the signal and the desired destination of the signal. It should be remembered, however, that any signal placed on a bus is carried by the entire bus and not only by that portion of the bus that has been darkened for the purpose of illustration.

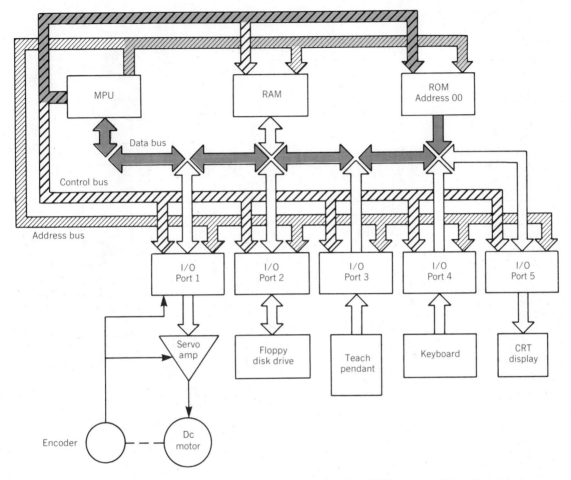

Figure 8.31 ROM outputs data stored at ROM address 00 on the data bus.

Programming the Control

To begin the process of programming, the power is turned on. The control automatically goes through a reset cycle. During the reset cycle all the flip-flops, latches, and counters are reset. The MPU is also loaded with the address of the first instruction that is to be executed. The MPU then outputs the first instruction.

1. The MPU outputs address 00. (Note: the addresses used by the control are actually in binary notation, but to simplify this program all addresses have been written in decimal notation.)

2. The MPU outputs a control signal on the control bus to read the ROM (Figure 8.30).

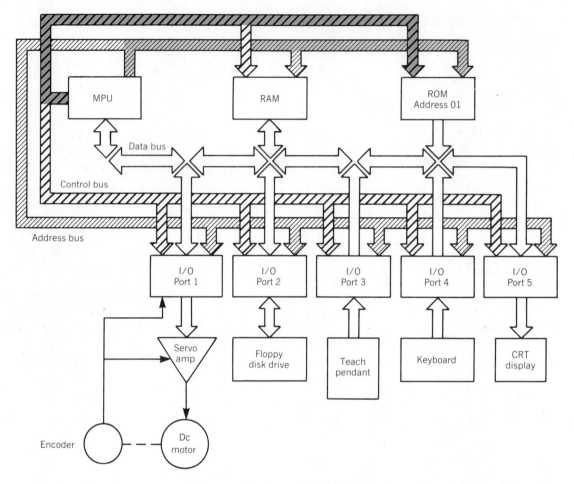

Figure 8.32   MPU addresses ROM 01 and outputs the read control signal on the control bus.

3. The ROM outputs an instruction on the data bus (Figure 8.31).

4. The MPU reads the instruction on the data bus.

5. The MPU decodes the instruction. The instruction directs the ROM to output data stored at address 01.

6. The MPU addresses ROM 01 and sends a control signal on the control bus to read data (Figure 8.32).

7. The data is placed on the data bus and read by the MPU (Figure 8.33).

8. The data is temporarily stored in the MPU.

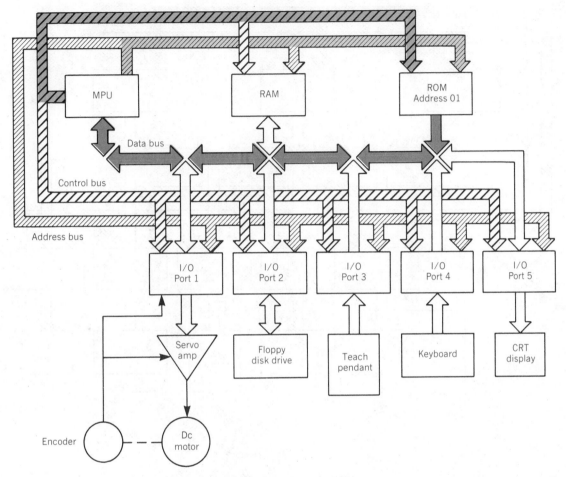

Figure 8.33 ROM outputs data stored at address 01 on the data bus.

9. The MPU outputs address 02 on the address bus and a control signal on the control bus to read data (Figure 8.34).

10. The ROM places the data on the data bus (Figure 8.35).

11. The data on the data bus is read and decoded by the MPU. The instruction directs the MPU to turn on port 5 and write data on the data bus.

12. The MPU outputs the address of port 5 on the address bus and sends a write control signal on the control bus. The data temporarily stored in the MPU is outputted on the data bus (Figure 8.36).

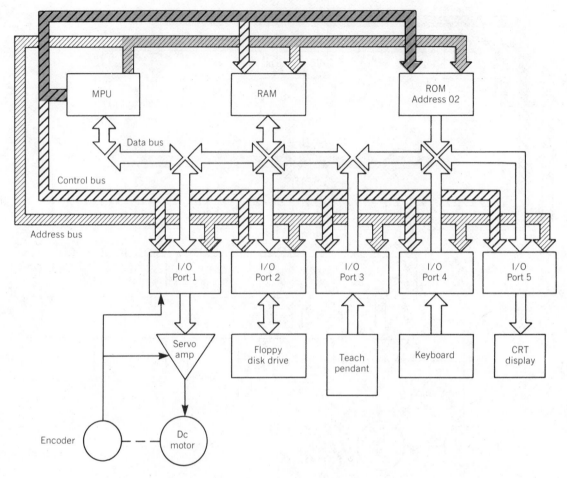

Figure 8.34 MPU addresses ROM 02 and outputs the read control signal on the control bus.

13. Port 5 reads and decodes the data on the data bus. "READY" appears on the CRT display.

14. The MPU addresses the ROM to output an instruction stored at address 03 and sends a signal on the control bus to read data.

15. The ROM outputs the instruction on the data bus (Figure 8.37).

16. The MPU reads and decodes the instruction on the data bus. The instruction directs the MPU to turn on port 4 (keyboard) and read data from the keyboard. The MPU addresses port 4 and the control bus switches port 4 to output data.

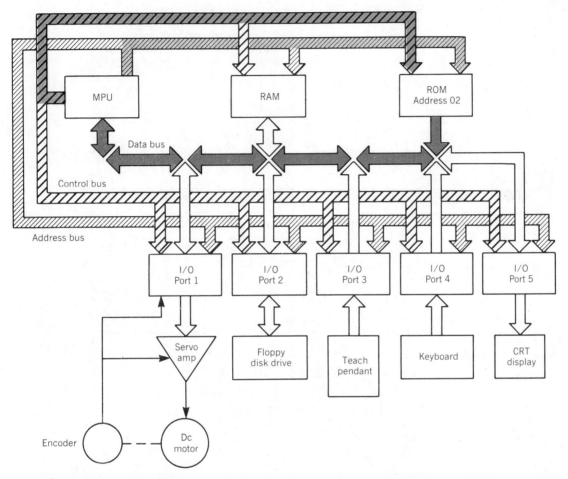

Figure 8.35 ROM outputs data stored at address 02 on the data bus.

17. The programmer types in the word "program" and hits the RETURN key.

18. The data is placed on the data bus one character at a time and is read by the MPU (Figure 8.38). The characters that are being transmitted are in a special binary code. This code also includes information that tells the MPU when the complete character has been sent. A description of this code (the ASCII code) will be presented later in this chapter.

19. The MPU decodes the word "program" by "fetching" data from the ROM and comparing it, one character at a time, with the

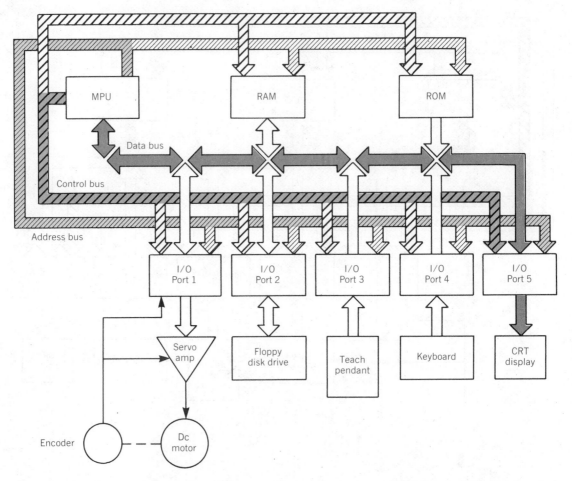

Figure 8.36 MPU outputs port address 5 on the address bus, the write control signal on the control bus, and data on the data bus.

> word "program." This process of fetching and comparing data is similar to that described in program steps 8 through 11.

20. The decoded word "program" tells the MPU to branch to a set of instructions beginning at ROM memory address 100.

21. The MPU outputs address 100 on the address bus and the READ command on the control bus.

22. The ROM outputs the instruction on the data bus (Figure 8.39).

23. The instruction is received and decoded by the MPU. The instruction directs the MPU to turn on port 3.

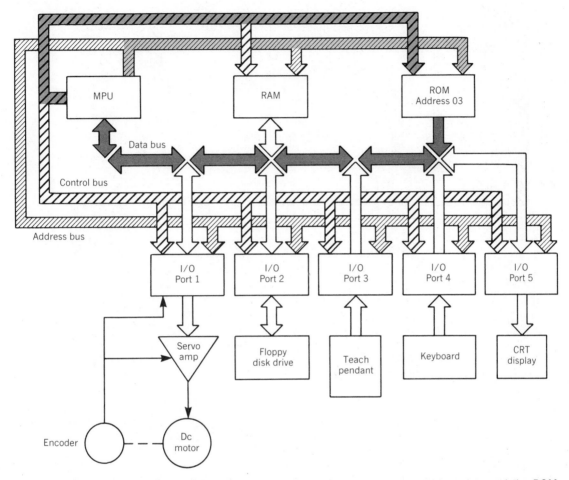

Figure 8.37 MPU outputs ROM address 03 on the address bus and the ROM read control signal on the control bus. ROM outputs data on the data bus.

24. The MPU addresses port 3 (teach pendant) and the control bus switches port 3 to output data.

25. The programmer selects a button on the teach pendant that corresponds to the desired movement of the robot (for example, the programmer might push a button labeled "shoulder up"). When the button is pushed, coded data is placed on the data bus (Figure 8.40).

26. The MPU reads the data and temporarily stores the data in one of its registers.

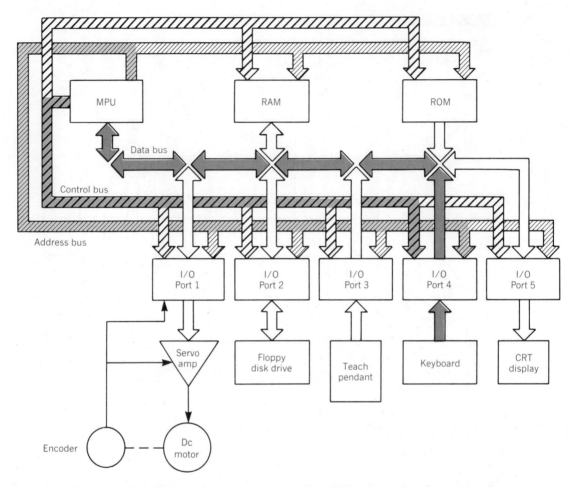

Figure 8.38 MPU outputs port address 4 on the address bus and the data read command on the control bus. The data is placed on the data bus by port 4 and read by the MPU.

27. The MPU outputs address 101 on the address bus and the ROM read command on the control bus.

28. The ROM outputs the instruction stored at address 101 on the data bus (Figure 8.41).

29. The MPU reads and decodes the instruction on the data bus. The instruction directs the MPU to address port 1 and the control bus switches to read the data on the data bus.

30. The MPU outputs the data in temporary storage onto the data bus.

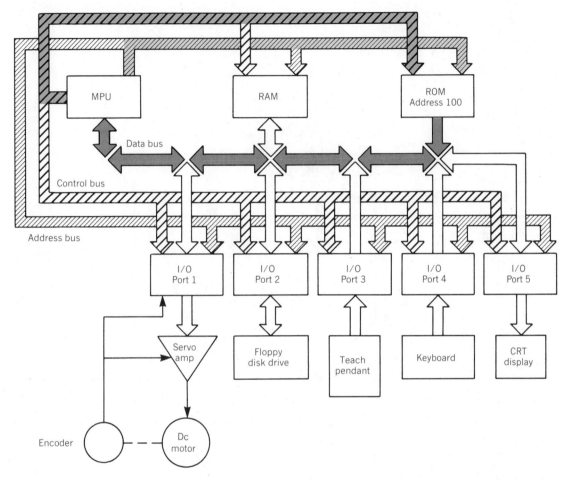

Figure 8.39 MPU outputs ROM address 100 on the address bus and the ROM read control signal on the control bus. The ROM places the instructions on the data bus.

31. Port 1 receives the data and turns on the servo amp. The robot's arm begins to move. Port 1 is latched on. Port 1 will remain on, and the robot's arm will continue to move, until port 1 receives data which switches it off (Figure 8.42).

32. The MPU outputs address 102 on the address bus and the ROM read command on the control bus. The ROM places data stored at memory address 102 on the data bus (Figure 8.43).

33. The MPU reads the instruction on the data bus and decodes it. The MPU is directed to address port 3.

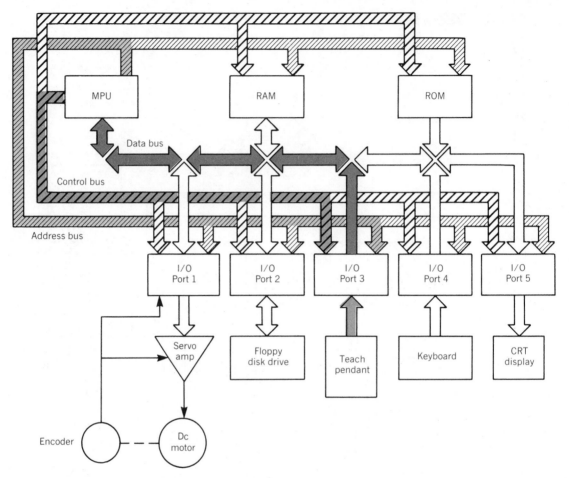

Figure 8.40 MPU addresses port 3 on the address bus and places a read control signal on the control bus. The teach pendant outputs data on the data bus, and the data is read by the MPU.

34. The MPU addresses port 3 and the control bus switches the port to read data. If the programmer is still holding the button down, data is present. As long as data is present on the data bus, the MPU "halts." The instant the programmer releases the button, data is no longer present on the data bus. When data is no longer present on the data bus, the MPU addresses the ROM for a new instruction (Figure 8.44).

35. The MPU outputs address 103 on the address bus and the control bus directs the ROM to place the instruction on the data bus (Figure 8.45).

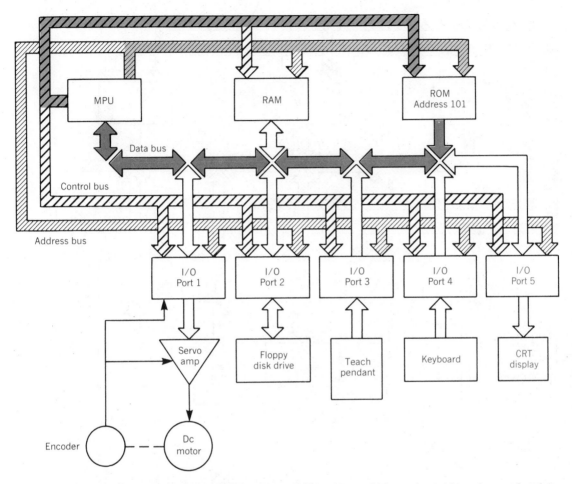

Figure 8.41 MPU outputs ROM address 101 on the address bus and a data read command on the control bus. ROM outputs data on the data bus, and the data is read by the MPU.

36. The MPU reads the instruction and decodes it. The MPU addresses port 1 and the control bus switches port 1 to accept data.

37. The MPU outputs data to turn off the servo amp (Figure 8.46).

38. Port 1 reads the data and switches the servo amp off. The process of reading the stop instruction from the ROM, decoding it, and outputting the instruction to port 1 happens so quickly that the robot's arm appears to stop moving the instant the programmer releases the button.

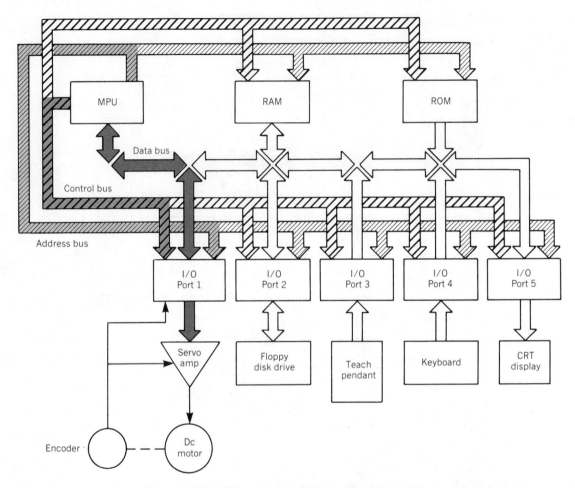

Figure 8.42 MPU addresses port 1 on the address bus and places a write command on the control bus. MPU writes data on the data bus, and the data is read by port 1, which turns on the servo motor.

39. The MPU puts address 104 on the address bus and the control bus directs the ROM to output data (a "read" command).

40. The data is placed on the data bus. The MPU reads and decodes the data (Figure 8.47).

41. The MPU addresses port 3 and directs the port to place data on the data bus. If the robot programmer again presses the button to move the robot's arm, the MPU will branch to step 23 in this program and the robot's arm will move. If the programmer

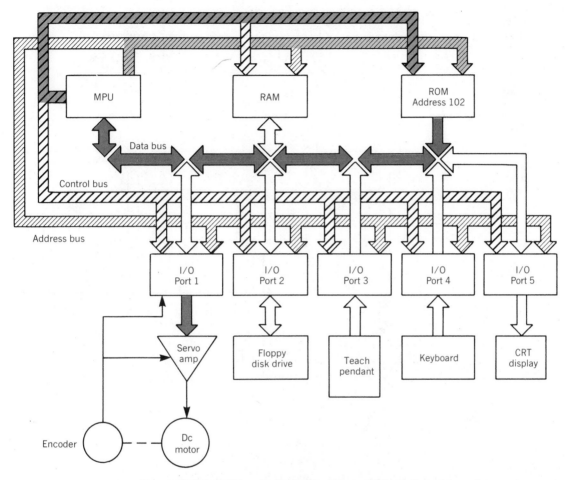

Figure 8.43 MPU outputs ROM address 102 on the address bus and a read command on the control bus. The ROM outputs data stored at address 102, and the data is read by the MPU.

pushes the "record" button on the teach pendant the data will be read and decoded by the MPU (Figure 8.48).

42. Assuming that the programmer has pushed the record button, the MPU puts address 105 on the address bus and a read command on the control bus.

43. The ROM outputs the data on the data bus. The MPU reads and decodes the data. The data directs the MPU to open port 1 and the control bus directs port 1 to output the robot's arm posi-

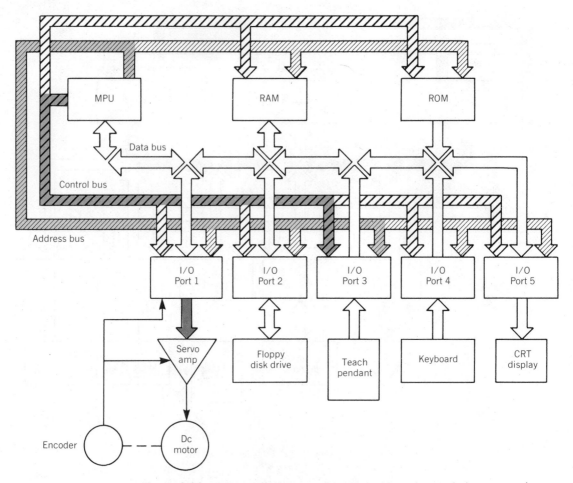

Figure 8.44 MPU addresses port 3 on the address bus and places a read command on the control bus. Port 3 places data on the data bus, and the data is read by the MPU.

tion. Notice that port 1 has an "encoder" connected to it. The encoder is attached to the robot's arm and outputs a binary number that represents the position of the arm.

44. The MPU reads the binary number that represents the position of the robot's arm and temporarily stores it (Figure 8.49).

45. The MPU outputs address 106 on the address bus and a ROM read command on the control bus.

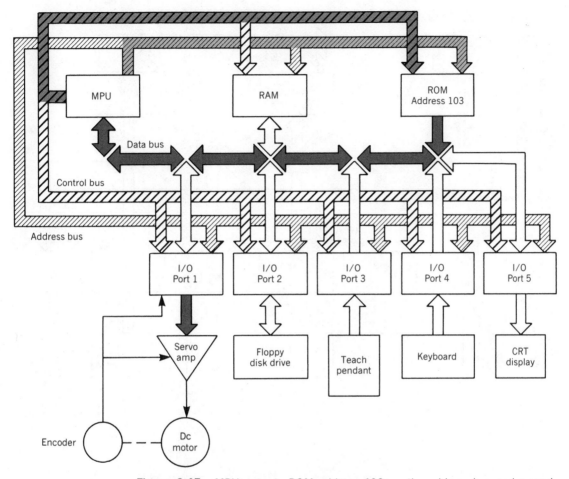

Figure 8.45 MPU outputs ROM address 103 on the address bus and a read control signal on the control bus. The ROM outputs data on the data bus, and the data is read by the MPU.

46. The ROM outputs the instruction on the data bus (Figure 8.50).

47. The MPU reads and decodes the data on the data bus. The instruction directs the MPU to move the robot-arm-position data which it holds in temporary storage into the RAM. The MPU addresses RAM memory 1000 and the control bus directs the RAM to write the data. The MPU outputs the data on the data bus, and the data is stored in RAM memory 1000 (Figure 8.51).

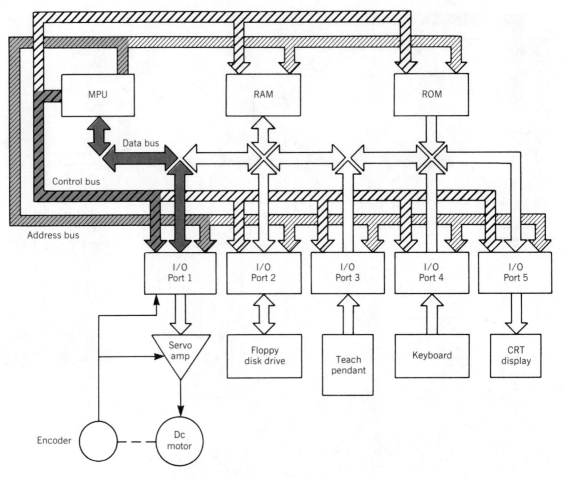

Figure 8.46 MPU addresses port 1 on the address bus and places a write command on the control bus. The MPU writes data on the data bus, which turns off the servo motor.

THE MICROPROCESSOR UNIT (MPU)

The microprocessor unit is the heart of the control. As you saw in the previous section, the microprocessor controls the operation of the RAM, the ROM, and the I/O ports. A complete study of the microprocessor is beyond the scope of this text, but a brief examination will demonstrate how the microprocessor exercises its control.

Figure 8.52 is a block diagram of a simple microprocessor. The microprocessor includes an **arithmetic logic unit (ALU)**, an **accumu-**

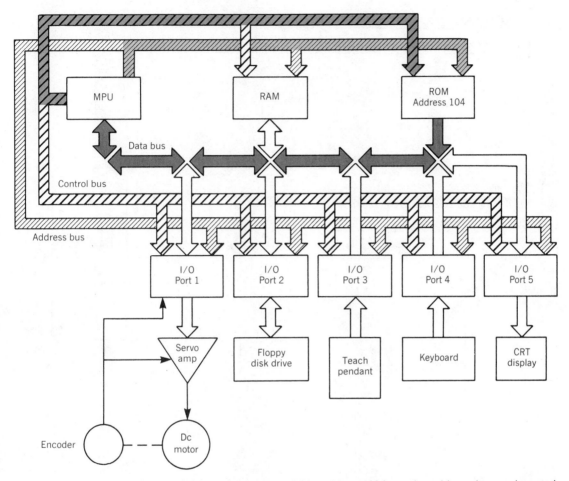

Figure 8.47 MPU outputs ROM address 104 on the address bus and a read command on the control bus. The ROM places the data on the data bus, and the data is read by the MPU.

lator, a **data register**, an **address register**, a **program counter**, an **instruction decoder**, and a **stack pointer**.

Arithmetic Logic Unit (ALU)

The **Arithmetic Logic Unit (ALU)** performs all the mathematical and logic operations of the microprocessor. The ALU can add and subtract, as well as perform AND, OR, XOR, and NOT operations on the data. All the logic and mathematical operations studied in this text can be performed by the ALU.

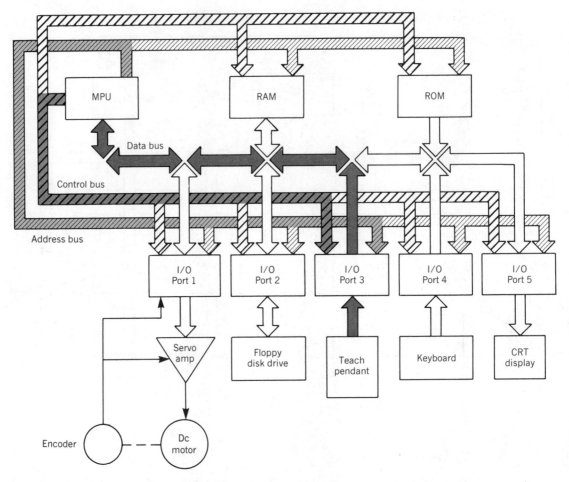

Figure 8.48 MPU addresses port 3 on the address bus and places a read command on the control bus. Port 3 places data on the data bus, and the data is read by the MPU.

The Accumulator

The **accumulator** is a register. It is the primary holding area for data before the data enters the ALU and the recipient of data after the data has been processed by the ALU.

If we wish to add two numbers, the first number is "fetched" from a memory location and placed in the accumulator. The other number is also fetched from memory and placed in a data register. The contents of the accumulator and the data register are summed by the ALU and the result of this addition is placed in the accumula-

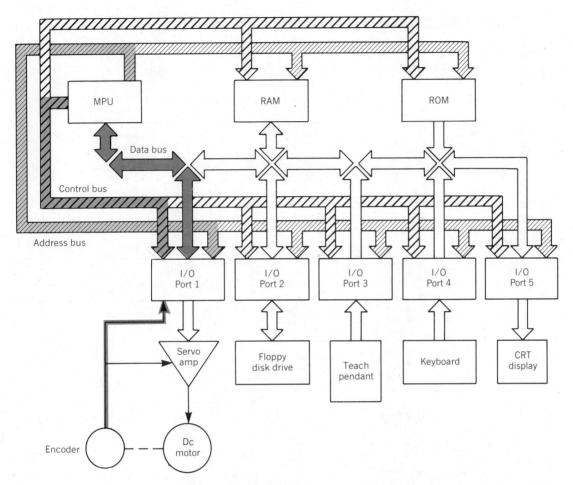

Figure 8.49 MPU addresses port 1 on the address bus and places a read command on the control bus. Port 1 places the robot's arm position on the data bus, and the data is read by the MPU.

tor. Recall that the accumulator was used to hold the data before the addition. When the result of the addition exits the ALU, it is stored in the accumulator, and the previous contents are lost.

Data Register

The **data register** is a temporary holding area for data. One example of the use of the data register is given above, in our description of the accumulator. The data register can also be used to store the contents of the accumulator before sending the data to memory.

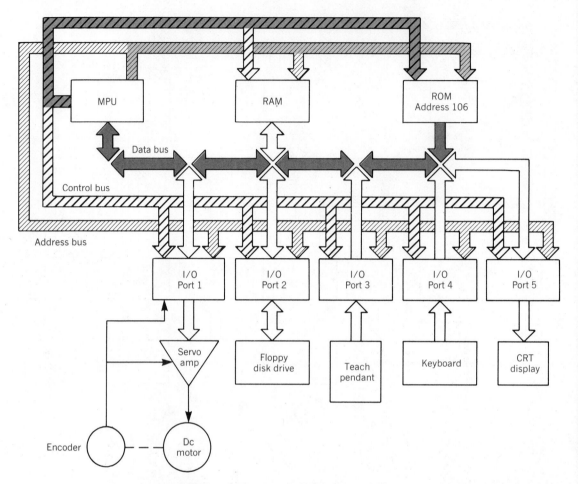

Figure 8.50 MPU outputs ROM address 106 on the address bus and a read command on the control bus. The ROM places data on the data bus, and the data is read by the MPU.

Address Register

The **address register** is used to temporarily hold a memory address before it is used to address a memory location. Many microprocessors allow for programming of the address register. If the address register is programmable, the contents of the register can be modified by a program instruction.

Program Counter

The **program counter** holds the address of the next instruction or data to be accessed. Recall from the previous discussion that the

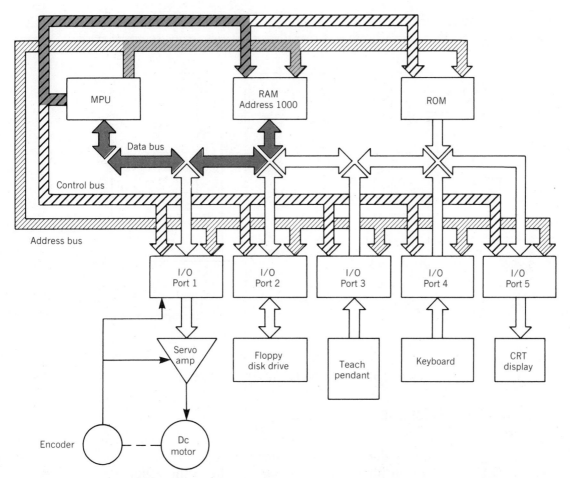

Figure 8.51 MPU addresses RAM address 1000 on the address bus and places a write command on the control bus. The MPU places data on the data bus, and the data is written into RAM address 1000.

addresses of the instructions and data were in sequence. The program counter was placing these addresses on the address bus.

The address held in the program counter is the next address in memory. The program counter outputs the address of the next instruction or data. The program counter can be modified so that the next address is not simply 1 greater than the previous address. Recall from the previous section that the MPU had accessed address 02, then the programmer typed in the word "program." The word "program" caused the program counter to "jump" to address 100.

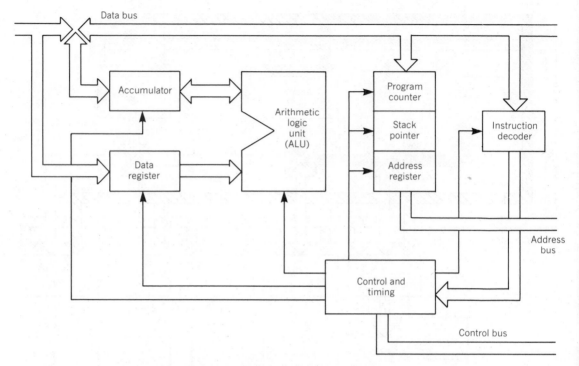

Figure 8.52   Block diagram of a microprocessor

Instruction Decoder

The **instruction decoder** does just as its name implies—it decodes the program instructions. Note that the program instructions are in binary. Recall from Chapter 6 the robot that was designed with a program panel. The program panel was designed using binary codes and was operating as an instruction decoder. The program instructions in the microprocessor are divided into two parts: the **op code** and the **operand**.

The op code (an abbreviation for "operation code") is the instruction that is to be executed. For example, the program instruction "ADD B" will add the contents of the B register to the contents of the accumulator. The op code is the "ADD" instruction. The "B" in the "ADD B" instruction is called the operand because this is the data to be "operated on."

The instruction decoder reads the op code "ADD" and prepares the ALU for an add operation. It then interprets the B as the B register and activates the appropriate control lines to execute the instruction.

Stack Pointer

The **stack pointer** is similar to the program counter. The stack pointer holds the address of data that has been temporarily stored in a section of the RAM called the **stack**.

Programs often simply step from one address to the next under the control of the program counter. At times, the result of an operation directs the program to jump out of sequence and execute another program called a **subroutine**. The problem is that data stored in the various registers may be lost while the subroutine is being executed. To preserve the data in the temporary registers, the data is put on a stack in the RAM and the stack pointer keeps track of the memory location. The stack pointer is also used to temporarily store the address in the program counter before the program is commanded to jump to a memory address that is out of the normal sequence.

One unusual aspect of the stack is the manner in which data is stored in it. Rather than being stored in sequence, the data is "stacked." The data from one register is entered into the stack, followed by the data from another register, which is placed on top of the data from the first register.

The stack is often referred to as **Last In First Out,** or **LIFO**. The last data put on the stack is the first data retrieved from the stack. The op code often used to place data on the stack is "PUSH." The data is "pushed" onto the top of the stack. The op code used to retrieve data is "POP." The data is "popped" off the top of the stack.

It is the job of the stack pointer to keep track of the RAM addresses of the data stored in the stack.

MICROPROCESSOR PROGRAMMING

The process of **programming** a microprocessor is time-consuming and requires a great deal of skill. Most microprocessors have more than 200 different instructions that can be used in writing the program. There are some larger microprocessors that have as many as 2000 instructions.

In programming a microprocessor it is the responsibility of the programmer to direct the microprocessor's every step. For example, after an ADD instruction, the result of the addition is stored in the accumulator. If the programmer forgets to store this result in memory and continues on to another operation (such as a subtraction), the result of the addition will be lost.

The microprocessor operates using binary numbers. Programming instructions such as ADD and POP are meaningless to the microprocessor. When programming, programmers use these three- and four-letter words, call **mnemonics**, to write their programs. Mnemonics are used because they are easier to remember than the binary codes actually used by the microprocessor. Writing programs using mnemonics is called "assembly language programming." Assembly language programs must be converted into binary codes. The process of converting mnemonics into their respective binary codes is called **assembling**. Assembling a program is time-consuming, and the probability of making an error is high. Many computers have a special program that allows the programmer to enter the mnemonics and operands. The assembler program converts the mnemonics into binary codes and stores the program in the RAM.

High-level Programming Languages

Assembly language programming is difficult, and becoming skilled at it requires years of practice. For this reason, other program languages have been developed, such as BASIC, FORTRAN, and COBOL. These program languages are more like the English language. Since computers can operate only with machine code, a special program called a "compiler" translates the high-level languages. Below is a sample BASIC program:

```
10 PRINT "ENTER A NUMBER BETWEEN 0 AND 10"
20 INPUT A
30 PRINT "ENTER ANOTHER NUMBER BETWEEN 0 AND 10"
40 INPUT B
50 IF A<B THEN PRINT "THE FIRST NUMBER IS SMALLER THAN
   THE SECOND"
60 IF A>B THEN PRINT "THE FIRST NUMBER IS GREATER THAN
   THE SECOND"
70 IF A=B THEN PRINT "THE NUMBERS ARE EQUAL"
80 END
```

This sample program asks the operator to enter two numbers between 0 and 10. It then compares the two numbers and prints the result of the comparison. Even if you are not familiar with programming you can read these instructions and interpret the program.

If written in assembly language, this program would normally require more than 30 instructions. In fact, when this program is run on a computer, the computer converts the program into all the steps necessary for execution of the program by the microprocessor.

Most computer operators and robot programmers will never have to write an assembly language program, because the programming that they will perform will be in high-level languages.

In addition to the high-level language programs mentioned above, there are specialized robot programming languages such as VAL, RAIL, and AML. These program languages use English-language statements such as those found in BASIC or COBOL.

We will write a sample program, in VAL, for a point-to-point robot. The robot is going to pick up a part at one location and move it to another. Before the program is written, the robot must be taught the points at which the part will be picked up and placed.

The robot's arm is moved to the point where the part will be picked up, and the programmer types "HERE PICK." The robot's arm is then moved to the point where the part will be deposited, and the programmer types "HERE PLACE" (Figure 8.53). The following program is typed in:

```
10  Appro Pick, 100
20  Move Pick
30  CloseI
40  Depart 100
50  Appro Place, 100
60  Move Place
70  OpenI
80  Depart 100
90  End
```

This program commands the robot to "Approach" the point identified as "Pick" within 100 mm. The robot is then commanded to move to the point called "Pick". The third instruction, "CloseI," tells the robot to close the gripper immediately on arrival at "Pick." Step 40 directs the robot to "Depart" point "Pick" by 100 mm. The

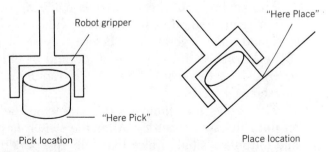

Figure 8.53 Illustration of a VAL program with locations "Pick" and "Place" identified

robot is then commanded to "Approach" the point identified as "Place" within 100 mm. The next instruction commands the robot to "Move" to the point called "Place." Step 70 commands the robot to open the gripper immediately. The next step commands the robot to "Depart" 100 mm. The final step ends the program.

MICROPROCESSOR INTERFACING

To allow the microprocessor to communicate with the outside world, it is connected to "ports." The ports are used to accept data from, and send data to, the outside world. In this section we will examine how these ports can be used to input information to the robot controller and receive information from the controller.

ASCII Code

Many robots have a typewriter keyboard which is used by the robot programmer to communicate with the microprocessor. When the keys of the typewriter keyboard are pushed, switches are closed. The switches are connected to a special decoder which generates a special binary code. The binary code that is most often used is called **ASCII** (American Standard Code for Information Interchange).

The ASCII code is made up of seven binary bits. A seven-bit code allows for 128 (2^7) possible combinations of 1's and 0's. With 128 possible combinations, all the letters of the alphabet, both upper- and lower-case, can be represented, as well as the numbers 0 through 9 and several special codes that include punctuation and machine control information.

Many keyboards transmit ASCII in parallel format. There are seven wires that carry the code. An eighth wire, called the "strobe line," is also connected between the keyboard and the port.

The keyboard of a controller is made up of switches. When a switch is pushed, its contacts may "bounce"—that is, they may close and open several times before they settle and remain closed. If the keyboard did not have a method of eliminating bouncing, several characters would be sent very rapidly rather than the single character desired.

The strobe line eliminates this problem. The strobe line remains off until the switch settles. After the switch has settled, the strobe line outputs a single pulse that tells the port that the data on the seven data lines are valid.

The ASCII code is also used to transmit data in a serial format. At times it is inconvenient to run eight wires between the keyboard or other peripheral device and the port. The serial format allows the data to be sent over a pair of wires. One of the wires carries the data; the other is a ground connection.

The data signals can be transmitted as changes in voltage or in current. There are four standards in use today. The two voltage standards are **TTL** and **RS232C**, and the two current standards are the **20mA current loop** and the **60mA current loop.**

TTL Standard

The **TTL (Transistor Transistor Logic) standard** is directly compatible with all TTL logic. The TTL standard specifies that 5 V represents a logic 1 and 0 V represents a logic 0.

The advantage of the TTL standard is that it directly interfaces with TTL logic. Its disadvantage is that it is susceptible to noise and voltage drops. Noise can cause inaccurate data transmission. The resistance of the transmission line causes a voltage drop. If the voltage drops more than ½ V along the transmission line, inaccurate data may be received.

RS232C Standard

The **RS232C standard** was developed by the Electronic Industry Association (EIA). The RS232C standard states that a voltage between −3 and −25 V represents a logic 1 and that a voltage between +3 and +25 V represents a logic 0 (Figure 8.54).

The advantages of the RS232C are its immunity to noise and the fact that a small voltage drop along the transmission line will not distort the data. Its major disadvantage is that a converter is required to convert the voltage levels to the TTL standard at the port of the computer.

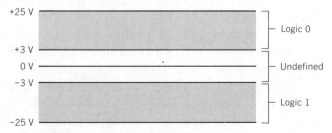

Figure 8.54   RS232C standard

20mA and 60mA Standards

The **20mA standard** specifies that a current of 20 mA represents a logic 1 and that a current of 0 represents a logic 0. The **60mA standard** specifies that a current of 60 mA represents a logic 1 and that a current of 0 represents a logic 0.

The advantages of these current standards are their high noise immunity and their ability to transmit data over long distances. The disadvantage of the current standards is that they must be converted to voltage levels if they are to be used as inputs at the computer port.

SYNCHRONOUS AND ASYNCHRONOUS DATA TRANSMISSION

During transmission of data, the rate at which the data is being transmitted must match the timing of the receiver. There are two different methods used to ensure that the transmitter and the receiver are operating at the same speed. These two methods are **synchronous data transmission** and **asynchronous data transmission.**

Synchronous Data Transmission

In **synchronous data transmission,** the clock that controls the rate of transmission is common to both the receiver and the transmitter. This is referred to as a "common clock." The advantage of the common clock is that it ensures that the receiver and the transmitter are operating at exactly the same speed, but this requires extra leads.

When data is transmitted using the synchronous method, the beginning of each word transmitted is indicated by the clock.

Asynchronous Data Transmission

In **asynchronous data transmission,** there is no common clock. The asynchronous method is far more common than the synchronous method.

Even though there is no common clock between the transmitter and the receiver, it is critical that both the transmitter and receiver operate at the same speed. The transmitter and the receiver have their own clocks. These clocks must be set at the same speed.

To ensure that both the transmitter and the receiver are operating at the same speed, standards have been established. The standard speeds are referred to in terms of "Baud rate." The Baud rate is generally defined as the number of bits of data that are being transmitted per second or, more specifically, as 1 divided by the time of one bit, also called "one bit-time." For example, if a bit is high for 0.833 ms (0.000833 sec), its Baud rate is 1/0.000833, or 1200.

The standard Baud rates in use today are 75, 110, 150, 300, 600, 1200, 2400, 4800, 9600, and 19200.

When the asynchronous method of data transmission is used, the beginning and end of the word being transmitted must be indicated as part of the transmission. To indicate the beginning of the transmission, a 0 is sent for 1 bit-time. This is called a "space" (Figure 8.55). After all the data bits have been sent, a "stop" bit called a "mark" is transmitted. The stop bit is always a 1, and can be transmitted for 1, $1\frac{1}{2}$, or 2 bit-time periods. Since the transmission has a start bit and one or more stop bits, the data is said to be "framed." When no data is being transmitted, a "mark" is continuously sent. The system is said to be "marking time."

Parity. To ensure that data is being received accurately, a **parity** bit is added to each series of bits. Recall that the ASCII code uses seven bits to transmit a single character or symbol. When parity is included, an eighth bit is transmitted. There are two parity systems in use: **odd parity** and **even parity.**

The **odd parity** system requires that an odd number of high bits be transmitted and received. Before the character is transmitted, the high bits are counted. If the number of high bits is odd the parity bit remains low, but if the number of high bits is even the parity bit is set high. When the character is received, the high bits are again counted. If an odd number of high bits are received it is assumed that the data transmission is accurate, but if an even number of high bits are received it is known that an error has occurred in the transmission (Figure 8.56).

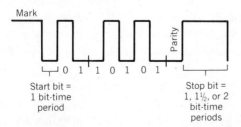

Figure 8.55 Framed data

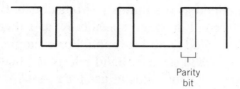

Parity
bit

Figure 8.56　Odd parity

The parity system assumes that there will never be more than one error per transmission. One error changes the count of high bits from odd to even. If, however, two errors occur in the transmission, the number of high bits received will be odd and it will be assumed that the data transmission is accurate (but this is very unlikely).

The **even parity** system requires that an even number of high bits be transmitted and received. The number of high bits is determined. If the number of high bits is even the parity bit remains low, but if the number of high bits is odd the parity bit is set high.

The receiver is always looking for an even number of high bits. When the receiver counts the high bits and the count is even, the character is assumed to have been received correctly, but if the count should be odd it is assumed that the character has been incorrectly received.

It is very important when using parity to ensure that both the transmitter and the receiver are operating on the same standard. If the transmitter is set for even parity and the receiver is set for odd parity, all characters that are transmitted properly will be rejected.

MULTIPLEXING AND DEMULTIPLEXING

Often we wish to send many different signals over a single set of transmission lines. The method used to "share" a single set of lines is called **multiplexing and demultiplexing.** Multiplexing can be done on the basis of frequency or time. Frequency multiplexing is used by telephone companies to carry thousands of conversations on a single pair of transmission lines. Here we will discuss only the time multiplexing method.

Figure 8.57 is a simple illustration of a time multiplexer/demultiplexer. The multiplexer/demultiplexer is made of two rotary switches. The wipers of the rotary switches are connected in series. All inputs are connected to rotary switch A, and the input port of the computer is connected to rotary switch B. If rotary switch A is

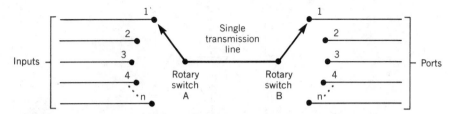

Figure 8.57 Time multiplexer-demultiplexer

set at position 1 and rotary switch B is set at position 1, data present at input 1 will be transmitted to input port 1. If both rotary switches are set at position 2, data from input 2 will be transmitted to computer port 2. Using multiplexing/demultiplexing allows any number of inputs to be transmitted along a single pair of transmission lines.

When multiplexing/demultiplexing is used, data from any particular input is not transmitted continuously. The data from input 1 is transmitted and received only when both rotary switches are in the same position. The data is said to be "sampled."

Figure 8.58 shows a solid-state eight-input multiplexer. It is built using eight four-input AND gates, one eight-input OR gate, and six NOT gates. The "select" lines are used to choose which input line is active. If all select lines (S1, S2, and S3) are low, the output of the first group of inverts is high, enabling AND gate 1. Any data present on input line 1 will be transferred to the output of AND gate 1. The data is passed through the OR gate and appears on the output line.

Figure 8.59 shows the same multiplexer with select line S1 high and select lines S2 and S3 low. Three inputs of AND gate 2 are high. Any data present at data input D2 will pass through AND gate 2 and through the OR gate and will appear at the output.

Any of the data input lines can be selected by using the correct combination of select lines.

Display Multiplexing

Figure 8.60 is a schematic diagram of an LED display which is being multiplexed. This display has four seven-segment LED's. All of the LED's are connected to a single data bus. A multiplexer is used to control which seven-segment display is to turn on.

Using a single data bus for all four seven-segment displays is accomplished by placing data on the data bus and turning on the first seven-segment display. The multiplexer then turns off the first display, places new data on the data bus, and turns on the second

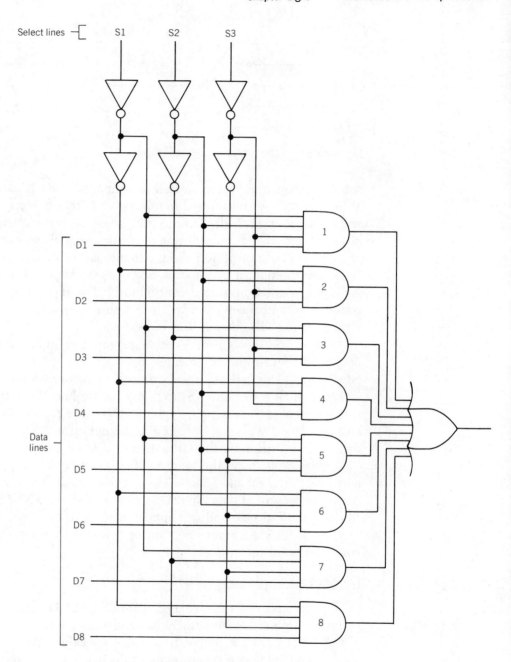

Figure 8.58   Solid-state eight-input multiplexer

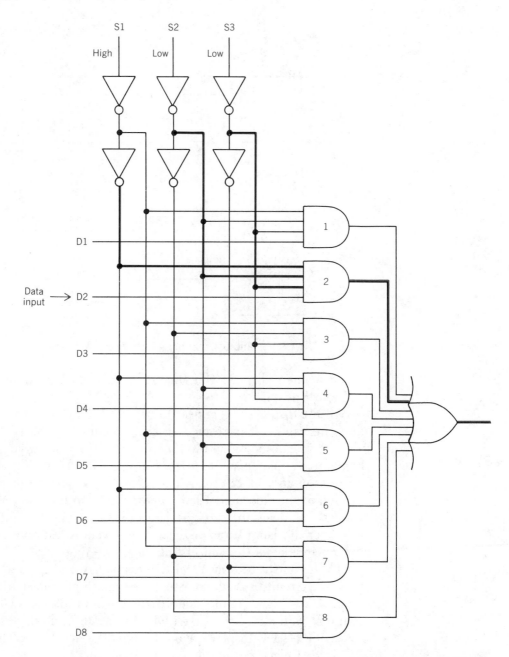

Figure 8.59 Multiplexer shown in Figure 8.58 with S1 high and S2 and S3 low

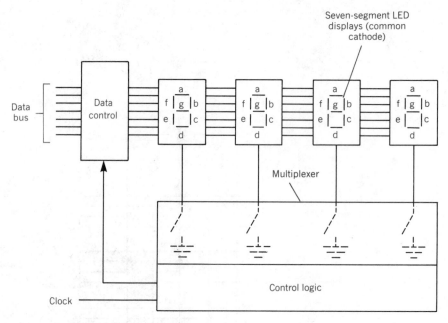

Figure 8.60 Schematic diagram of an LED display with multiplexer

seven-segment display. The same procedure is followed for the third and fourth displays.

All the displays will appear to be on all the time if the sequence described above is continued at a rate higher than 50 Hz. The displays appear to be on all the time just as a fluorescent lamp in your home or school appears to be on continuously although it actually is turning on and off 120 times per second. Each time the ac voltage applied to the fluorescent lamp rises, the lamp turns on. When the voltage falls back to zero, the lamp turns off. As the voltage goes negative, the lamp turns on again. The cycle is fast enough to give the appearance of the lamp being on all the time.

To demonstrate how the LED display is multiplexed, we will trace the circuit through one step at a time (Figure 8.61). Data lines b and c are high. At the same time, the multiplexer switches the common cathode of the first LED display to ground. Segments b and c of the LED display are on, forming the number 1. This occurs on the first clock pulse.

On the second clock pulse, the control logic places new data on the data bus. Data lines a, b, c, d, and g are high. At the same time, the common cathode of the second display is grounded. The second display lights, forming the number 3.

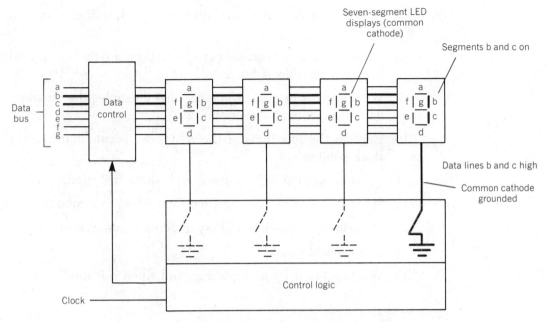

Figure 8.61 Same as Figure 8.60, with segments b and c on

On the next clock pulse, new data is placed on the data bus and the common cathode of the third LED is grounded. The third display lights. On the fourth clock pulse, the last bits of data are placed on the data bus and the common cathode of the fourth display is grounded. The fourth display lights. If this sequence is repeated very rapidly, all the LED displays will appear to be on continuously.

QUESTIONS FOR CHAPTER 8

1. Draw schematic and timing diagrams for a PISO shift register, and explain the operation of this register.

2. What are some of the applications for a PISO shift register?

3. Define the terms RAM, ROM, PROM, EPROM, and EEPROM.

4. Explain two-dimensional addressing.

5. What is meant by $\overline{\text{CS}}$? Explain its use in memory expansion.

6. Explain the difference between static and dynamic RAM's.

7. Define volatile and nonvolatile memories, and give examples of both.

8. What types of information are carried on a data bus, a control bus, and an address bus?

9. Write a short program sequence that illustrates the functions of data, control, and address buses.

10. Explain the functions of the ALU, the accumulator, and the stack pointer.

11. What is the RS232C standard, and where is it used?

12. What is meant by ASCII? Give some of its applications.

13. Describe synchronous and asynchronous data transmission.

14. Explain odd parity.

15. What is meant by multiplexing, and when is it used?

CHAPTER NINE

Servo Systems

In Chapter 1, robots were categorized as being either servo controlled or nonservo controlled. Recall that the nonservo controlled robot moves from position to position by "banging" into end stops, whereas the servo controlled robot can be taught to stop at any point within its work envelope.

There are two basic types of servo systems: the **open-loop servo system** and the **closed-loop servo system.**

OPEN-LOOP SERVO SYSTEM

The **open-loop servo system** (Figure 9.1) is the simpler of the two servo systems. In an open-loop servo system, a controller sends out a signal directing the robot's arm to move. In response to this signal, the power system moves the arm. There is no way for the controller to know whether or not the power system has done precisely what it was commanded to do. If the arm has not moved far enough or if it has moved too far, its position is not corrected. We can say that an **error** exists. When the arm is commanded to move again, it starts from its current position (which is not the correct

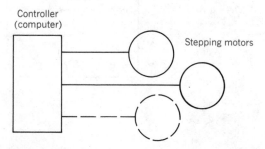

Figure 9.1   Open-loop servo system (no feedback)

position) and moves toward the new position. The error is not eliminated. If the arm should again not move the distance commanded by
the controller, another error is introduced. The first error and the
second error are added together. The errors will continue to accu-
mulate throughout the entire program.

This accumulation of errors can cause significant problems. The
robot may not be close enough to its commanded position to pick up
a part, or, if the robot is loading a part into a machine, the robot's
arm may crash into the machine. For this reason, open-loop servo
systems are not commonly used in industrial robots.

There are, however, many situations in industry where the open-
loop servo system is adequate. Figure 9.2 shows a simple open-loop
servo system comprising an inlet pipe, a control valve, a vat, and an
outlet pipe. Oil flows through the inlet pipe and into the vat. If the
amount of oil entering the vat through the inlet pipe is the same as
the amount exiting the vat through the outlet pipe, the oil level in
the vat will remain the same. The problem is that something may
upset the balance of this process. Suppose, for instance, that the
temperature of the room goes down. The oil in the vat will cool, will
become thicker, and will not flow out of the vat as fast as it did when
it was warmer. Since there will be less oil exiting the vat than entering the vat, the oil level in the vat will rise. If this situation is not
corrected, the vat will eventually overflow.

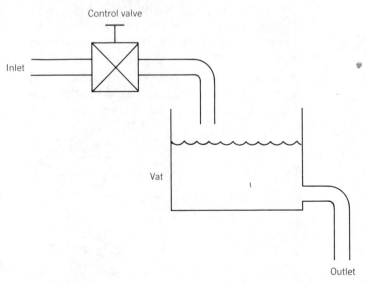

Figure 9.2   Open-loop servo vat system

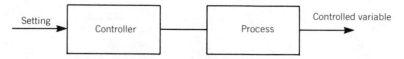

Figure 9.3 Block diagram of open-loop servo vat system

Figure 9.3 is a block diagram of the vat system. The label **setting** represents the adjustment of the inlet valve, and the block labeled **controller** represents the valve itself. The operator of the system chooses the setting by turning the handle of the valve (controller). The block labeled **process** represents the vat and the inlet and outlet pipes as well as the "process" of the oil entering and exiting the vat. The result of the process, which is labeled **controlled variable,** is the level of oil in the vat. The controller controls the controlled variable. It is obvious that the "setting" (adjustment of the valve) affects the "controlled variable" (the level of oil in the vat).

The diagram shown in Figure 9.4 is identical to that shown in Figure 9.3 with the exception of the additional term **disturbances**. Disturbances are factors that upset the process and as a result affect the "controlled variable." For our simple vat system, a decrease in the temperature of the room would be a disturbance. As the temperature of the room goes down, the oil becomes thicker and thus exits the vat more slowly, and the level of oil in the vat begins to rise. The "disturbance" (change in room temperature) has affected the "controlled variable" (the level of oil in the vat). If the operator is watching the vat, the "setting" can be changed to compensate for the effect of the "disturbance," and the "controlled variable" will remain the same.

CLOSED-LOOP SERVO SYSTEM

The simple vat system discussed above can be changed to a **closed-loop servo system** to eliminate the human operator and allow

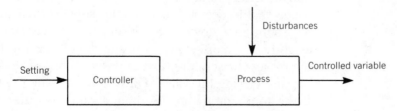

Figure 9.4 Block diagram of open-loop servo vat system, showing "disturbances"

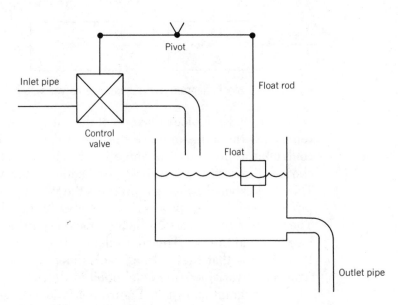

Figure 9.5 Closed-loop servo vat system

for automatic control of the process (Figure 9.5). The objective of the system is the same: to maintain the oil level in the vat. The changes that are made in the system are the removal of the manually controlled valve and the addition of an automatic valve controlled by a float.

As the level of oil in the vat goes up the float is pushed upward, and as the level in the vat goes down the float moves downward. The float is connected to the inlet valve through a lever. If the oil level rises, the float moves upward, pushing on the lever and closing the valve, and thus reducing the flow of oil into the vat. If the oil level goes down, the float moves downward, pulling on the lever and opening the valve, and thus allowing more oil to enter the vat.

To adjust the system for the proper level of oil in the vat, the float is moved up or down the float rod. With the float properly positioned, the system will automatically adjust itself even if a disturbance should occur. For example, suppose that the temperature in the room goes down. The oil in the vat becomes thicker and thus exits the vat more slowly, causing the oil level in the vat to rise. As the oil level rises, the float is pushed upward, pushing on the linkage and closing the valve. Oil is now entering the tank more slowly. If the temperature in the room should rise, the oil in the vat will become thinner and will exit the tank more quickly. In this case the float will follow the oil level down, pulling on the linkage and opening the valve. The oil in the vat will return to the desired level.

Figure 9.6 is a block diagram of this closed-loop servo system. Notice that the only difference between Figure 9.6 and Figure 9.4 is the addition of a **comparator** and a **feedback loop.**

When used in reference to servo systems, the term "comparator" refers to a circuit or other device that compares the setting and the controlled variable. The difference between the setting and the controlled variable is the output of the comparator. In the systems studied in this chapter, a differencing Op Amp (operational amplifier) will be used as the comparator.

In Chapter 7, a comparator was described as an Op Amp that compares two inputs. If the inputs are equal the output of the Op Amp is zero, and if the inputs are not equal the Op Amp saturates and the output voltage is at a maximum. Although such comparators exist and are used in many digital circuits, they are not the same as, and should not be confused with, the comparators used in servo systems.

The comparator in a servo system must always find the difference between the setting and the controlled variable. It is unfortunate that the same term is used for both digital comparators and the comparators used in servo systems. In this chapter, the term "comparator" will always denote the circuit that compares the setting and the controlled variable, and the output of the comparator will be the difference between the two.

In the system illustrated in Figure 9.6, the setting is controlled by the location of the float on the control rod. The comparator is the float itself. The float "knows" the desired level of oil in the vat and compares the desired level with the actual level. The controller is

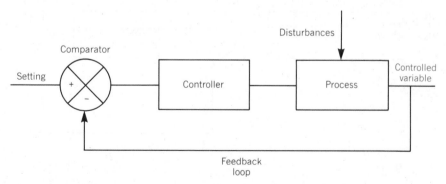

Figure 9.6 Block diagram of closed-loop servo vat system

the valve. If a disturbance (such as a change in temperature) should occur, the controlled variable will change (the level of oil in the vat will rise or fall). The float (comparator) will monitor this change and will make whatever adjustment is needed. In this system, the desired level of oil in the vat is maintained automatically.

CLOSED-LOOP SERVO ROBOT

A **closed-loop servo robot** operates much the same as the closed-loop vat system discussed above. Recall the method of teaching a robot. The operator leads the robot to the desired position by pushing buttons on a teach pendant. When the robot's arm is in the desired position, the operator pushes the "record" button on the teach pendant. A measurement device mounted on the robot's arm sends a coded electrical signal, which represents the position of the arm, to the robot's computer, and the coded electrical signal is recorded in the computer's memory. The operator moves the robot's arm to the next position and again pushes the record button. The signal from the measurement device sends a coded signal to the computer, which is recorded at the next memory location. When the robot is run it looks up in the computer's memory the first arm position that the operator recorded and moves to that position.

Figure 9.7 shows a block diagram of a closed-loop servo robot. The computer contains the recorded positions. The desired position of the robot's arm is outputted from the computer. The output from the computer is a digital signal. The digital signal is fed to a D/A con-

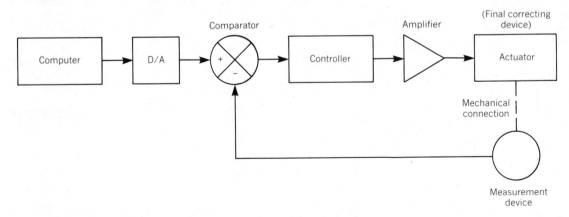

Figure 9.7   Block diagram of closed-loop servo robot

verter. The output voltage is the desired position of the arm (setting). This voltage is fed to one input of the comparator. The output from the comparator is fed to the controller. The output of the controller is fed through a power amplifier to the **actuator**. The actuator is an electric motor or a hydraulic servo valve and cylinder. The actuator is also referred to as a **final correcting device.** We will use the term "actuator" throughout this text.

A measurement device, such as a potentiometer, a resolver, or an encoder, is mechanically connected to the actuator. Each of these devices produces a voltage or a digital signal that represents the position of the shaft of the device. To indicate the mechanical connection of the measurement device to the actuator, we will use a dotted line. A full explanation of the operation of the resolver and the encoder is included later in this chapter.

The measurement device generates an electrical signal that represents the position of the arm. This is the same device that was used during the teach sequence to generate the positional information that was recorded in the computer's memory. The electrical signal from the measurement device is connected to the negative input of the comparator.

Servo Operation

The desired position of the robot's arm is outputted from the robot's computer. This positional information is in the form of a digital signal. The digital signal is converted to an analog signal by the D/A converter, which feeds this signal (the "setting") to the comparator in the form of a voltage.

For example, refer to Figure 9.8. In this case the desired position of the arm is converted by the D/A converter into a 5-V signal. At the same time, the measurement device outputs its measured position. Its output is 2 V. This is fed to the negative input of the comparator. The 5-V "setting" from the D/A and the 2-V output from the measurement device are compared. The difference between the setting and the measured position is 3 V. An error exists (the robot's arm is out of position). The 3-V error is amplified and the actuator begins to move. The measurement device continues to monitor the position of the arm.

As the arm moves, the output voltage of the measurement device changes. When the output of the measurement device is the same as the setting voltage, the output from the comparator is 0 V and the arm stops moving. The robot's arm is in the desired position.

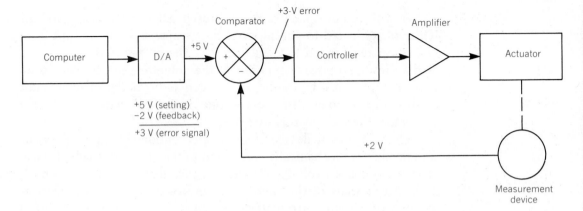

Figure 9.8 Closed-loop servo robot. Setting (5 V) − feedback signal (2 V) = error signal (3 V).

The robot's computer now outputs the next desired position (Figure 9.9). The digital output of the computer is again converted to a voltage by the D/A and fed to the comparator. The output voltage of the D/A is 2 V. Recall that in the first example the output voltage of the measurement device was 2 V. The computer is telling the robot to return to its original position. The output of the measurement device is 5 V, and the command voltage (setting) is 2 V. The output of the comparator is −3 V. This −3-V output is amplified, and the actuator moves in the opposite direction. The measurement device continues to monitor the position of the robot's

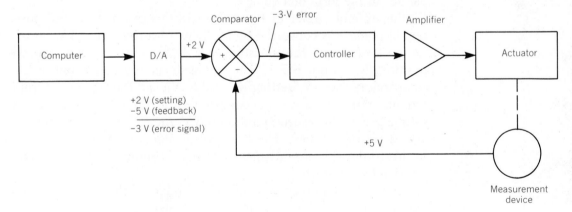

Figure 9.9 Closed-loop servo robot. Setting (2 V) − feedback signal (5 V) = error signal (−3 V).

arm. When the output of the comparator is again 0, the arm has re-turned to its original position and stops moving.

SERVO CONTROL MODES

All closed-loop servo systems react to errors. When an error is detected, the closed-loop servo system responds to the error and attempts to correct it. It is generally agreed that there are five control modes by which closed-loop servo systems react to errors: **on-off control**, **proportional control**, **proportional plus integral control**, **proportional plus derivative control**, and **proportional plus integral plus derivative control.**

Although the proportional plus integral plus derivative (PID) mode is the only control mode used in servo robots, it is important to understand the other modes of control since these control systems increase in complexity from on-off control to PID control.

All five modes of control respond to errors. They differ in the speed and accuracy with which they eliminate the error between the setting and the controlled variable.

It is also worthy of note that closed-loop servo systems are not limited to robotics. Many forms of automation use the closed-loop control method. The simplest of all control modes is on-off control, and the majority of automated systems in industry use this mode.

On-off Control

When an error exists and the system is in the **on-off control** mode, the actuator will be either fully on or fully off in attempting to correct the error. To help illustrate this mode of control we will examine a common home heating system (Figure 9.10). A home heating system has been chosen because of its simplicity. After we have explained the principles of the various control modes using the

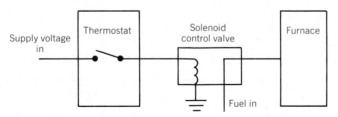

Figure 9.10   Home heating system

home heating system, we will apply these principles to a particular
type of robot.

The thermostat is the measurement device. When the temperature in the room is too low the thermostat turns the furnace fully on,
and when the room becomes too warm the thermostat turns the
furnace fully off.

The thermostat is an on-off device. When the room temperature
falls below the setting, the thermostat closes a switch. The switch is
connected to a fuel valve in the furnace. When the switch closes,
the valve is fully opened, and the furnace turns on and begins to
generate heat.

The graph in Figure 9.11 shows the operation of the thermostat
and the fuel valve in the furnace. As you know, the temperature in
the room does not rise the instant the furnace turns on. It takes time
for the temperature in the room to rise to the point at which the
furnace turns off. Even after the thermostat turns the furnace off,
the furnace contains enough heat so that the temperature continues
to rise for a while. With the furnace off, the temperature in the
room eventually begins to fall. When the temperature has gone low
enough, the thermostat again closes and the furnace turns on.

The graph in Figure 9.12 shows the rise and fall of the temperature and the operation of the fuel valve. The fuel valve is either fully
closed or fully open. The temperature rises and falls about the set
point. The temperature is said to oscillate. The oscillation is consistent.

Differential Gap.   The length of time between oscillations is controlled by the **differential gap.** The differential gap is defined as the
smallest change in the controlled variable that causes the valve to
shift from off to on or from on to off.

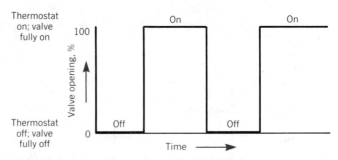

Figure 9.11   Graph illustrating operation of thermostat and furnace fuel valve

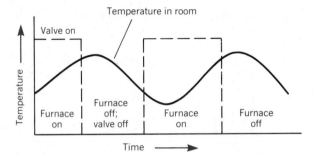

Figure 9.12 Graph illustrating rise and fall (oscillation) of temperature as fuel valve turns on and off

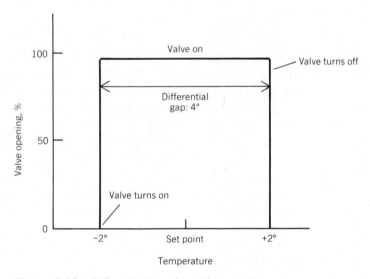

Figure 9.13 Differential gap for a thermostat

Figure 9.13 illustrates the differential gap for a thermostat. The temperature must rise 2° above the set point before the furnace turns off and must fall 2° below the set point before the furnace turns back on. The smallest possible change in temperature that will turn the furnace on or off is 4°. The differential gap for this thermostat is 4°.

The differential gap can also be expressed as a percentage of the full range of the controller. For example, assume that the thermostat can control the temperature between 40 and 90°. The control range of the thermostat is 50° (90 − 40 = 50). To find the differential gap in percentage, divide the differential gap by the total control range:

$$\% \text{ Differential gap} = \frac{\text{Differential gap}}{\text{Total control range}}$$

$$\% \text{ Differential gap} = 4/50$$

$$\% \text{ Differential gap} = 0.08$$

$$\% \text{ Differential gap} = 8\%$$

The differential gap is both an advantage and a disadvantage. The wider the differential gap, the slower the oscillation but the larger the amplitude. The narrower the differential gap, the faster the oscillation but the smaller the amplitude.

When the differential gap is very narrow, the controlled variable remains very close to the set point. The disadvantage of maintaining a narrow differential gap is the number of times the valve is turned on and off. The rapid oscillation wears equipment and shortens its life.

When the differential gap is wide, the controlled variable does not remain very close to the desired set point; however, the equipment does not turn on and off very often and its life is greatly extended.

Figure 9.14 demonstrates the difference between a system with a narrow differential gap and a system with a wide differential gap.

Since the on-off control mode is always in oscillation whether the differential gap is wide or narrow, it is never used in robot design. It should, however, be emphasized that the on-off control mode is the most common feedback system due to its simplicity and inherent reliability.

Proportional Control

Proportional control does not operate in the full on-off mode but rather operates in a continuum from fully off to fully on. To obtain this type of control, the simple solenoid valve in the furnace must be replaced with a servo valve. Recall that the servo valve is powered by a torque motor. As voltage rises the motor turns, opening the valve; the higher the voltage, the further the valve opens. At some point the voltage will drive the valve to its fully open position.

To help illustrate the operation of proportional control we will again examine the operation of a furnace. The standard solenoid valve is removed and a servo valve is substituted. The standard on-off thermostat is also removed and replaced with a thermostat that

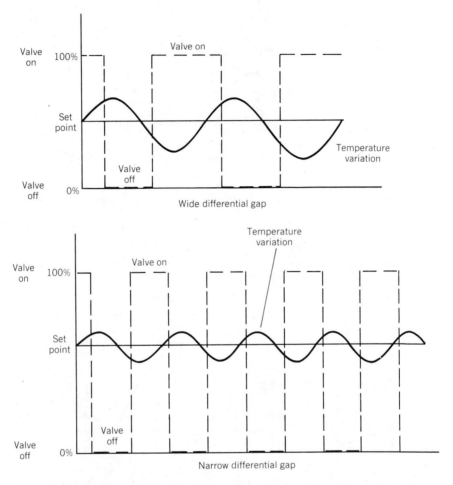

Figure 9.14 Comparison of systems with wide and narrow differential gaps

varies the output voltage in response to changes in room temperature (Figure 9.15).

The graph in Figure 9.16 plots the percentage of valve opening versus room temperature. We will choose 70° as the set point for our system. When the room temperature is 70°, the valve is 50% open. The 50% valve opening is the neutral position of the valve. When no voltage is applied to the valve it remains exactly 50% open. When a positive voltage is applied to the valve it opens more than 50%, and when a negative voltage is applied to the valve it closes more than 50%. If the temperature in the room suddenly drops to 60°, a positive voltage is applied to the valve. The valve opens from 50 to 70%,

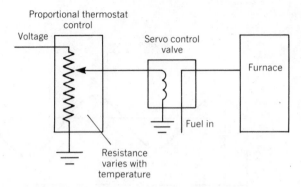

Figure 9.15 Proportional control heating system

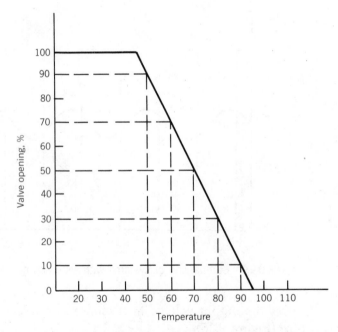

Figure 9.16 Percentage of valve opening vs. temperature

raising the temperature in the room back toward the set point. Note that the percentage of valve opening has increased in response to the decrease in room temperature, but that the valve is still less than fully open.

You can also see that the percentage of change in the valve is not equal to the percentage change in room temperature. For every 10° change in temperature the valve opening changes 20%. A change in temperature of 20° will result in a 40% change in valve opening.

The term "proportional control" is appropriate since the amount of correction is "proportional" to the error. The ratio between correction and error is constant. The valve shifts 20% to correct an error of 10° and shifts 40% to correct an error of 20°.

Proportional Band.   Look again at Figure 9.16. The valve is fully open when a disturbance causes the temperature to fall to 45° and is fully closed when a disturbance causes the temperature to rise to 95°. The valve shifts from fully open to fully closed over a 50° range in temperature (95 − 45 = 50). This range over which the valve shifts from fully open to fully closed is called the **proportional band.**

The proportional band is often expressed as a percentage of the total adjustment possible for the set point. In the system shown in Figure 9.16, the full control range is 110° minus 0°, or 110°. We have already determined that the proportional band is 50°. To find the percentage value of the proportional band, divide the proportional band by the full control range and multiply by 100:

$$\text{\% Proportional band} = \frac{\text{Proportional band}}{\text{Full control range}} \times 100$$

$$\text{\% Proportional band} = \frac{50}{110} \times 100$$

$$\text{\% Proportional band} = 0.4545 \times 100$$

$$\text{\% Proportional band} = 45.45\%$$

Notice that the formula for computing the proportional band is the same as the formula for computing the differential gap. Outside either the differential gap or the proportional band, the valve is fully open or fully closed. The difference between the proportional band and the differential gap is the manner in which the system responds within the band or gap (between the points at which the valve is fully open or fully closed). In an on-off system the valve opening does not change as the measured value passes through the differential gap, whereas in a proportional system the valve opening changes proportionally with the error as the measured value passes through the proportional band.

Figure 9.17 shows four graphs of percentage of valve opening versus temperature. The full control range is the same for each graph, but the graphs differ in the width of the proportional band (the slope of the line decreases as the proportional band becomes wider).

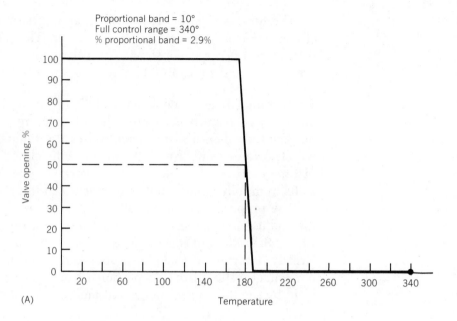

(A)

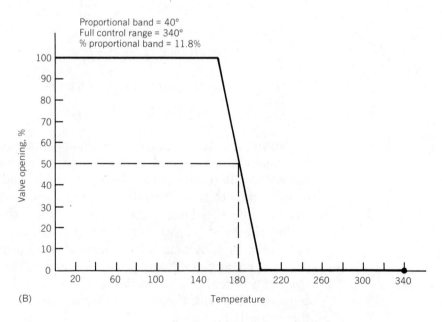

(B)

Figure 9.17 Four graphs (continued on facing page) of percentage of valve opening vs. temperature, illustrating proportional bands of four different widths

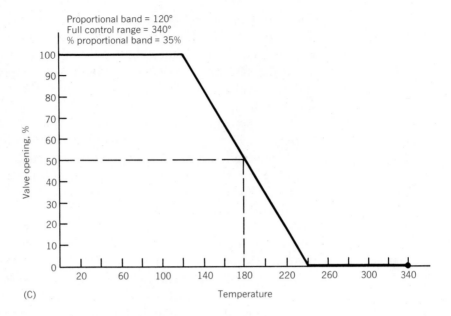

(C)

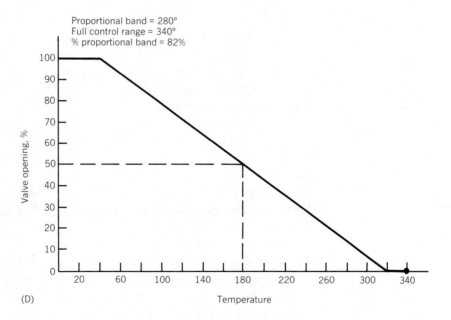

(D)

Figure 9.17 (Continued)

The narrower the proportional band, the more quickly the system responds to a disturbance. For all of the systems illustrated in Figure 9.17, the set point is 180° and the control valve is 50% open.

In the system represented by graph A, a temperature drop of 5° opens the valve to 100%. The furnace warms the air quickly, and the set point temperature is restored.

The proportional band in graph B is wider than that in graph A. If there is a 5° drop in temperature, the valve opens to 62.5%. More fuel is injected into the furnace, and the set point temperature is restored. Since the valve opens to only 62.5% instead of to 100%, the temperature is not returned to the set point as quickly as in the system illustrated by graph A.

The proportional band in graph C is wider still. In this system, a 5° drop in temperature causes the valve to open to 54%. The set point temperature is restored, but it takes even longer.

The proportional band in graph D is very wide. If a 5° temperature drop occurs in this system, the fuel valve opens — but to only 51.8%. This slight opening of the valve restores the temperature to the set point, but only after a long period of time.

From this discussion it seems that a narrow proportional band would always be an advantage: the set point temperature would be restored more quickly, and that is the objective of a closed-loop servo system. However, the problem with a narrow proportional band is that the narrower the proportional band, the higher the tendency for the system to go into oscillation.

The graph at left in Figure 9.18 represents a system with a proportional band of zero width. This graph should look familiar: it is similar to the graph of an on-off control system (the differential gap has been omitted). Recall that all on-off control systems oscillate. By the time the valve shifts, the system has already overshot the set point, and the valve shifts in the other direction (see graph at right in Figure 9.18).

The same problem exists for a proportional control system with a narrow proportional band. With a narrow proportional band, the system tends to overshoot the set point. The system responds by shifting the valve in the opposite direction, but the system overshoots again in the opposite direction. These oscillations normally die out after a period, and the system is said to be **stable**. At times, however, the overshoot is so large and the system overcorrects so far that the amplitude of the oscillations increases. These increasing oscillations, if left uncorrected, will destroy the system. When a system will not settle but continues to oscillate, it is said to be **unstable**.

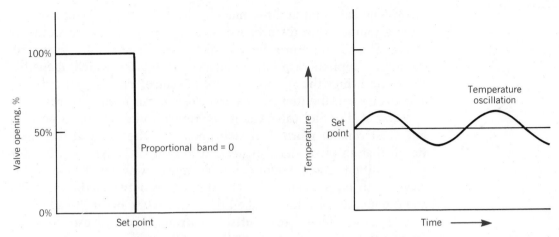

Figure 9.18 Oscillation in a proportional control system with a proportional band width of zero

There are methods for controlling the stability problems of servo systems with narrow proportional bands. We will cover these methods later in this chapter.

Steady-State Error. Up to this point we have assumed that when a disturbance occurs the measured value and the set point eventually become equal and the error is reduced to zero. Unfortunately, this is not the case.

Figure 9.19 illustrates what really happens. We will assume that the fuel valve is 50% open. There is no error voltage applied to the

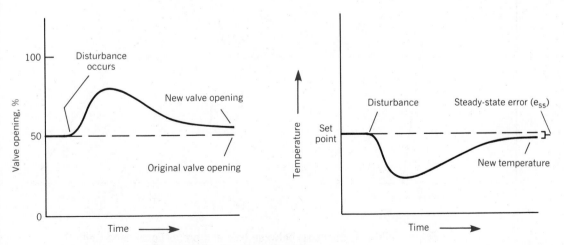

Figure 9.19 Steady-state error

valve. The set point and the measured value are in perfect balance. Then a temperature disturbance occurs. The temperature suddenly drops. The temperature disturbance is permanent. The system responds by applying a positive voltage to the valve which opens the valve and brings the system back into balance. The temperature begins to rise and the fuel valve starts closing. Since the disturbance is permanent, the fuel valve can never return to its 50% open position but must remain more than 50% open to "offset" the disturbance. Recall that an error signal is needed to shift the valve from its 50% open position, and therefore an error must always exist to keep the valve open more than 50%. The difference between the set point and the measured value is called the **steady-state error.** Steady-state error is also referred to as **offset** or **droop.** We will use the term "steady-state error" throughout this text.

Recall from the previous discussion of the proportional band that the narrower the proportional band the more the valve shifts in response to an error. The proportional band is often referred to as "gain" and is the slope of the ramp. The proportional band is really the amount by which an error signal is amplified. Figure 9.20 illustrates this point. The "gain" of system A is higher than that of system B, and the proportional band of system A is narrower than that of system B.

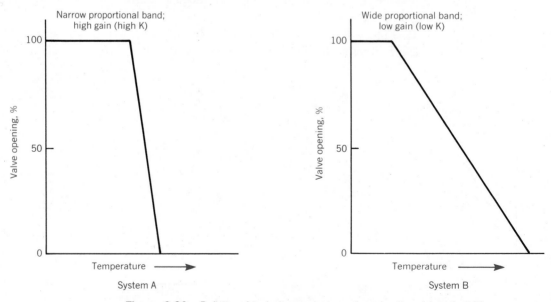

Figure 9.20 Relationship between gain and proportional band width

Figure 9.21 illustrates the relationship between gain and steady-state error, where e_{ss} is the steady-state error, A is the set point, and K is the gain. As the value of K increases, both the proportional band and the steady-state error are reduced.

Recall that the narrower the proportional band the higher the probability that the servo system will become unstable. To state it differently, the higher the gain (K) the higher the probability that the system will become unstable. On the other hand, the narrower the proportional band (the higher the gain) the smaller the steady-state error (e_{ss}) and the faster the system will respond to a disturbance.

Disturbances are not the only causes of steady-state error. Up to this point we have assumed that the set point results in a 50% valve opening. This is a textbook example. In the real world it is very unlikely that this situation will ever occur. It is true that there is one unique set point that will result in a 50% valve opening, but we will probably not choose that exact set point. If we do not choose the exact set point that results in the valve being in its neutral position (50% valve opening), a steady-state error will exist. A steady-state error may also exist as a result of losses in the system.

The farther the set point from the 50% valve opening, the larger the steady-state error (Figure 9.22). With these kinds of problems, the simple proportional control system is not used very much for industrial control systems and is never used for robots. By adding integral control to the simple proportional control, the system response is greatly improved.

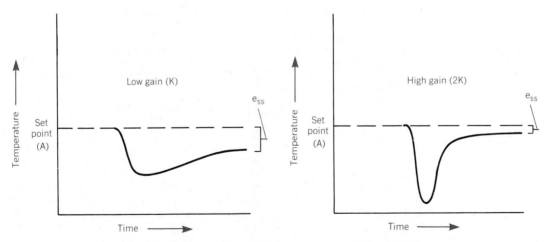

Figure 9.21   Relationship between gain and steady-state error

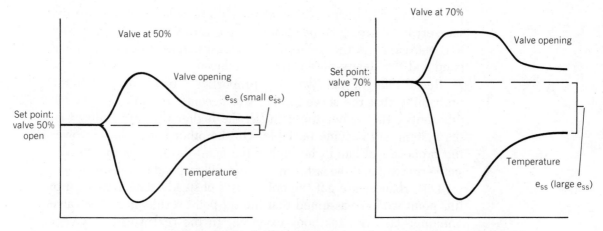

Figure 9.22   Relationship between steady-state error and set point

Schematic Diagram for Proportional Control.   Figure 9.23 is the schematic diagram of a simple proportional control system. The control system is connected to the arm of a robot. Although the simple proportional control is not used in robotics, the more complex proportional plus integral plus derivative system is used, and it is

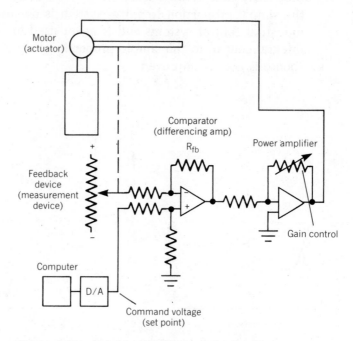

Figure 9.23   Simple proportional control system

important that you understand the operation of a simple system before attempting to understand the more complex system.

The robot's arm is connected to the wiper of a potentiometer. As the robot's arm moves up and down, the wiper on the potentiometer is moved, which causes the output voltage of the potentiometer to vary. The potentiometer is the "feedback" device in this system and is connected to the negative input of an Op Amp.

A computer is used as the control for the robot. The output of the robot is connected to a D/A converter. The output of the D/A converter is connected to the positive input of the Op Amp.

The Op Amp is connected so as to detect the difference between the command voltage output (set point) of the D/A converter and the output voltage of the feedback potentiometer.

If the output of the computer commands the robot's arm to move upward, the output of the D/A converter will be positive. The feedback voltage of the potentiometer is 0. The command voltage (set point) and the feedback voltage are compared by the Op Amp, and the result (difference) is the output of the Op Amp. The positive output voltage of the Op Amp is further amplified by the power amplifier. The output of the power amplifier is connected to a motor mounted on the robot's arm. With a positive voltage being fed to the motor, the robot's arm moves upward. When the arm approaches the set point, it stops moving. The arm will not stop at the precise point commanded by the computer because of the steady-state error.

The speed at which the robot's arm moves is determined by the "gain" of the power amplifier. The more the power amplifier amplifies the output of the Op Amp, the faster the arm moves. If the gain of the power amplifier is too high, the system will become unstable. The arm of the robot will not come to rest but rather will oscillate around the set point. If the gain of the power amplifier is set still higher, the oscillations will increase and the robot will eventually destroy itself.

Proportional Plus Integral Control

To correct the steady-state error, an **integral amplifier** is added to the proportional system (Figure 9.24).

Recall how an integral amplifier operates (Figure 9.25). When a voltage is applied to the input of an integral amplifier, the output of the amplifier is zero. If the input voltage remains unchanged, the capacitor forming the feedback loop begins to charge. As the capaci-

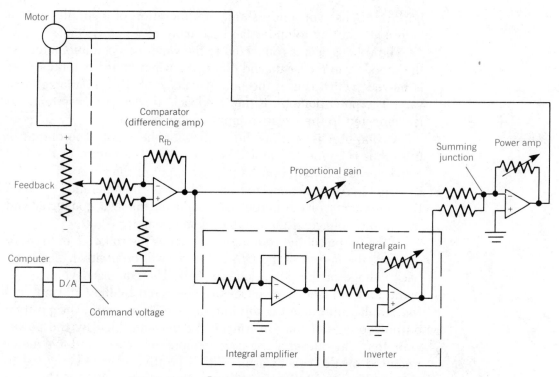

Figure 9.24 Proportional plus integral control system

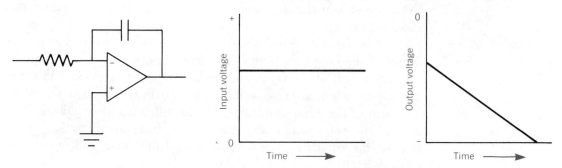

Figure 9.25 Integral amplifier

tor charges, the feedback voltage falls and the output of the amplifier begins to rise. After a period of time the capacitor becomes fully charged, and the output of the amplifier is at a maximum.

Note in Figure 9.24 that the output of the differencing amplifier is connected to the inverting input of an integral amplifier. The output of the integral amplifier is connected to the inverting input of another Op Amp.

In order to make the system operate properly, the output of the integral amplifier must follow the input. If the input to the integral amplifier is positive then the output should also be positive, and if the input is negative the output should also be negative. Unfortunately, the integral amplifier inverts the signal. If a positive voltage is applied to the integral amplifier the output will be negative, and if the input of the integral amplifier is negative the output will be positive. This problem is corrected by connecting the output of the integral amplifier to the inverting input of an Op Amp. Adding the second Op Amp allows the output to follow the input (Figure 9.26).

The output of the inverting Op Amp is connected to one input of a summing junction (Figure 9.27). The output of the differencing amplifier is connected to the other input of the summing junction. Finally, the output of the summing junction is connected to a power amplifier, and the output of the power amplifier is connected to the robot's motor.

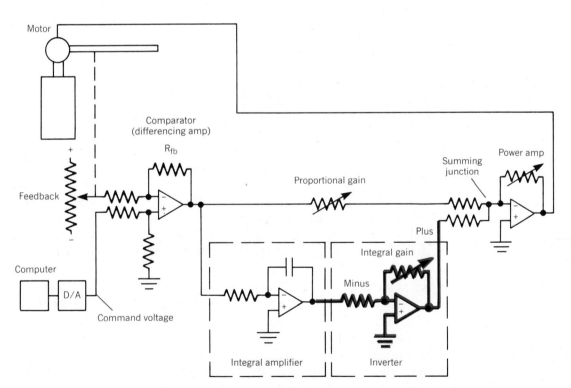

Figure 9.26 Proportional plus integral control system, showing inversion of the output of the integral amplifier so that the output follows the input. As the input of the integral amplifier goes positive, the output of the inverter goes positive.

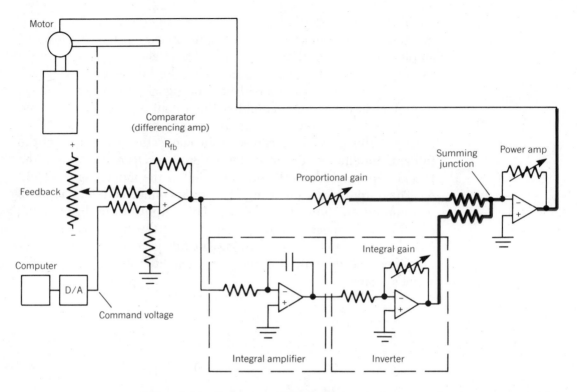

Figure 9.27 Proportional plus integral control system, showing summing of the proportional signal and the integral signal

    When the robot's computer outputs a command (set point), the D/A converter transforms the digital signal into an analog voltage. The voltage from the position potentiometer is compared with the set point voltage from the D/A converter, and the difference is outputted by the differencing amplifier (Figure 9.28). This error signal drives the power amplifier through the summing amplifier; the power amplifier drives the motor, correcting most of the error. As we saw earlier, the total error will not be eliminated. The error voltage that exists is also being fed to the integral amplifier (Figure 9.29). The longer the error signal exists, the higher the output voltage of the integral amplifier. Remember that the integral amplifier is inverting the error signal. Since we want the output of the integral amplifier to be added to the primary error signal, it must be inverted by a second Op Amp before being summed by the summing

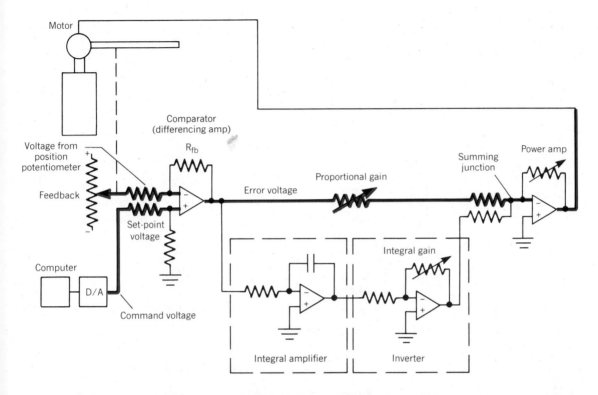

Figure 9.28 Proportional plus integral control system, showing the proportional error voltage resulting from the difference between the set point and the actual position

amplifier (Figure 9.30). If this is not done, the integrated signal will be subtracted, causing an even larger error.

The integrated error signal is summed with the primary error signal, driving the power amplifier. The robot's motor turns, eliminating the steady-state error (Figure 9.31).

Figure 9.32 presents graphs of the system's response showing the input error at the summing point, the response of the proportional amplifier, the response of the integral amplifier, and the summed response of the proportional and integral amplifiers.

This robot system has one major problem: stability. Recall that proportional systems alone may oscillate about the set point and exhibit steady-state error. Addition of the integral function eliminates the steady-state error but does not solve the oscillation problem. In

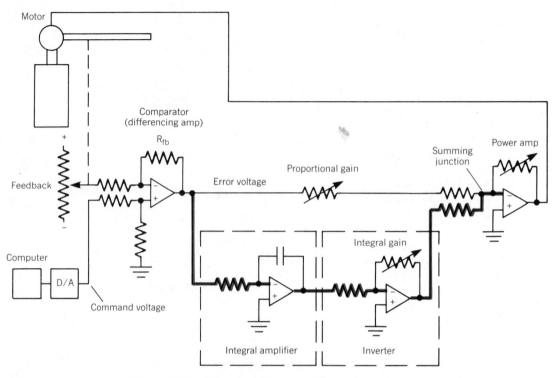

Figure 9.29 Proportional plus integral control system, showing error voltage being applied to the integral amplifier

the next section we will see how a derivative amplifier can be incorporated to eliminate the stability problem.

Proportional Plus Derivative Control

To eliminate the stability problem in the system, the derivative function is added. The **derivative amplifier** will perform the derivative function. The derivative amplifier responds to the rate of change of an error signal. If the error signal is changing very rapidly, the output of the derivative amplifier will be high. On the other hand, if the error signal is stable, the output of the derivative amplifier will be zero.

Figure 9.33 shows the schematic diagram of a derivative amplifier along with graphs illustrating the response of the amplifier to various types of inputs. You know that a capacitor blocks direct current but passes alternating current. If a constant dc voltage is applied to the derivative amplifier, the capacitor will block the voltage

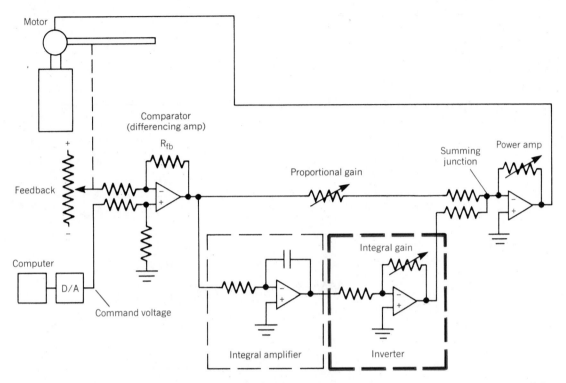

Figure 9.30 Proportional plus integral control system, showing inversion of the output of the integral amplifier

after a period of time. Since the capacitor is blocking the voltage, the input to the amplifier itself will be zero and the output will be zero. If, on the other hand, the input to the derivative amplifier is a square wave, the output will be a spike. If the input to the amplifier is a ramp, the output will be a dc voltage proportional to the ramp's rate of change.

Figure 9.34 shows a robot control system with a proportional plus derivative feedback system. The set point is controlled by the computer. The output of the computer is connected to a D/A converter, and the output of the D/A converter is connected to the non-inverting input of a differencing amplifier. A position potentiometer is connected to the inverting input of the differencing amplifier. This is the same as the proportional system described earlier.

The output of the differencing amplifier is connected to one input of a summing amplifier. The output of the derivative amplifier is connected to the other input of the summing amplifier

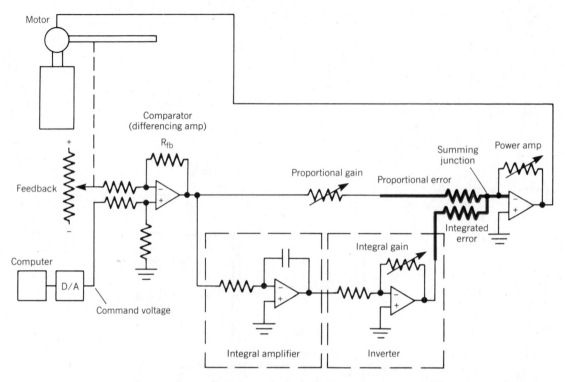

Figure 9.31 Proportional plus integral control system. The input to the power amplifier is the sum of the proportional error and the integral error.

(Figure 9.35). Notice that the derivative amplifier is an inverting amplifier. As the input of the derivative amplifier goes more positive, the output of the amplifier goes more negative. Since the output of the derivative amplifier is connected to the summing junction, the output of the derivative amplifier will be subtracted from the error signal being produced by the proportional function. Figure 9.36 illustrates this point.

Suppose that the computer is outputting a string of numbers such as 1, 2, 3, 4, etc. The output of the D/A is a rising voltage. The rising voltage is being compared with the voltage from the position potentiometer. Since the arm of the robot is not moving at this time, the output of the position potentiometer is stable and the output of the differencing amplifier is rising at the same rate as the input. The rising error signal is feeding one input of the summing amplifier. The error signal is also feeding the input of the derivative amplifier. The output of the derivative amplifier is a negative voltage. The

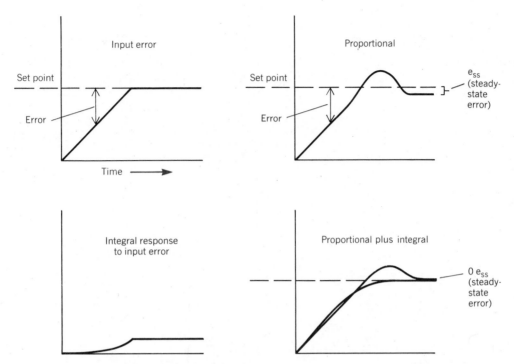

Figure 9.32 Graphs of input error, response of proportional amplifier, response of integral amplifier, and summed response of proportional and integral amplifiers

rising voltage from the differencing amplifier and the negative voltage from the derivative amplifier are summed, and the output of the summing amplifier is reduced (Figure 9.37).

The derivative amplifier is slowing the response of the system. This is called "damping." Damping a system slows the response of the system to a quickly changing error and prevents the oscillation associated with a simple proportional system.

Figure 9.38 presents graphs of the input error, the response of the proportional amplifier, the response of the derivative amplifier, and the summed response of the proportional and derivative amplifiers.

Damping. The gain of the derivative amplifier affects the degree to which the system is damped. Figure 9.39 shows the response of a proportional plus derivative system to three different levels of damping.

An **undamped** system responds very quickly to a change in the set point. The problem is that the system overshoots and oscillates

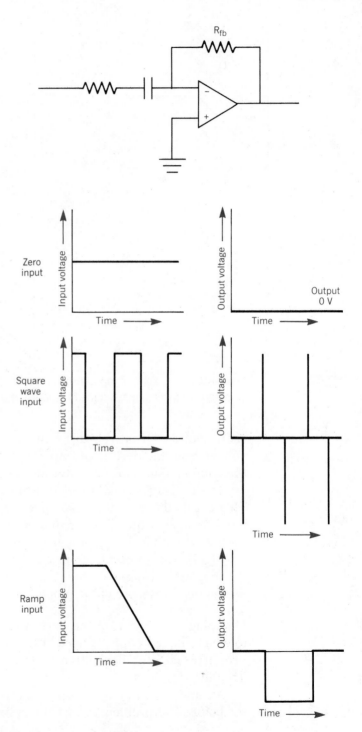

Figure 9.33   Derivative amplifier

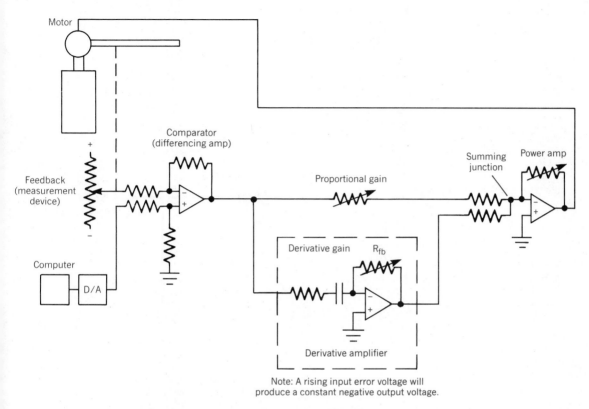

Figure 9.34 Proportional plus derivative control system

before settling down. In a robot, this type of system allows the robot's arm to oscillate before coming to rest. This oscillation can damage the robot and possibly even cause the robot's arm to crash into nearby machinery.

In an **overdamped** system, the gain of the derivative amplifier is high. This significantly slows the response of the robot. Since we want the robot to move as quickly as possible, overdamping is also undesirable.

Finally, in a **critically damped** system, the response is slowed just enough to prevent oscillation while still allowing the robot to respond to a command at near maximum speed. A critically damped system is the ideal situation but is very difficult to accomplish. An engineer begins the design of a damping system using complex mathematical modeling, but the final optimum combination of a proportional band width that allows the robot to respond very quickly to error signals and sufficient damping to prevent oscillations with-

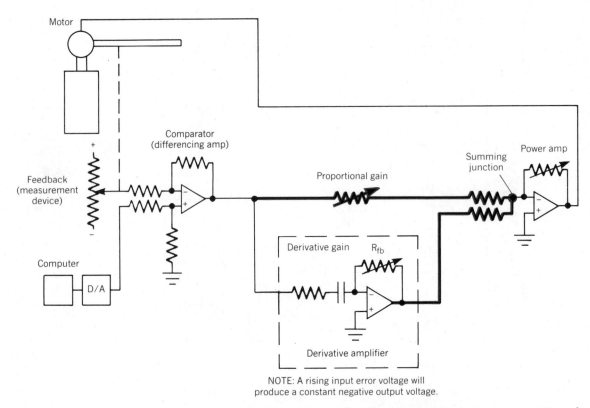

Figure 9.35 Proportional plus derivative control system, showing summing of the proportional error and the output of the derivative amplifier

out overdamping is adjusted through experimentation after the system has been built.

Recall that a proportional system contains a steady-state error. Adding the derivative function to the proportional system helps eliminate oscillations but has no effect on the steady-state error. To correct for the steady-state error we must also include the integral function.

Proportional Plus Integral Plus Derivative Control

In order to control the oscillation problem associated with a simple proportional controller as well as eliminate steady-state error, it is necessary to combine both the integral and derivative functions with proportional control.

Figure 9.40 is a schematic diagram of a **proportional plus integral plus derivative (PID) system.** We will examine the operation of such a system.

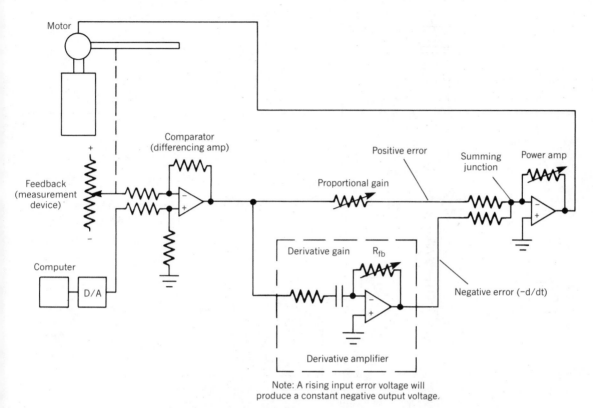

Motor

Comparator
(differencing amp)

Positive error

Summing
junction

Power amp

Feedback
(measurement
device)

Proportional gain

+

−

Computer

D/A

Derivative gain R_{fb}

Negative error ($-d/dt$)

Derivative amplifier

Note: A rising input error voltage will
produce a constant negative output voltage.

Figure 9.36 Proportional plus derivative control system, showing that the output of the derivative amplifier is negative with respect to the proportional error. The output of the derivative amplifier is subtracted from the proportional error.

Assume that the robot is stable and is waiting for a command to move. The computer outputs a command. The output of the computer is a string of increasing numbers telling the robot's arm to move upward. The D/A immediately converts the string of digits into a rising voltage. The rising voltage from the D/A is compared with the voltage from the position potentiometer, and the difference is outputted (Figure 9.41).

The output of the error amplifier is fed to the input of the summing amplifier, and the robot's drive motor begins to turn. At the same time, the integral amplifier is receiving the error voltage. The output voltage of the integrator circuit is summed with the error voltage to correct the error.

The output of the error amplifier is also feeding the derivative amplifier. The derivative amplifier responds by creating a negative voltage. The negative voltage of the derivative amplifier is added to

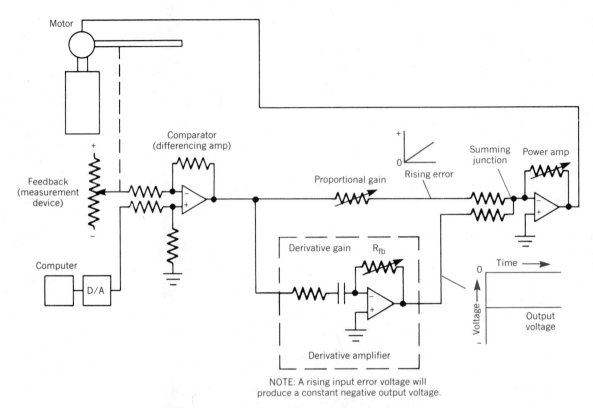

Figure 9.37 Proportional plus derivative control system. The output of the summing junction is the proportional error minus the output of the derivative amplifier.

the combined voltage of the error amplifier and the integral amplifier, and the result is a controlled speed.

We have described the events that occur as if they were sequential events. In reality, all the amplifiers are outputting voltages at the same time. In addition, the robot's arm is in motion, which produces a changing voltage from the position potentiometer. An understanding of the system's response to all of the amplifiers at the same time and the changing error voltage requires the use of calculus or Laplace transforms. A full description of the combining of signals and the study of Laplace transforms are beyond the scope of this text. If you simply remember that the proportional function affects the speed with which the system responds to an error, that the integral function eliminates steady-state error, and that the derivative function damps the system to prevent oscillations, you will have a sound basic understanding of a servo system.

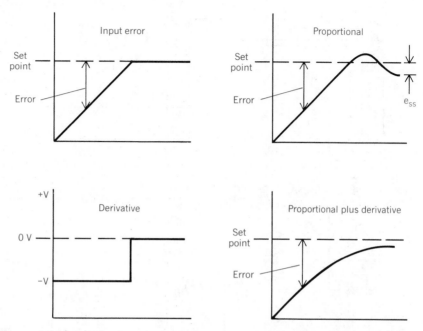

Figure 9.38 Graphs of error signal, response of proportional amplifier, response of derivative amplifier, and summed response of proportional and derivative amplifiers

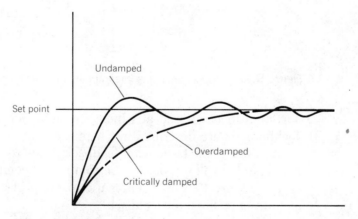

Figure 9.39 Proportional plus derivative control system, showing undamped, overdamped, and critically damped system responses

Tachometer Plus Proportional Plus Integral Plus Derivative Control

In addition to PID control, most robots also have a **tachometer feedback.** Recall that the tachometer generates a voltage that is pro-

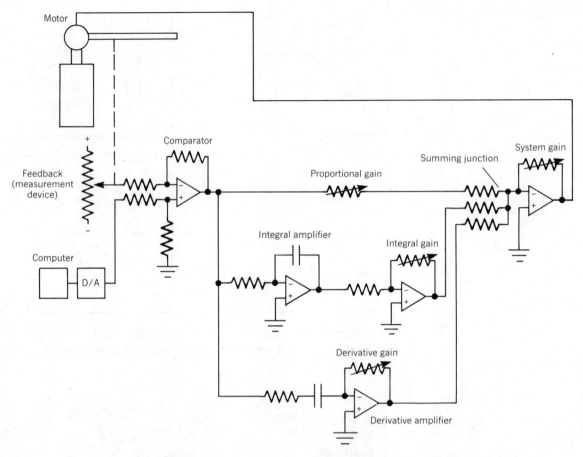

Figure 9.40   Proportional plus integral plus derivative (PID) control system

portional to the speed at which the tachometer's shaft is turned. Tachometers are described in Chapter 4.

Addition of the tachometer feedback to the robot's system ensures that the robot will respond to the computer's command at the correct speed. Figure 9.42 is a schematic diagram of a robot control that includes a tachometer feedback.

The computer outputs the command position as it did in the previous example. The computer does not simply output the final desired position but rather sends out a series of command positions. For example, to command the robot to move to a position represented by the number 10, the robot's computer outputs the numbers 0, 2, 4, 6, 8, and 10 (Figure 9.43, left). This series of numbers represents

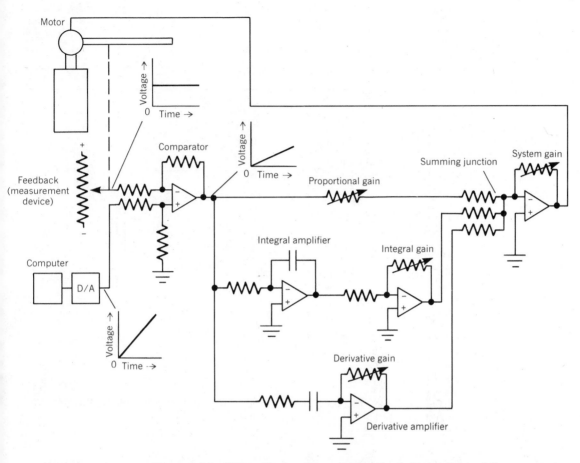

Figure 9.41 PID control system, showing the output voltage of the comparator which is feeding the proportional, integral, and derivative amplifiers

not only the positions of the robot's arm but also its velocity. If the computer outputs the numbers 0, 1, 2, 3, 4, 5, 6, 7, 8, 9, and 10 (Figure 9.43, right), it will take the robot's arm twice as long to move to the final position, and thus its speed will be only half as great.

The numbers outputted by the computer are converted to voltages by the D/A converter and are compared with the voltage from the robot's feedback potentiometer. The error voltage created by this comparison is sent to another comparator. The second comparator is also connected to the tachometer. The tachometer outputs a voltage that is proportional to the speed at which the robot's arm is moving. Since the robot's arm is initially not moving, the output error

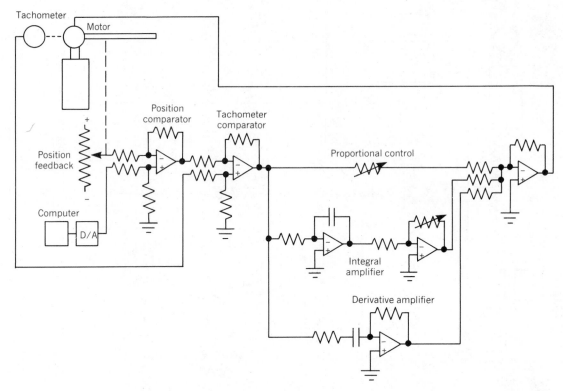

Figure 9.42 Tachometer plus proportional plus integral plus derivative control system

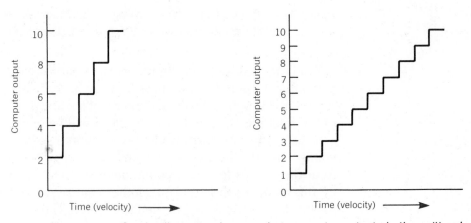

Figure 9.43 Graphs illustrating how a robot computer outputs both positional information and velocity information

signal of the second comparator is large. This error is processed through the PID portion of the control, and a high output voltage is applied to the robot's motor, creating a large starting torque.

The robot's arm begins to move. The tachometer outputs a voltage which is being compared by the second comparator. As the robot's velocity increases, the output of the second comparator is reduced and the applied voltage on the robot's motor goes down. The torque is automatically reduced to the level needed to keep the robot's arm moving at the command velocity. If the load on the robot should increase, the robot's arm will slow down, the output voltage of the second comparator will go up, and the torque will increase, returning the robot to the command velocity. If the speed of the robot should exceed the command speed, the output voltage of the second comparator will drop, the torque will decrease, and the arm will slow down. The robot continues at the controlled speed until it approaches the command position. As the error voltage of the first comparator goes down, the output voltage of the second comparator also goes down, and the robot's arm comes to rest.

Since the controller includes the PID functions, the robot's arm will come to rest at the command position without steady-state error or oscillation.

Following Error.   It may have occurred to you that if the robot ever accelerated to a rate that exactly matched the position error coming from the first comparator, the output voltage of the second compara-

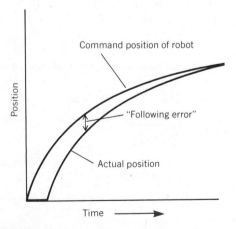

Figure 9.44 Following error

tor would be zero and the robot's arm would stop moving. This, however, is never the case, because the robot never accelerates to the full command velocity due to friction and the moment of inertia. The robot always lags behind the computer command. This lagging is called **following error.** If the following error were not present, the robot would accelerate, the output of the second comparator would drop to zero, the robot would slow down, and, since the robot is not at the final position, the error from the first comparator would increase and the robot would accelerate again. This process would create a jerky movement of the robot.

Figure 9.44 is a graph comparing the robot's command position with its actual position as it moves from one point to another. The following error is clearly illustrated.

DIGITAL SERVO SYSTEMS

Up to this point we have studied **analog servo systems,** in which the comparison between the set point and the controlled variable is done with an operational amplifier. Another type of system that is becoming more popular is the **digital servo system.**

Digital servo systems compare the set point with the controlled variable by adding and subtracting numbers rather than voltages. Figure 9.45 shows a block diagram of a digital servo system. The position potentiometer is connected to an A/D converter. The input voltage from the position potentiometer is converted to a number that represents the position of the robot's arm. The computer reads the next set point from its memory and compares that point with the

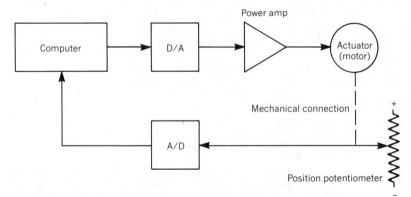

Figure 9.45   Block diagram of digital servo system

output of the A/D converter. The difference between the set point and the position of the robot's arm is outputted by the computer to a D/A converter. The output of the D/A converter is amplified by the power amplifier, and the robot's drive motor turns.

This simple system is a proportional system and suffers from the same problems that plague the analog proportional system: steady-state error and oscillation. The digital system adds integral and derivative functions through a computer program. The computer computes the integral and the derivative of the error and combines this information with the proportional information before outputting a number to be converted by the D/A to a voltage that will drive the motor.

The major problem with this type of system is speed. It takes time for the computer to compute the integral and the derivative. Until recently, computers took too much time to complete the calculations. With the advent of faster computers, more robot companies are turning to the digital system, since the "tuning" of the system can be done through changes in the computer program rather than by adjustments of the rather sensitive amplifiers.

FEEDBACK DEVICES

In our descriptions of servo systems we have used the potentiometer as the feedback device. The potentiometer is inexpensive and has been a popular choice by engineers for many years. There are, however, problems associated with the use of the potentiometer as a feedback device. The potentiometer has a wiper that is moved across a wire or a piece of carbon. As the wiper moves across the wire or carbon, the wiper and the wire or carbon both wear, making the potentiometer unreliable.

Potentiometers also become dirty. You have probably experienced a dirty potentiometer at one time or another. The volume controls in radios and TV's are potentiometers. When they become dirty you hear a "scratchy" sound when you increase or decrease the volume. The scratchy sound that you hear is caused by the wiper not making good contact with the wire or carbon and causing an erratic output voltage. When a volume control (potentiometer) is used as a position feedback device in a servo system, the system will respond erratically when the potentiometer becomes dirty.

The feedback devices that are commonly being used to replace potentiometers in robots are the **optical encoder** and the **resolver**.

Optical Encoders

An **optical encoder** (Figure 9.46) has four major elements: a light source, an optical disk, a light sensor, and some signal processing electronics.

The light source may be either an incandescent lamp or a light emitting diode (LED). The light source must be powered by direct current since alternating current will cause the light to fluctuate and may result in an error. The incandescent lamp is not as tolerant of vibration as the LED, but the LED is more sensitive to temperature and has a much lower light output.

The optical disk is usually made of plastic or glass. It has both opaque and transparent areas. The disk is placed between the light source and the light sensor. When the disk is rotated, light passes through the transparent portions and is blocked by the opaque portions of the disk.

Two types of disks are available: the incremental disk and the absolute disk (Figure 9.47). We will cover the operation of incremental and absolute disks in detail a little later in this chapter.

The light sensor is usually a phototransistor (Figure 9.48), but on occasion a photovoltaic cell is used. The phototransistor has a collector, a base, and an emitter. When light strikes the base of the transistor, the transistor begins to conduct. The brighter the light, the more the transistor conducts.

The signal processing electronics in an incremental encoder counts the number of times the phototransistor is struck by light and in some cases determines in which direction the disk is rotating (this method of determining direction is described in the next section). The signal processing electronics in an absolute encoder interprets the digital information stored on the disk.

Incremental Optical Encoder.   The incremental optical encoder has one or two tracks. A single-track encoder simply outputs a series

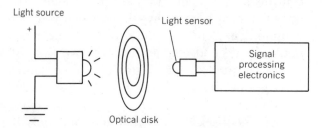

Figure 9.46   Optical encoder

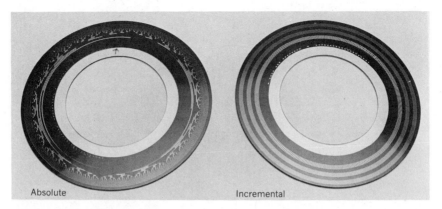

Figure 9.47 Incremental and absolute disks for optical encoders

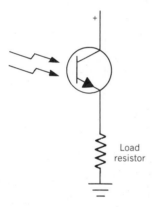

Figure 9.48 Phototransistor (light sensor)

of pulses as the disk is rotated and cannot determine the direction of rotation. Its output is the same for both directions of rotation.

The more pulses that are generated per revolution, the better the **resolution**. The term "resolution" means the smallest distance the disk must rotate to produce new positional information. The more opaque and transparent "windows" in the disk, the better the resolution. There are incremental optical encoders available with resolutions of 5000 increments per revolution.

The two-track incremental optical encoder can be used to determine the direction of rotation. The tracks are in quadrature, meaning that they are 90° out of phase. Each track has its own phototransistor. The outputs of the transistors are connected to the signal processing electronics (Figure 9.49). The signal processing electronics determines which of the two transistors is first to change from off

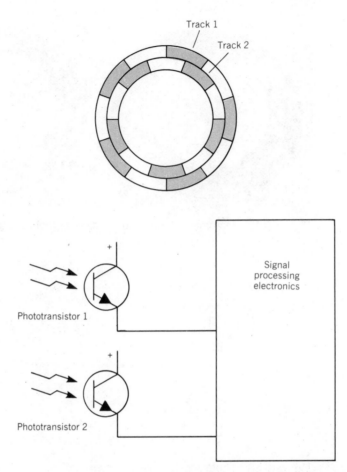

Figure 9.49   Two-track incremental optical encoder, showing phase quadrature

to on, which in turn determines the direction of disk rotation (Figure 9.50).

In addition to the two primary tracks, many incremental optical encoders have a track that outputs one pulse for each revolution of the disk. This is called the **zero reference pulse** or the **index pulse.**

The output pulses of an incremental encoder are fed to a counter. The count stored in the counter represents the position of the encoder shaft. If the power to the counter is lost, the count will be lost, and we will no longer be certain of the position of the encoder shaft.

The output of the counter is digital. Since the feedback systems of most robots require analog rather than digital information, the output of the incremental encoder is connected to a D/A converter.

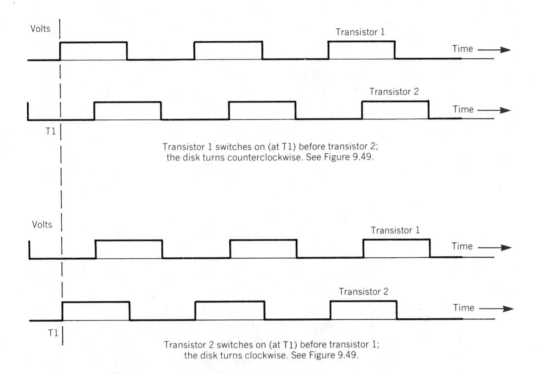

Figure 9.50 Strings of pulses outputted by the transistors of a two-track incremental optical encoder. Order in which transistors switch on determines direction of disk rotation.

Absolute Optical Encoder. The absolute optical encoder differs from the incremental optical encoder in the makeup of the disk. The disk of an absolute encoder has many tracks, rather than only one or two. The tracks on the disk of an absolute encoder produce a specific code for every position of the encoder. The output of the encoder is some form of binary code. The output may be in BCD or in standard binary, or may be in **Gray code** (see below). Whatever the output code of the encoder, the fact remains that each position outputs a number that is unique to that position.

Since the absolute optical encoder can produce a specific code for each position, it does not rely on a counter to produce positional information. Loss of power is not a problem, because when the power is restored the exact position of the encoder shaft can be read from the disk.

Gray Code. The Gray code is a specialized form of the binary system. Figure 9.51 presents a comparison between the standard

Decimal	Binary	Gray code
0	0000	0000
1	0001	0001
2	0010	0011
3	0011	0010
4	0100	0110
5	0101	0111
6	0110	0101
7	0111	0100
8	1000	1100
9	1001	1101
10	1010	1111
11	1011	1110
12	1100	1010
13	1101	1011
14	1110	1001
15	1111	1000

Figure 9.51   Comparison of the standard binary system and the Gray code

binary system and the Gray code. Notice that the two codes have identical equivalents for the decimal numbers 0 and 1 but different equivalents for decimal numbers greater than 1.

When counting is done in the standard binary system, it is not uncommon for more than one digit to change at a time. For example, changing the binary equivalent for 7_{10} (0111) to the binary equivalent for 8_{10} (1000) requires that the MSB be changed from 0 to 1 and that all other bits be changed from 1 to 0. Now examine the Gray code equivalents for the same numbers. When the Gray code equivalent for 7_{10} (0100) is changed to the Gray code equivalent for 8_{10} (1100), the MSB changes from 0 to 1 but all other bits remain the same. The Gray code has been designed so that only one bit changes at a time.

Use of the Gray code for optical encoders significantly reduces the magnitudes of the errors that may occur. For example, if a standard binary encoder is read at the instant of transition from 0111 (7_{10}) to 1000 (8_{10}), the first three digits from the right may be read as 0's and the MSB may be read as a 0 also because the disk may not have rotated far enough to register the change in the MSB. In this instance the number read would be 0000 rather than 1000, and the resulting error would be equivalent to the difference between 8_{10} and 0_{10}. When the standard binary system is used, it is possible for an error to be produced that is as large as the "weight" of the most significant digit. Now examine what would happen in the same situation if the encoder were built using the Gray code. The transition from 7_{10} to 8_{10} in the Gray code would be the transition from 0100 to 1100. Since this transition requires a change in only the digit at the far left, the output of the encoder will be either 0100 (7_{10}) or 1100 (8_{10}), and thus the only error that can occur is equivalent to the decimal number 1. When the Gray code is used, the largest error that can be produced is equivalent to the "weight" of the least significant digit.

Obviously, an error as large as the most significant digit could cause severe control problems in a robot, and for this reason most robot manufacturers have chosen to use Gray code encoders rather than standard binary encoders.

RELATIVE AND ABSOLUTE POSITIONING

Robots can be built using either relative or absolute positioning. Relative positioning is the less expensive of the two systems but suffers from amnesia when power is lost. Absolute positioning is more expensive than relative positioning. We will examine both systems since both are in common use.

Relative Positioning

Whenever **relative positioning** is used, every arm position is measured from a starting position. The starting position is called the "home position." The first arm position stored in the robot's program is the home position. As the robot's arm is moved to the next desired position by the robot programmer, the encoders feed positional information to the robot's computer. When the arm is in the desired position, the "record" button is pushed. The information stored in the robot's memory is the distance the arm has moved from the home position. We can say that it is "relative" to the home position.

Both absolute and incremental encoders can be used for relative positioning. We will examine the operation of the incremental encoder first.

Incremental Encoders in Relative Positioning.   The incremental encoder outputs a string of pulses as it is turned. The number of pulses in the string is stored in a counter. When the incremental encoder is used, the home position resets the counter to zero. As the robot's arm is moved, the output pulses are counted. When the arm is in the desired position, the "record" button is pushed and the count stored in the counter of the incremental encoder is entered into the robot's memory.

To play the program back, the robot's arm is brought to the home position and the encoder's counter is reset to zero. The robot's computer outputs the desired position and compares the desired position with the count stored in the encoder's counter. The count stored in the computer is compared with the count stored in the encoder's counter. This comparison shows that an error exists, and the robot moves to correct the error. As the robot moves, the robot's computer continues to compare the two counts. When the count stored in the encoder's counter is identical to the computer's stored position, the robot's arm is in the correct position and stops moving. (This description assumes that the robot is using a digital servo system. If the robot is using an analog servo system, the command position is converted to a voltage by a D/A converter and the output of the counter is converted to a voltage by another D/A converter. The two voltages are compared, and the difference is amplified. This was discussed in detail in a previous section.)

Since all positions are measured in relation to the home position, it is very important that the robot start from the exact home position each time. If the operator starts the robot from a position different from that used as the home position when the program was recorded, the robot will miss each point in the program by the exact distance between the home position and the actual starting position.

To ensure that the robot is always started from the same home position, many industrial robots have a "hard home." The hard home is established by the robot manufacturer. The common practice is to use small switches to establish the hard home. When the robot is commanded to go to its hard home position, it moves until it bumps into the small switches. If the robot is always taught from the hard home, and all programs are executed from the hard home position, the robot will always move to the desired points.

Absolute Encoders in Relative Positioning. The absolute optical encoder can also be used for relative positioning. This may sound like a contradiction in terms, but in fact many industrial robots use this design.

The absolute optical encoder is normally connected to the robot's drive motor if the robot is electric, or to the joint of the robot, through a gear train, if the robot is hydraulically powered.

When the robot's arm is moved through its work envelope, the encoder is turned many times (possibly hundreds of times). Recall from the discussion of the absolute optical encoder that the output of the encoder is absolute only if the encoder is not allowed to rotate more than one full turn. Since in reality the encoder may be rotated many times, the output of the encoder is not absolute.

Figure 9.52, which is an elevation view of a spherical robot, will help illustrate this point. The arc shows the total possible movement of the robot's arm through the vertical axis. The arc is segmented into 12 equal parts that are labeled 0 through 12. Each segment indicates when the encoder has completed a full revolution. For example, assume that the robot's arm is located at the 0 position. As the robot's arm is moved, the optical encoder is turned. When the arm reaches position 1 the encoder will have made one full revolution. As the arm is moved from position 1 to position 2, the encoder outputs the identical positional information that it did while moving from position 0 to position 1. The robot's computer is not certain whether the arm is located between positions 0 and 1 or between positions 1 and 2.

Since this type of ambiguity cannot be tolerated, robot manufacturers have added a counting circuit. Every time the encoder makes one full revolution it outputs a pulse that is stored in a counter. For example, if the robot's arm is between positions 0 and 1, the counter

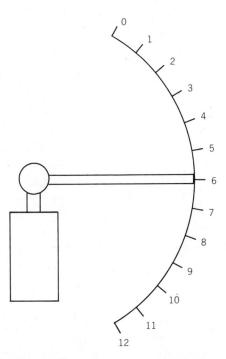

Figure 9.52 Elevation view of a spherical robot. Each number represents one full revolution of the optical encoder.

will be at 0. We simply read the output from the encoder and we are certain of the robot's position. When the robot's arm arrives at position 1, the encoder outputs a pulse to the counter. If the robot's arm is stopped between positions 1 and 2, we can determine its exact location by first reading the count stored in the counter and then reading the encoder itself. With the robot's arm stopped between positions 1 and 2, the count in the counter will be 1. We know that the robot's arm is someplace between positions 1 and 2. By reading the encoder we can determine its exact position.

This process of storing counts occurs every time the robot's arm is moved from one segment to the next. When the arm is moved to position 2, two counts are stored in the counter; and when the arm is moved to position 3, three counts are stored in the counter.

When the robot is first started, the counter is set to 0. The robot's arm can be in any position, and that position is established as the home position. Since the counter is incremented once for each revolution of the encoder relative to the home position, this system is also a relative positioning system. To ensure that the robot is always started from the same position, a hard home is normally used.

Absolute Positioning

Absolute positioning does not require a home position. Even if power is lost, the true position of the robot's arm is not lost.

Absolute positioning requires the use of two absolute optical encoders for each axis. One of the optical encoders is connected to the robot as described earlier. This encoder will turn many full revolutions throughout the robot's work envelope. The second optical encoder is connected through a gear train to the first encoder. The second encoder is geared in such a way that it never turns more than one full revolution throughout the work envelope of the robot.

The second optical encoder, called the "coarse" encoder, identifies the approximate position of the arm, and the first encoder, called the "fine" encoder, identifies its exact position.

This system of positioning works like a clock. The coarse encoder is like the hour hand on a clock, and the fine encoder is like the minute hand. The hour hand tells us the approximate time. If the hour hand is between 4 and 5 we know that the time is between 4 o'clock and 5 o'clock. By looking at the minute hand in combination with the hour hand we can determine the exact time.

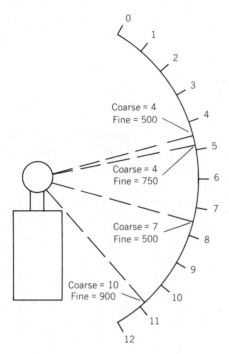

Figure 9.53 Absolute positioning

Figure 9.53 illustrates the operation of the absolute positioning system. The output of the coarse optical encoder is 4. We know that the arm is located between positions 4 and 5. The fine encoder outputs 1000 absolute positions per revolution. If the output of the fine encoder is 500, we know that the robot's arm is exactly halfway between positions 4 and 5. If the output of the fine encoder is 750, we know that the robot's arm is three-quarters of the way from position 4 to position 5.

Since the absolute positioning system does not rely on a counter for indication of the robot's arm location, the arm's positional information is not lost when the power is turned off. After a power failure, the computer simply reads the outputs of the coarse and fine encoders, and the position is established.

Since this system has the advantage of never getting lost due to power failure or power fluctuation, you may wonder why not all robot manufacturers have gone to the absolute system. The absolute system is far more expensive to implement than the simpler relative system. Many manufacturers feel that power losses are rare enough so that the additional cost is not justified. They argue that when power is lost it is a simple matter of returning the robot to the home position and starting the program over.

RESOLVERS

Another device that is commonly used for feedback information is the **resolver**. The resolver differs from the encoder in that it resembles a motor (Figure 9.54).

The positional information created by the resolver is generated through the relationship of an armature coil to the field coils. Figure 9.55 is an illustration of a resolver. The field coils are mounted 90° apart. One of the field coils is fed with a sine wave, and the other is fed with a cosine wave.

The output of the armature depends on the position of the armature coil. If the armature coil is parallel to the sine coil, the output of the armature will be a sine wave (Figure 9.56); and if the armature coil is parallel to the cosine coil, the armature's output will be

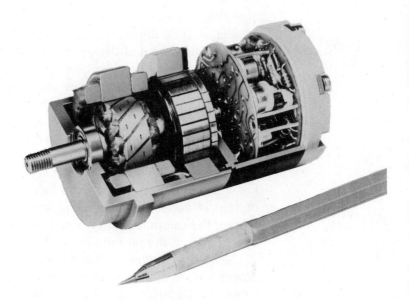

Figure 9.54 Cutaway view of a resolver

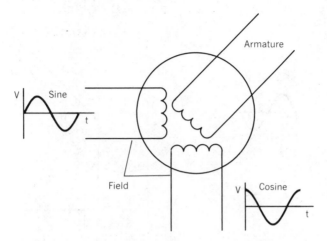

Figure 9.55   Resolver, showing armature and field coils

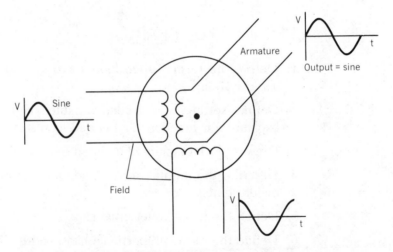

Figure 9.56   Sine wave output results when armature coil is parallel to sine coil

a cosine wave (Figure 9.57). At every position of the armature the output will have a unique phase relative to one of its inputs. By comparison of the output of the armature with the sine input, the exact position of the armature can be determined.

The resolver can be used for both relative and absolute positioning. A single resolver is used for relative positioning, and two resolvers geared together are used for absolute positioning.

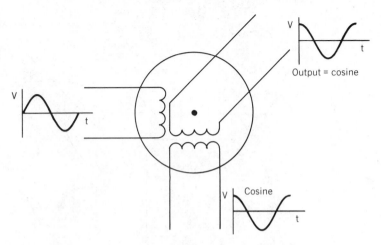

Figure 9.57   Cosine wave output results when armature coil is parallel to cosine coil

QUESTIONS FOR CHAPTER 9

1. Define the terms "open loop servo system" and "closed loop servo system."

2. Define "setting," "controlled variable," and "disturbances," and explain their functions in a closed loop servo system.

3. Draw a block diagram of a closed loop servo system.

4. Describe the on-off control mode (use graphs to illustrate the major points).

5. Define the term "differential gap."

6. Define the term "proportional band width."

7. What control problems are associated with the simple proportional control mode?

8. Define "steady-state error," and explain what can be done to eliminate it.

9. What can be done to eliminate oscillations in a servo system?

10. Draw the schematic diagram of a PID control system, and explain the operation of such a system.

11. Explain the operation and purpose of tachometer feedback.

12. Define the term "following error."

13. Explain the operation of an optical encoder.

14. What is the advantage of an optical encoder?

15. Describe absolute positioning and relative positioning, and give the advantages and disadvantages of each.

CHAPTER TEN

Robot Interfacing

A group of machines connected together to perform an operation is referred to as a **work cell.** When a robot is installed as part of a work cell, it becomes the central control of the work cell. The other machines send information to the robot, and the robot processes the information, makes decisions regarding this information on the basis of its program, and sends signals back to the other machines to control their processes.

Figure 10.1 shows a work cell for manufacture of shafts. The work cell includes a robot, an infeed conveyor to deliver parts to the robot, a Computer Numeric Controlled (CNC) lathe to machine the shafts, and an outfeed conveyor to carry the finished parts out of the

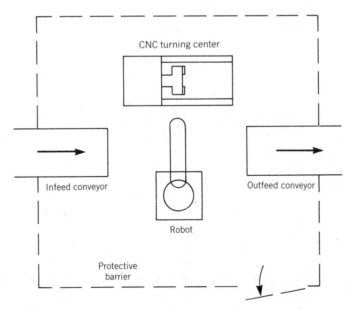

Figure 10.1 Work cell for manufacture of shafts

work cell. Notice that the work cell is surrounded by a protective barrier. The protective barrier keeps workers from accidentally walking into the work cell while the robot is operating.

Before examining the design of a work cell, we will discuss the various switches, controls, and sensors which are used to send information to the robot.

PUSH-BUTTON SWITCHES

Push-button switches are used to turn the work cell on and off and to provide an emergency stop should something fail during the operation of the work cell.

The push-button switches that are used in industry are heavier duty than those found in the home. In general they are sealed to prevent oil from leaking into the switch and thereby causing high resistance at the contacts. These switches are normally rated for 120-V ac and continuous currents of approximately 6 A, but they can withstand momentary currents of up to 10 times their rated current. This is necessary because, as discussed in Chapter 4, the starting current can be quite high.

Push-button switches are available with either normally open (NO) or normally closed (NC) contacts. A normally open contact is open when the button is not being pushed. A normally closed contact is closed when the button is not being pushed. The **combination push-button switch** has both NO and NC contacts in the same enclosure. When the button is not being pushed, the NC contact is closed and the NO contact is open. When the button is being pushed, the NC contact is open and the NO contact is closed. The schematic symbols for NO, NC, and combination switches are shown in Figure 10.2.

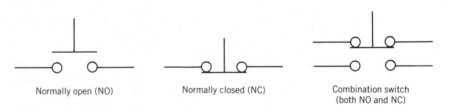

Figure 10.2 Schematic symbols for NO, NC, and combination push-button switches

There are many different designs of switches in use. The most common types are the **flush head switch,** the **extended head switch,** the **mushroom head switch,** and the **keylock switch.**

Figure 10.3 shows a **flush head switch.** The button is flush with a bezel which prevents the button from being accidentally bumped and activated. To activate the switch the operator must press the button straight down. The flush head switch has a return spring. After the button is pushed, the spring returns the button to the unactivated position.

In the **extended head switch** (Figure 10.4), the button extends beyond the mounting bezel. This type of switch is used in applications where accidental activation is not a major concern. The extended head switch also has a return spring.

The **mushroom head switch** (Figure 10.5) is normally used as an emergency stop switch. The large button is easy to hit in an emergency situation. Some mushroom head switches are spring returned to the unactivated position, but most are not. If a switch maintains its contact after it has been pushed it is called a "maintain contact" switch. When the button of a maintain contact mushroom head switch is pushed it remains in the "in" position until the operator pulls it out. When a mushroom head switch is used to disconnect the power from the work cell it is called an "Emergency Power Off" (EPO) switch.

Bezel

Figure 10.3   Flush head switch

Figure 10.4   Extended head switch

Figure 10.5   Mushroom head switch

A standard color code has been established for switches used in control panels. A red switch is used for stop or emergency stop, a yellow switch is used for return to neutral or emergency return to neutral, and a black switch is used for motor start or cycle start.

A **keylock switch** is shown in Figure 10.6. A keylock switch is normally included in the control panel of a work cell. When the keylock switch is off, the machinery in the work cell cannot be started. This prevents unauthorized personnel from operating the

Figure 10.6 Keylock switch

work cell. It is also an important safety feature. When maintenance must be performed on the work cell, the system can be locked out so that it cannot be activated by accident while it is being serviced.

The schematic symbols for the various switches discussed above are shown in Figure 10.7.

SELECTOR SWITCHES

A **selector switch** is shown in Figure 10.8. The selector switch can be turned to a selected position so as to activate a specific set of contacts. The most common selector switches are either two- or

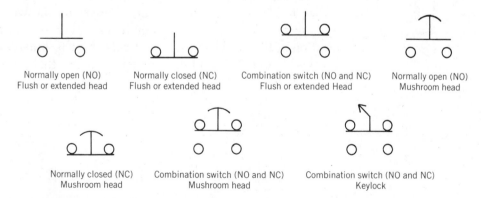

Figure 10.7 Schematic symbols for flush head, extended head, mushroom head, and keylock switches

Figure 10.8 Selector switch

three-position switches, although selector switches with as many as eight positions are available.

Selector switches may be of either the "spring return" type or the "maintain contact" type. For example, a three-position selector switch with the center position off may have contacts of either the spring return type or the maintain contact type at both the left and right positions or may have one contact of each type.

The schematic symbol for a three-position selector switch is shown in Figure 10.9. Notice the asterisks on the wires. The asterisks indicate which lines have inputs and outputs connected at each position of the switch. In this particular switch, lines A and C are asterisked under the number 1, which indicates that lines A and C are

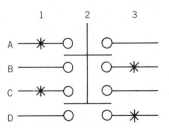

Figure 10.9 Schematic symbol for a three-position selector switch

on when the selector is at position 1. The location of the number 2 above the center of the switch means that no lines are on at selector position 2. Lines B and D are asterisked under the number 3, indicating that lines B and D are on when the selector is at position 3.

Figure 10.10 shows a four-position selector switch. Lines A, B, and C are on when the selector is at position 1; lines B and C are on at position 2; lines B and D are on at position 3; and line D is on at position 4.

INDICATOR LAMPS

In addition to switches, most control panels have an array of **indicator lamps** that inform the operator of what conditions exist in the work cell. An indicator lamp is shown in Figure 10.11.

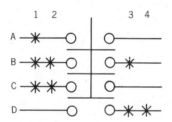

Figure 10.10   Schematic symbol for a four-position selector switch

Figure 10.11   Indicator lamp

Indicator lamps come in four different lens colors. The standard colors and their uses are as follows:

Red: danger; stopped; alarm
Amber: attention; warning
Green: safe; running
White: normal condition or special information.

The schematic symbol for an indicator lamp is shown in Figure 10.12. The color of the lens is always indicated by a letter inside the symbol—R for red, A for amber, G for green, or W for white.

In addition to the standard indicator lamps, "press to test" indicator lamps are available. When the lens of the lamp is pushed, a switch in the lamp closes and the lamp lights. The advantage of the press to test indicator lamp is that it can be easily tested, thus eliminating from the operator's mind the question of whether a system is malfunctioning or the indicator lamp is simply burned out.

A press to test lamp and its schematic symbol are shown in Figure 10.13.

CONTROL RELAYS

In addition to switches and indicator lamps, many work cell control consoles include **control relays.** Control relays are shown in Figure 10.14. When a voltage is applied to the coil of the relay, a magnetic field forms. The magnetic field pulls the armature toward the relay and closes the contacts.

Control relays normally have more than one set of contacts. Most control relays have at least one set of normally open (NO) contacts and one set of normally closed (NC) contacts. Many relays used in industry contain multiple NO and NC contacts.

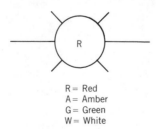

R = Red
A = Amber
G = Green
W = White

Figure 10.12 Schematic symbol for an indicator lamp

Figure 10.13 Press to test lamp and schematic symbol

Figure 10.14 Control relays

The schematic symbol for a control relay is shown in Figure 10.15. The coil of the control relay is normally labeled "CR." The symbols for NO and NC contacts are also shown.

Relay coil   Normally open (NO) contact   Normally closed (NC) contact

Figure 10.15 Schematic symbol for a control relay

LADDER DIAGRAMS

Figure 10.16 shows a simple control diagram. This diagram is called a **ladder diagram** because it has the appearance of a ladder. The vertical lines are referred to as "rails" and the horizontal lines are referred to as "rungs."

When the start button is pushed, a voltage is applied to the control relay coil (CR). The normally open (NO) contacts of the relay close. Notice that NO contact CR-1 is connected around the start switch. When the start button is pushed, CR-1 closes. When the start button is released, it returns to its NO position, but voltage is still applied to the relay coil through relay contact CR-1. To turn the relay off, the stop button is pushed, which breaks the current path to the control relay coil. With relay coil CR no longer activated, all contacts return to their normal positions. Releasing the NC stop button does not reactivate CR; CR remains off until the start button is pushed again.

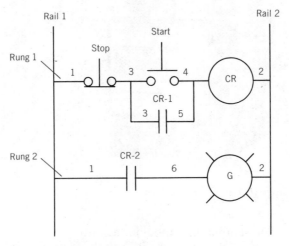

Figure 10.16 Ladder diagram

Notice that there is another relay contact (CR-2) on rung 2. When the start button is pushed, CR-2 closes and the indicator lamp lights.

Ladder diagrams are common in industry, and you should become familiar with them. Ladder diagrams will be used throughout this chapter to illustrate control circuits.

TIME DELAY RELAYS

In addition to simple control relays, **time delay relays** are also available. There are two types of time delay relays: on-delay relays and off-delay relays.

The time delay of the time delay relay can be controlled through the use of a pneumatic chamber with variable orifice or by means of an electronic circuit. Figure 10.17 shows a pneumatic time delay relay. When a voltage is applied to the relay coil, the armature pushes against the diaphragm and air is forced out of the chamber.

Figure 10.17 Pneumatic time delay relay

The speed at which the air exits the chamber controls the length of the time delay.

An electronic time delay relay is shown in Figure 10.18. The electronic time delay relay may be set for either on-delay or off-delay.

On-delay Relays

When power is applied to the relay coil of an **on-delay relay,** a period of time elapses before the relay contacts change state. Figure 10.19 shows the schematic symbol for an on-delay relay along with the symbols for its NO and NC contacts.

Figure 10.20 is a ladder diagram of a control circuit with an on-delay relay. When the start button is pushed, the control relay (CR) is activated. All CR contacts close. When the start button is released, CR remains activated. Relay contacts CR-1 and CR-2 are closed, and a voltage is applied to the coil of the time delay relay (TR). Even though a voltage is applied to the timing relay coil, contact TR-1

Figure 10.18   Electronic time delay relay

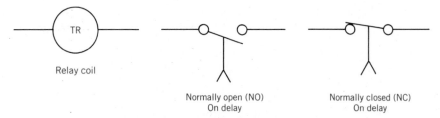

Figure 10.19 Schematic symbols for an on-delay relay and its NO and NC contacts

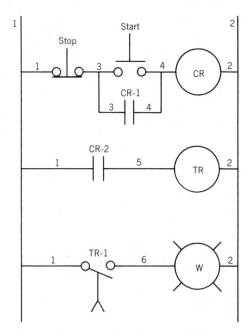

Figure 10.20 Ladder diagram of a control circuit with an on-delay relay

does not close immediately. After a period of time, TR-1 closes and the white indicator lamp turns on.

This relay is called an "on-delay" relay because a period of time elapses before the contact is turned on. If the circuit also includes an NC contact, the NC contact is also delayed for a period of time before it opens. When the circuit is turned off, all contacts immediately return to their normal positions.

Off-delay Relays

Figure 10.21 shows the symbols for an **off-delay relay** and its NO and NC contacts. When voltage is applied to the coil of an off-delay

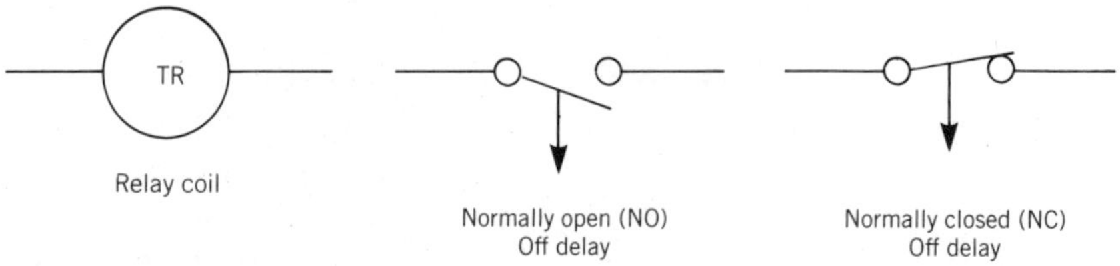

Figure 10.21   Schematic symbols for an off-delay relay and its NO and NC contacts

relay, the contacts change state immediately as they do in a normal control relay. The NO contact closes and remains closed as long as a voltage is applied to the coil. When the voltage is removed from the coil, the contacts remain in their activated states for a preset period of time. After the relay "times out," the contacts return to their normal states.

Figure 10.22 is a ladder diagram of a control circuit with an off-delay relay. When the start button is pushed, a voltage is applied to relay coil CR. When the start button is released, CR-1 closes, maintaining voltage to coil CR. Contact CR-2 also closes, and a voltage is applied to coil TR. Contact TR-1 closes immediately and the indicator lamp turns on. When the stop button is pushed, voltage is removed from coil CR, and all CR contacts open. Voltage is also removed from coil TR, but contact TR-1 remains closed for a preset period of time and the indicator lamp remains on. After the relay "times out," contact TR-1 opens and the light goes out.

LIMIT SWITCHES

Limit switches are the most common sensors used in work cells to provide signals to the robot. The limit switch is simple and relatively inexpensive. There are many different types of limit switches on the market today. Limit switches are divided into three major categories: mechanical switches, proximity switches, and vane-type switches.

Mechanical Limit Switches

The most common type of limit switch used in industry is the **mechanical limit switch.** The mechanical limit switch is generally less expensive than the other types. Figure 10.23 shows some common mechanical limit switches. The limit switches shown at the top

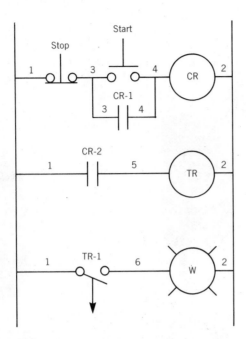

Figure 10.22 Ladder diagram of a control circuit with an off-delay relay

of the illustration are actuated by pushing a lever. The limit switches shown at the bottom of the illustration are actuated by pushing a roller or a button. In addition to these actuators (lever, roller, button), many other types of actuators are available. Figure 10.24 shows a few of the other common types of actuators used on mechanical limit switches. Generally, the limit switch and the actuator are purchased separately. The actuator is simply screwed onto the limit switch.

Mechanical limit switches are available with normally open contacts and normally closed contacts. A normally open contact on a limit switch is open when pressure is not being placed on the actuator. A normally closed contact is closed, meaning that current can pass through the switch, when it is in the unactuated position.

The schematic symbol for a normally open contact limit switch is shown in Figure 10.25. When pressure is applied to the actuator, the switch is closed and current can pass through the switch.

At times the normally open contacts are held in the closed position when the system is not operating and are open when the system is operating. The schematic symbol used to indicate that a normally

Figure 10.23   Mechanical limit switches

open contact is being held in the closed or actuated position is shown in Figure 10.26.

The schematic symbol for a normally closed contact is shown in Figure 10.27. If no pressure is applied to the actuator, the contacts remain closed and current passes through the switch. At times the normally closed switch is held open when the machine is not operating. The schematic symbol of a normally closed switch held open is shown in Figure 10.28.

Figure 10.24 Actuators for mechanical limit switches

Figure 10.25 Schematic symbol for a normally open contact limit switch in the unoperated position

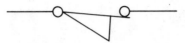

Figure 10.26 Schematic symbol for a normally open contact limit switch in the operated position

Figure 10.27 Schematic symbol for a normally closed contact limit switch in the unoperated position

Figure 10.28 Schematic symbol for a normally closed contact limit switch in the operated position

Limit switches are available not only with various types of actuators but also with actuators that require various amounts of pressure. The operating force needed to actuate a limit switch may be as little as 1 lb. To reduce even further the force needed to actuate the switch, a lever actuator can be attached to the switch. The longer the lever, the less pressure is needed to actuate the switch but the farther the lever must travel before the switch actuates.

Figure 10.29 is an illustration of a lever-type limit switch. There are several terms that you should be familiar with when specifying

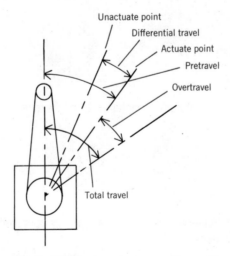

Figure 10.29 Lever-type mechanical limit switch

limit switches. "Pretravel" is the distance that the lever must travel before the switch is actuated. While the lever is moving from its unactuated position to the end of the pretravel, the limit switch remains in its normal condition. For example, a normally open limit switch remains open throughout the pretravel. When the actuator reaches the end of the pretravel, the switch closes.

"Overtravel" is the distance from the point at which the switch actuates to the safe limit of travel before the switch is damaged. The overtravel allows for additional travel of the actuator to ensure that the limit switch can be actuated.

"Total travel" is defined as the sum of the pretravel and the overtravel.

When the switch is released, the lever moves toward the at rest position and passes through the actuate point. However, the limit switch remains actuated until it has moved closer to the at rest position. The distance between the point at which the switch actuated and the point at which it returns to its unactuated state is called "differential travel."

Often more than one limit switch is mounted inside the same enclosure. For example, you may find one set of normally open (NO) contacts and another set of normally closed (NC) contacts. The schematic symbol which illustrates that two switches are in the same enclosure is shown in Figure 10.30. Both switches are drawn and are connected with a dotted line.

Vane-type Limit Switches

The basic operating principle of the **vane-type limit switch** is illustrated in Figure 10.31. A vane-type limit switch is operated by passing a soft piece of steel between the main magnet and the switch. The magnetic lines of force from the main magnet are concentrated in the ferrous vane. Since the magnetic field is now con-

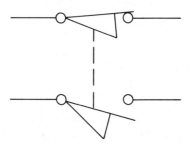

Figure 10.30 Schematic symbol for two limit switches in the same enclosure

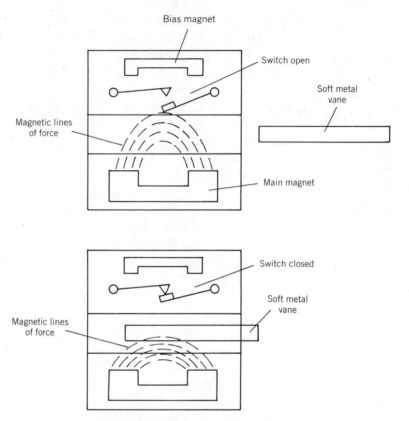

Figure 10.31   Vane-type limit switch

centrated in the vane, the field is no longer strong enough to hold the switch open.

Proximity Switches

The **proximity switch** (Figure 10.32) does not have to be contacted in order to be actuated. The proximity switch is actuated when an object passes in close "proximity" to it.

There are two basic types of proximity switches: **ferrous** and **nonferrous**. The schematic symbols for normally open (NO) and normally closed (NC) proximity switches are shown in Figure 10.33 (the same symbols are used for both ferrous and nonferrous switches).

Ferrous Proximity Switches.   The ferrous proximity switch senses when any metal that can be magnetized (ferrous metal) passes close to it. The ferrous proximity switch operates through magnetic induction. A magnetic field is created by a transmitter. When ferrous

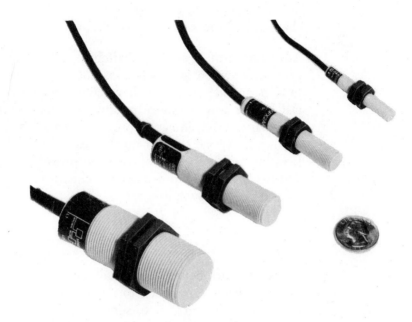

Figure 10.32   Proximity switches

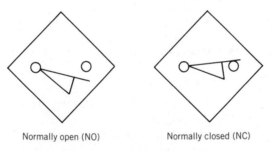

Normally open (NO)   Normally closed (NC)

Figure 10.33   Schematic symbols for NO and NC proximity switches

metal passes near the switch, the magnetic field is disturbed by the metal. The disturbance in the magnetic field is sensed, and the switch is actuated.

Nonferrous Proximity Switches.   The nonferrous proximity switch can sense nonferrous metals such as aluminum, nickel, brass, and copper as well as most other materials such as plastics, cardboard, wood, and ceramics. The nonferrous proximity switch senses the presence of a material through a change in capacitance. When a material comes in close proximity to the switch, the capacitance changes and the switch is closed.

Advantages of Proximity Switches. The major advantage of the proximity switch is that it does not have to be contacted in order to be actuated. Since they have no moving parts, proximity switches last longer and are more reliable than mechanical limit switches. The major disadvantage of proximity switches is that they generally are more expensive than mechanical limit switches. However, the long life and high reliability of the proximity switch are worth the additional cost, and proximity switches are replacing mechanical and vane-type limit switches.

PHOTODETECTOR SENSORS

Many work cells use **photodetector sensors** to sense the presence or absence of parts, to sense whether a machine is ready to accept another part, or to sense whether the robot is clear of the machine before the process is started.

There are two basic types of photodetector sensors: **photoconductive cells** and **photovoltaic cells.**

Photoconductive Cells

The resistance of a **photoconductive cell** changes in response to light. As the light level increases, the resistance of the photoconductive cell decreases. Light is often thought of as being comprised of waves similar to radio waves. Photoconductive cells are sensitive to wavelengths between 0.3 and 0.8 μm. This range includes light that is visible to humans as well as infrared light, which is not. The particular material used in the design of the photoconductive cell determines the range of wavelengths to which the cell is the most sensitive. Some photoconductive cells are sensitive in the visible range whereas others are more sensitive in the infrared range. Photoconductive cells also vary in the degree to which the resistance changes in response to a change in light level. The most sensitive photoconductive cells vary from a resistance of 1 MΩ in the absence of light to a resistance of as little as 100 Ω when the cell is being struck by light.

There are three types of photoconductive cells: **photoresistors, photodiodes,** and **phototransistors.**

Photoresistors. A photoresistor can be thought of as a variable resistor. In fact, the schematic symbol used for a photoresistor is similar to that used for a variable resistor. Figure 10.34 shows these symbols. Notice that there are two symbols for the photoresistor. One of the symbols has arrows pointing toward the photoresistor to indicate that the photoresistor changes its resistance in response to light. The other symbol uses the Greek letter λ (lambda) to indicate wavelength.

Figure 10.35 shows a photoresistor. This cell is made from cadmium sulfide (CdS). Other photoresistors are made from cadmium selenide (CdSe). Figure 10.36 shows the range of sensitivity to light for each type of cell. In industry, both CdS and CdSe cells are normally called "Cad cells" ("Cad" stands for "cadmium").

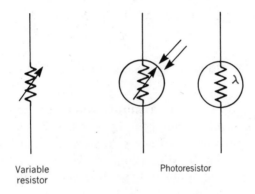

Variable
resistor

Photoresistor

Figure 10.34 Schematic symbols for a variable resistor and a photoresistor

Figure 10.35 Photoresistor (Cad cell)

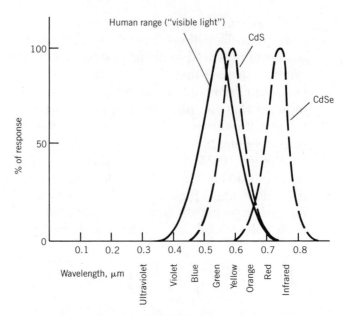

Figure 10.36   Ranges of light sensitivity for CdS and CdSe photoresistors

Figure 10.37 shows a simple voltage divider circuit in which a Cad cell is used as one of the resistors. With the Cad cell dark, very little voltage is dropped across resistor R1 and the output voltage is approximately 10 V. When light strikes the Cad cell, the resistance drops to approximately 100 Ω. Most of the voltage is dropped across resistor R1, and the output voltage is 0.91 V.

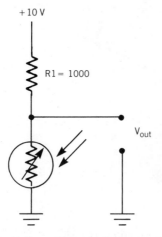

Figure 10.37   Voltage divider circuit with Cad cell

Photodiodes. The photodiode is another form of photoconductive cell. When light strikes the PN junction of a photodiode (Figure 10.38), more current can flow through the diode. The change in the amount of current in response to light is quite small, and amplification is needed. Figure 10.39 shows a photodiode connected to the noninverting input of an Op Amp. A small change in the amount of current flowing through the diode is amplified and appears at the output of the amplifier.

Photodiodes are sensitive to a much wider range of wavelengths than Cad cells. Figure 10.40 shows the range of sensitivity for a photodiode. Compare this graph with that for the Cad cell (Figure 10.36). Figure 10.41 shows the schematic symbols for the photodiode (note that there are two symbols, just as there are for the photoresistor).

Phototransistors. The phototransistor (Figure 10.42) is similar to the photodiode. When the base of a phototransistor is exposed to light, it is forward biased and begins to conduct. The equivalent circuit is shown in Figure 10.43.

A graph showing the sensitivity of the phototransistor is presented in Figure 10.44. Note that the peak of the curve is in the infrared range. The advantage of having the sensitivity in the infrared range is that the phototransistor is not affected by ambient light.

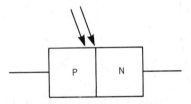

Figure 10.38   Schematic symbol for a photodiode

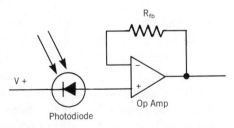

Figure 10.39   Photodiode connected to the noninverting input of an Op Amp

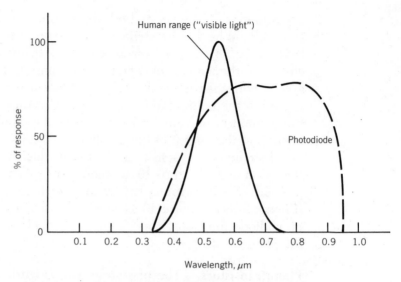

Figure 10.40 Range of light sensitivity for a photodiode

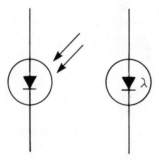

Figure 10.41 Schematic symbols for a photodiode

When Cad cells or photodiodes are used in industrial applications they must be shielded from ambient light, which otherwise would cause them to conduct.

Phototransistors amplify the small current caused by light striking the transistor. The output of the phototransistor is high enough to connect directly to a TTL logic circuit. However, the phototransistor is often connected to an Op Amp to amplify its output.

Phototransistors are available with two or three leads. The collector and the emitter are always brought out of the transistor, but the base is not always brought out. If the base is brought out, the phototransistor can be biased externally to control its sensitivity. Be careful not to confuse photodiodes with two-wire phototransistors, because they cannot be directly substituted in circuits.

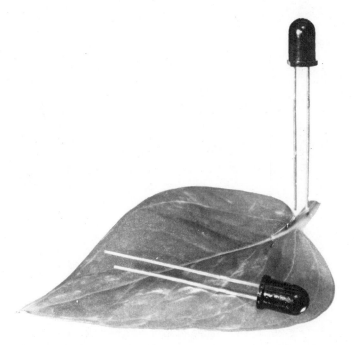

Figure 10.42 Phototransistor

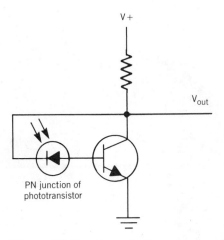

Figure 10.43 Equivalent circuit of a phototransistor. When struck by light, the PN junction forward biases the transistor, switching it on.

Photovoltaic Cells

The **photovoltaic cell** actually produces electricity when struck by light. A photovoltaic cell is a PN junction much like a photodiode (Figure 10.45). When light strikes the junction, a potential is created. If a load is connected across the cell, current flows.

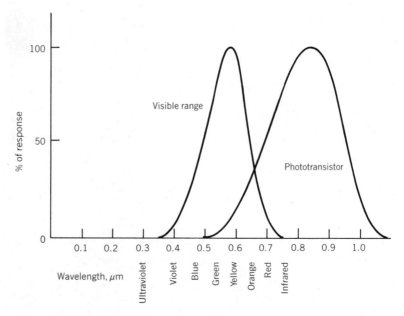

Figure 10.44 Range of light sensitivity for a phototransistor

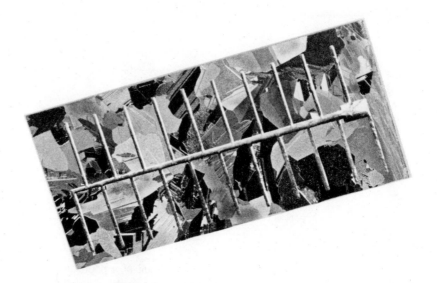

Figure 10.45 Photovoltaic cell

Photovoltaic cells can be made from selenium or from silicon. Figure 10.46 is a graph showing the sensitivities of the selenium cell and the silicon cell. Under high light levels the silicon cell can produce up to 0.5 V at 100 mA. The selenium cell is more sensitive to

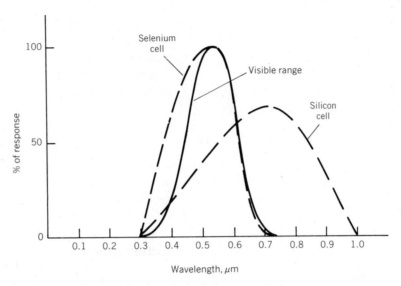

Figure 10.46 Ranges of light sensitivity for selenium and silicon photovoltaic cells

light in a narrow spectrum but is less efficient when exposed to light in a broad spectrum and is normally not used as a solar cell for generation of electricity. Silicon photovoltaic cells are used as the power sources for small hand-held calculators. Selenium photovoltaic cells are used in photographic light meters.

PHOTODETECTOR APPLICATIONS

Both photoconductive and photovoltaic cells have been used for many years in industry to identify parts on a conveyor (or simply to count the number of parts on the conveyor), to determine the level of fluid in a vat, and to perform a variety of quality control inspections in manufacturing. When photodetectors are used in industry they are normally connected to a circuit which determines the light level. If the light level is higher than a specified level the photodetector closes a normally open contact on a relay, and if the light level is below the specified level the normally open contact of the relay remains open.

Both photoconductive and photovoltaic cells can be used and are normally referred to simply as "receivers." The light source — whether it be an LED, focused light from an incandescent lamp, or general plant illumination — is referred to as the "transmitter." It is also referred to as an "emitter."

There are five different ways in which photodetectors and light sources can be arranged to accomplish these objectives. These transmitter/receiver arrangements are generally referred to as "scanning modes." The five scanning modes are **opposed scanning, retroreflective scanning, diffused scanning, convergent scanning,** and **specular scanning.**

Opposed Scanning

Opposed scanning, also called "direct scanning," is one of the most common scanning modes. If opposed scanning is used on a conveyor line, the transmitter (emitter) is mounted on one side of the conveyor and the receiver is mounted on the other side (Figure 10.47). As an object moves along the conveyor, the light beam from the transmitter is broken, signaling the presence of the object.

Opposed scanning is an excellent choice when the objects being detected are opaque (objects through which light cannot pass). The light from the transmitter is focused directly on the receiver. The

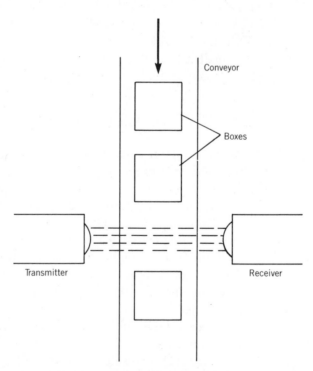

Figure 10.47   Opposed scanning

intense beam of light produces a high output voltage in the case of a photovoltaic cell or a low resistance in the case of a photoconductive cell. When an object breaks the light beam of the transmitter, the output of the photovoltaic cell goes down significantly, or the resistance of the photoconductive cell becomes very high.

Opposed scanning is not a very effective method of recognizing transparent objects (objects that pass light, such as those made of clear glass or plastic) or translucent objects (objects that diffuse but still pass light, such as those made of frosted glass). We will subsequently describe scanning methods that are effective when the objects to be sensed are transparent or translucent.

Retroreflective Scanning

Retroreflective scanning is similar to opposed scanning except that the receiver and the transmitter (emitter) are mounted on the same side of the conveyor (Figure 10.48). In retroreflective scanning, the transmitter sends out a beam of light. If there is no object to interrupt the light beam, the light beam bounces off the reflector and returns to the receiver. When an opaque object passes in front

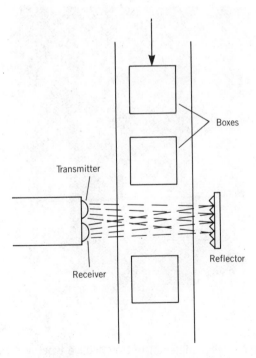

Figure 10.48 Retroreflective scanning

of the transmitter, the light beam cannot strike the reflector, and the receiver senses the presence of the object.

The transmitter and the receiver are normally mounted in the same enclosure (Figure 10.49). The retroreflective scanner is installed by simply mounting the transmitter/receiver package on one side of the conveyor and mounting a reflector similar to a bicycle reflector directly across from the transmitter/receiver.

Retroreflective scanning is simple to install and service, but a word of caution is in order. If the object to be detected is itself highly reflective, the light beam from the transmitter will bounce off the object back to the receiver and the light beam will not be interrupted. Since the beam of light is not interrupted, the receiver cannot sense the object passing in front of it.

Diffused Scanning

Diffused scanning is similar to retroreflective scanning. Both the transmitter and the receiver are mounted in the same enclosure.

Figure 10.49   Photodetector (retroreflective type) with transmitter and receiver in the same enclosure

The difference between retroreflective scanning and diffused scanning is that in diffused scanning the object passing in front of the transmitter is expected to reflect the light back to the receiver.

In Figure 10.50, there is no object in front of the transmitter and there is no reflector to bounce the light back to the receiver. If the receiver is a photovoltaic cell its output voltage is low, and if the receiver is a photoconductive cell its resistance remains high. When a reflective object passes in front of the transmitter (Figure 10.51), the light strikes the object and diffuses, meaning that the light beam is no longer a focused beam of light but rather is bouncing off the object in many directions. If enough of the diffused light is returned to the receiver, the receiver will sense the object.

In opposed and retroreflective scanning, the light beam is broken by an object passing in front of the transmitter. In diffused scanning, the beam is returned to the receiver when the object passes in front of the transmitter. In diffused scanning, the beam is said to be **made** (rather than "broken") when an object is sensed.

In diffused scanning we rely on the object to reflect enough light for the receiver to sense its presence. For this reason the transmitter/receiver must be mounted very close to where the object will pass to ensure that enough light is reflected.

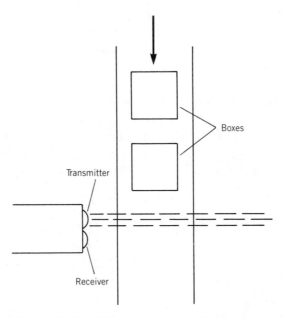

Figure 10.50 Diffused scanning (without box in position)

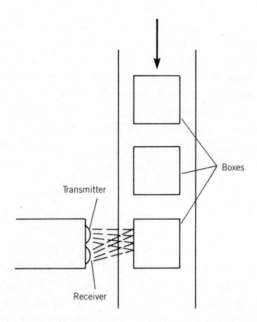

Figure 10.51   Diffused scanning (with box in position)

Convergent Scanning

Convergent scanning is similar to diffused scanning. The object passing in front of a convergent scanner "makes" (rather than "breaks") the light beam. The difference between convergent and diffused scanning is that the convergent scanner has a set of lenses in front of the transmitter. The purpose of the lenses is to focus the light at a particular distance (Figure 10.52). The light beam from the transmitter is the brightest at the point at which the light beams "converge." Recall that a magnifying glass can be used to start a fire. When light from the sun is passed through a magnifying glass and the magnifying glass is adjusted so that the light from the sun is focused at one point, the light is so intense at that point that a fire can be started.

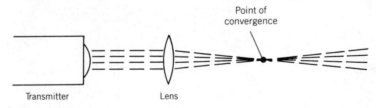

Figure 10.52   Convergent scanning. Light is focused at point of convergence.

The point of convergence is the brightest point. If the surface of an object is at the point of convergence, the light reflected to the receiver will be bright. If, on the other hand, the object is too close or too far away from the transmitter, the light striking the receiver will not be as intense and the light returned to the receiver will not be as bright.

Convergent scanning is used to detect objects that are close to another reflective object. When an object to be detected is not present, the light beam strikes the reflective surface and light is reflected to the receiver (Figure 10.53). The light returned to the receiver is not very intense since the light is not focused on the reflective surface, and the receiver ignores the light. When an object passes by the transmitter with its surface at the point of light convergence, high-intensity light is returned to the receiver and the object is detected (Figure 10.54).

Convergent scanners are often used to detect transparent or translucent objects. As a transparent or translucent object passes the transmitter at the point of convergence, the edge of the object reflects a high-intensity light back to the receiver.

Specular Scanning

Figure 10.55 shows the arrangement of the transmitter and the receiver for **specular scanning.** Note that the transmitter and receiver are not in the same enclosure. The transmitter and the receiver are mounted at equal angles to a perpendicular line drawn

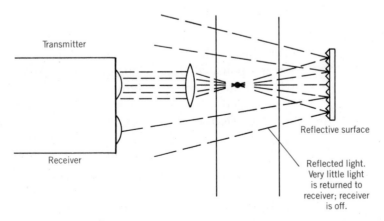

Figure 10.53 Convergent scanning. Reflective surface is not at point of convergence.

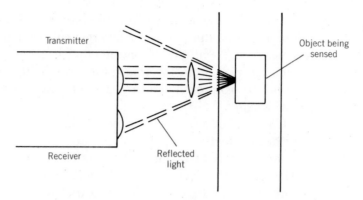

Figure 10.54   Convergent scanning. Reflective surface is at point of convergence.

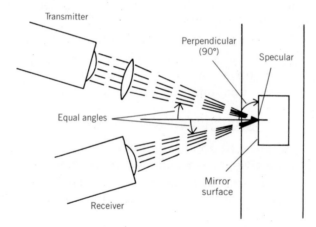

Figure 10.55   Specular scanning

to the object. The object being sensed has a highly reflective surface such as a polished steel or mirror surface. When the light strikes the surface of the object, an intense spot of light is reflected. The spot of light is called a "specular." You may see examples of speculars when you take a flash picture of someone wearing glasses. If the person is looking at the camera, bright spots of light show in their glasses.

The angle at which the light is reflected from the surface of the object is equal to the angle at which the light strikes the surface. We say that the angle of reflection is equal to the angle of incidence (Figure 10.56). If the object being sensed is not at the exact distance, the specular will not strike the receiver and the object will not be sensed (Figure 10.57).

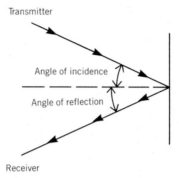

Transmitter

Angle of incidence

Angle of reflection

Receiver

Figure 10.56 Angle of reflection equals angle of incidence

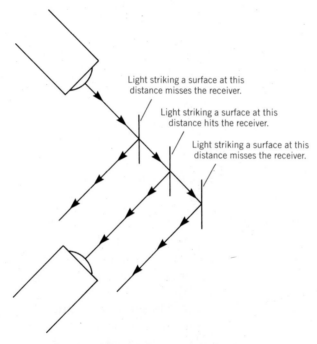

Light striking a surface at this distance misses the receiver.

Light striking a surface at this distance hits the receiver.

Light striking a surface at this distance misses the receiver.

Figure 10.57 Only an object at a specific distance is sensed by the receiver

It is possible to build a system that measures the distance from the object to the receiver (Figure 10.58). If the object is 1 ft away the specular will strike receiver A; if the object is 2 ft away the specular will strike receiver B; and if the object is 3 ft away the specular will strike receiver C. The more receivers, the more accurate the measurement.

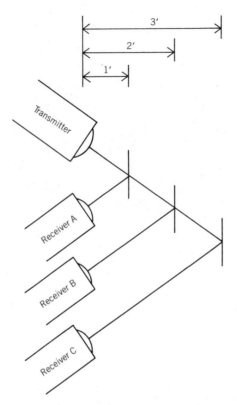

Figure 10.58   Measurement of distance by use of specular scanning

Modulated and Nonmodulated Light

In describing the various scanning systems in the previous section we have assumed that the transmitter is always turned on. If the transmitter is always illuminated, the system is said to be **nonmodulated**. If the transmitter is being turned on and off at a high frequency, the system is said to be **modulated**.

A common problem in the use of photodetectors in industry is the light in the plant. The general light in the plant is referred to as **ambient light.** The ambient light can confuse the photodetector system. For example, assume that we have installed a nonmodulated opposed scanning system. When an object breaks the light beam from the transmitter, the receiver does not sense the object because the ambient light level in the plant is so high that it keeps the photodetector switched on.

If the transmitted light is modulated (switched on and off) and the receiver has a circuit that responds only to the frequency of the transmitter, the receiver can differentiate between ambient light and the light from the transmitter. The receiver ignores the ambient light and responds only to the modulated light from the transmitter.

PRESSURE SWITCHES

Pressure switches are commonly used in industry to sense pressure in hydraulic and pneumatic systems. In addition, pressure switches can be used to sense the presence or absence of an object. More specific applications are given later in this chapter.

The symbols for NO and NC pressure switches are shown in Figure 10.59.

There are four basic types of pressure switches available today: the **Bourdon tube pressure switch**, the **piston-type pressure switch**, the **diaphragm-type pressure switch**, and the **bellows-type pressure switch.**

Bourdon Tube Pressure Switch

The **Bourdon tube** is the oldest of all pressure switches and still remains as one of the leading choices for many applications. The Bourdon tube was invented by Eugene Bourdon in 1851. Bourdon tube switches come in three styles: the C-type tube, the spiral tube, and the helical tube. Figure 10.60 illustrates the three designs.

The Bourdon tube is oblong in cross section (Figure 10.61). One end of the tube is open and is connected to pressure. The other end of the tube is blocked. All three styles of Bourdon tubes operate the same. When air pressure or hydraulic pressure is applied to the tube, the tube attempts to straighten. As the tube attempts to straighten, it closes or opens a precision switch depending on the design of the switch. Many Bourdon tubes have both normally closed (NC) and normally open (NO) contacts. Figure 10.62 shows a cutaway view of a C-type Bourdon tube.

Normally open (NO)
pressure switch

Normally closed (NC)
pressure switch

Figure 10.59 Schematic symbols for NO and NC pressure switches

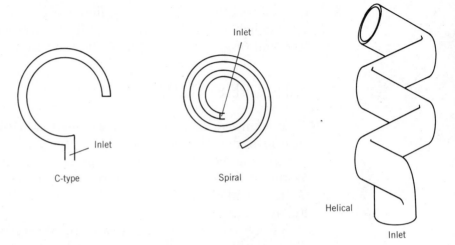

Figure 10.60   C-type, spiral, and helical Bourdon tubes

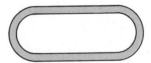

Figure 10.61   Cross-sectional view of a Bourdon tube

C-type Bourdon tube pressure switches are available in various pressure ranges within an over-all range of 50 to more than 15,000 psi. Caution should be taken to ensure that a switch that operates in the desired range is selected. A switch that is too small for the application will straighten under the pressure and will not return to its original shape when the pressure is released.

Piston-type Pressure Switch

Figure 10.63 is a cutaway view of a **piston-type pressure switch.** Pneumatic or hydraulic pressure is directed through the inlet. The pressure forces the piston upward against the spring pressure. When the pressure on the piston exceeds the spring tension, the piston moves and the switch is activated. Piston-type pressure switches are available with either normally closed (NC) or normally open (NO) contacts, and some models include both NC and NO contacts.

Piston-type pressure switches are available in various ranges from 15 to more than 10,000 psi.

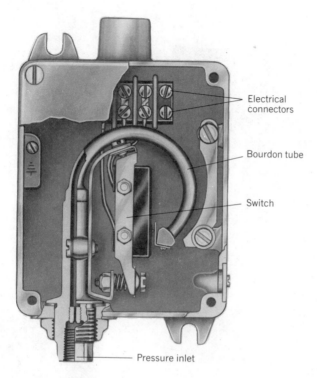

Figure 10.62   Cutaway view of a C-type Bourdon tube pressure switch

Diaphragm-type Pressure Switch

Figure 10.64 is a cutaway view of a **diaphragm-type pressure switch.** Air or hydraulic pressure is directed to the diaphragm. As the pressure builds, the diaphragm moves upward, activating the switch. Diaphragm-type pressure switches are available with NO contacts, with NC contacts, or with both NO and NC contacts.

Also available are diaphragm switches that respond to a vacuum. As air is removed from the chamber beneath the diaphragm, the diaphragm moves downward, activating the switch.

Diaphragm pressure switches can be purchased in various pressure ranges from vacuum to 150 psi.

Bellows-type Pressure Switch

Figure 10.65 is an illustration of a **bellows-type pressure switch.** As pneumatic or hydraulic pressure is directed to the inside of the bellows, the bellows expand. When the bellows have expanded far enough, a switch is activated.

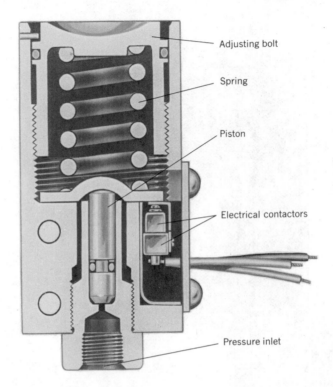

Adjusting bolt

Spring

Piston

Electrical contactors

Pressure inlet

Figure 10.63   Cutaway view of a piston-type pressure switch

Bellows type pressure switches are available with NO or NC contacts, or with both NO and NC contacts. The operating range of the bellows-type pressure switch is from vacuum to approximately 3000 psi.

Pressure Switch Applications

Pressure switches can be used to sense pressure in both hydraulic and pneumatic systems. It is common practice to include a pressure switch in a pneumatic system to sense the pressure in the receiver tank. The pressure switch turns the compressor on when the pressure in the receiver tank falls below a predetermined minimum pressure and turns it off when the pressure in the receiver tank rises above a predetermined maximum pressure.

Another common application of pressure switches is in hydraulic filters. An increase in the pressure differential between the input and the output of the filter indicates that the filter is becoming blocked. The pressure switch will either light a warning indicator on the control panel or shut the robot off.

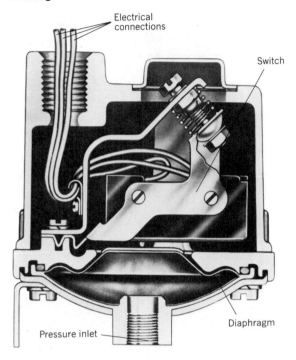

Electrical connections

Switch

Diaphragm

Pressure inlet

Figure 10.64 Cutaway view of a diaphragm-type pressure switch

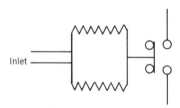

Inlet

Figure 10.65 Bellows-type pressure switch

A pressure switch can be used to sense when the arm of a nonservo robot bumps against its end stop. Figure 10.66 is a schematic drawing of such a system. The computer outputs a voltage that activates solenoid A. The solenoid opens the valve and air flows into the cylinder, moving the arm. As the arm is moving the pressure in the system goes down. When the arm bumps against the end stop, the pressure in the system builds, which closes pressure switch A. The pressure switch is connected to an input port on the computer. When the pressure switch closes, a voltage is applied to

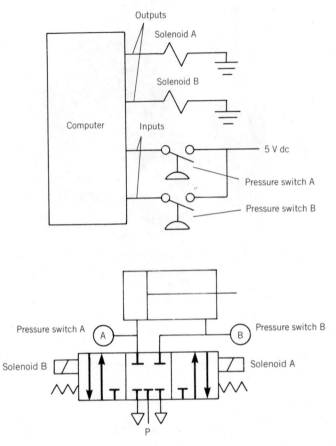

Figure 10.66 Schematic diagram of system in which a pressure switch is used to sense when the arm of a nonservo robot bumps into an end stop

the port. The computer senses the voltage on the port and steps to the next instruction in the program.

Pressure switches can also be used to sense when parts are properly located. For example, a robot places a part in a punch press. To ensure that the part is properly located in the die before cycling the press, a sensor can be mounted in the die. The life of proximity switches and limit switches is short due to the vibration and impact that occur during operation of the press. An alternative method for sensing the location of the part in the die is the use of a pressure switch. Small holes are bored in the die. Pressure lines are connected to the inlets of the holes, and a normally open pressure switch is connected. When the part is not properly seated in the die, air escapes from the bored holes and opens the pressure switch. When the part is properly located, the holes are blocked by the part and the pressure builds, closing the pressure switch. The pressure switch is connected to an input of the robot. When the pressure switch is closed the part is properly positioned in the die. A voltage is now applied to the input of the robot. The robot retracts its arm from the press and cycles the press.

VACUUM SWITCHES

Vacuum switches are similar in design to pressure switches. When a vacuum is applied to the inlet port of a vacuum switch, a normally open (NO) switch is closed and a normally closed (NC) switch is opened.

Whereas pressure switches are actuated by pressure and are rated in pounds per square inch (psi), vacuum switches can be actuated by various amounts of vacuum and are rated in inches of mercury (in. Hg).

The schematic symbols used for vacuum switches are identical to those used for pressure switches.

As you know, there is pressure all around us. The pressure at sea level is approximately 14.7 psi. When a pressure gauge is used at sea level, it is adjusted to read 0 psi rather than 14.7 psi. As pressure is increased by an air compressor, the gauge reads only the increase in pressure, ignoring the atmospheric pressure. This pressure is referred to as **gauge pressure.**

If the gauge were to read not only the increase in pressure due to the compressor but also the atmospheric pressure, the gauge would show the total pressure. The total pressure is called **absolute pressure.**

Gauges used for measuring vacuum are graduated in inches of mercury rather than in negative psi. Figure 10.67 shows a flask inverted in a bowl of mercury. The pressure inside the flask is 14.7 psi (14.7 psi absolute pressure or 0 psi gauge pressure) and the pressure on the mercury in the bowl is 14.7 psi (14.7 psi absolute pressure or 0 psi gauge pressure). The mercury does not rise in the flask. If air is drawn out of the flask (a vacuum is drawn), the level of mercury in the flask will rise. The more air that is drawn out of the flask the higher the level of mercury. In Figure 10.68 the pressure in the flask has been reduced to 9.8 psi absolute pressure (−4.9 psi

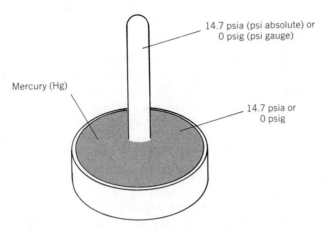

Figure 10.67 Flask inverted in bowl of mercury. Pressure in flask is 14.7 psi (absolute).

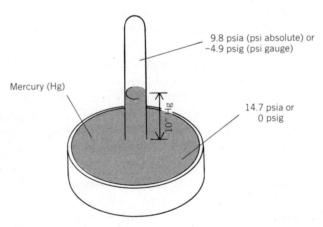

Figure 10.68 Flask inverted in bowl of mercury. Pressure in flask is 9.8 psi (absolute).

gauge pressure). The mercury in the flask will rise 10 in., and we say that we have a vacuum of 10 in. Hg. If all of the air were pumped out of the flask, the mercury would rise 29.92 in.; this would be a perfect vacuum.

It is more convenient to speak of inches of mercury than of absolute pressure or negative psi gauge pressure. For this reason, vacuum gauges and vacuum switches are normally calibrated in inches of mercury.

STRAIN GAUGES

The **strain gauge** is another type of sensor that can be used in robot applications. Before examining the applications of strain gauges, we will define the terms "stress" and "strain."

Stress

Stress is defined as force divided by area:

$$\text{Stress} = \frac{F}{A}$$

where F is force and A is the cross-sectional area to which this force is uniformly applied.

Figure 10.69 shows a rod with forces applied to the ends, attempting to stretch it. The result of such forces is called **tensile stress.**

Figure 10.70 shows a rod with forces pushing in on the ends, attempting to compress the rod. The result of these forces is called **compression stress.**

Figure 10.71 shows a rod connected to a wall. A downward force is applied to the free end of the rod, subjecting the rod to both a tensile stress and a compression stress. The top of the rod is stretched (tensile stress), and the bottom of the rod is compressed (compression stress).

Strain

When a sufficient tensile stress is applied to a rod, the rod elongates (Figure 10.72). **Strain** is defined as the change in length di-

Figure 10.69 Tensile stress

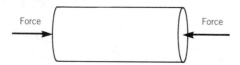

Figure 10.70 Compression stress

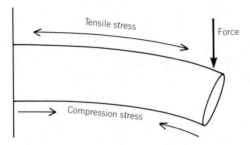

Figure 10.71   Tensile stress and compression stress

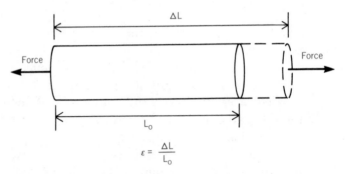

Figure 10.72   Strain equals change in length divided by original length

vided by the original length:

$$\varepsilon = \frac{\Delta L}{L_o}$$

where ε is strain, ΔL is the change in the length of the rod, and L_o is the original length of the rod.

Measurement of Strain

Strain can be measured through the use of a strain gauge. A strain gauge is simply a wire. The resistance of a wire is affected by three factors: the length of the wire, the cross-sectional area of the wire, and the resistance of the wire material. Different materials have different resistances. For example, the resistance of a copper-nickel alloy is much higher than that of copper alone. The resistance of a material is referred to as its **resistivity**.

When a tensile strain is applied to a wire, its length increases and its cross-sectional area decreases but the resistivity of the wire material remains unchanged. The increase in length and the reduction of the cross-sectional area increase the resistance of the wire. This is expressed mathematically as:

$$R = \frac{\rho L}{A}$$

where R is the resistance of the wire, ρ is the resistivity of the wire material, L is the length of the wire, and A is the cross-sectional area of the wire.

When the wire is stretched, its length increases and its cross-sectional area decreases, and as a result the resistance of the wire increases. When the wire is mounted on a bar and a force is applied to the bar, as shown in Figure 10.73, the bar elongates. Since the wire is mounted on the bar, it also elongates and its resistance increases. The change in the resistance of the wire is proportional to the strain on the bar.

Figure 10.74 shows commercially available strain gauges. These gauges consist of a zigzag wire mounted on a paper backing. (Commercial strain gauges are also made of thin foil mounted on paper backings.) Since the resistivity of the wire material is low, a long wire is needed to increase the resistance of the gauge and increase the change in resistance relative to the strain. The zigzag pattern allows a long wire to be mounted in a small area.

The strain gauge is sensitive to a strain along one axis. This axis is called the **longitudinal axis.** When a force is applied to the gauge as shown in Figure 10.75, the wire is stretched and its resistance increases. However, if a strain is applied to the gauge as shown in Figure 10.76, the gauge uncoils. The wire is stretched only where it turns back on itself. Since this area is so small, the resistance remains virtually unchanged. Commercial strain gauges normally have

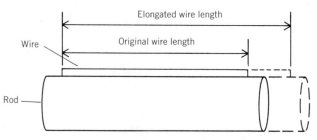

Figure 10.73   Measurement of tensile strain by use of a strain gauge

Figure 10.74   Commercial strain gauges

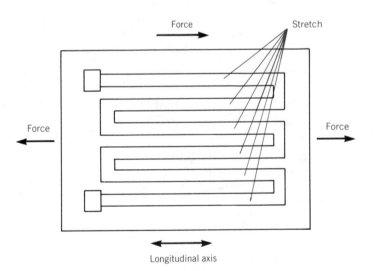

Figure 10.75   Force applied along longitudinal axis increases strain gauge resistance

arrows showing the direction in which the gauge should be mounted relative to the longitudinal axis and the force being measured. If the gauge is not mounted properly, the resistance of the gauge will not change proportionally with the strain.

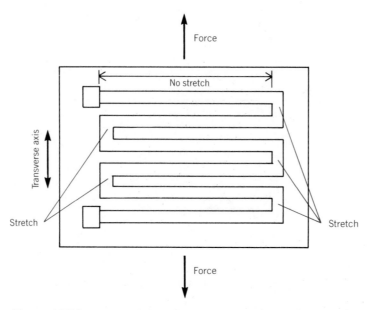

Figure 10.76 Force applied along transverse axis produces virtually no change in strain gauge resistance

The material on which the gauge is to be mounted should be dry and free from oil and dirt. Special adhesives are used to mount the gauge. These adhesives, which are available from strain gauge manufacturers, are formulated so that the deformation of the material is transferred to the gauge. If the gauge is properly mounted, its length will change proportionately with strain. There are also special gauges which can be attached by spot welding or by use of ceramic bonding agents. For certain special applications, strain gauges are embedded in plastic resins or concrete without special adhesives.

Up to this point we have been describing strain gauges for use in measuring tensile strain. The strain gauge also works for measuring compression strain (Figure 10.77). When a compression force is applied to the rod, the length of the rod is reduced, the length of the strain gauge wire is reduced, and the resistance of the gauge goes down.

Gauge Factor

The amount by which the resistance of the gauge changes in response to a strain is called the **gauge factor.** The gauge factor is defined mathematically as:

$$GF = \frac{\Delta R/R_o}{\Delta L/L_o}$$

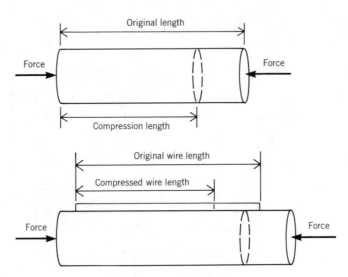

Figure 10.77   Measurement of compression strain by use of a strain gauge

where GF is the gauge factor, ΔR is the change in resistance of the gauge, R_o is the original resistance of the gauge, ΔL is the change in wire length due to a force, and L_o is the original wire length. Since the strain $\varepsilon = \Delta L/L_o$, then, by substitution:

$$GF = \frac{\Delta R/R_o}{\varepsilon}$$

The gauge factor is important in designing a strain gauge. The higher the gauge factor, the greater the change in resistance due to strain. The gauge factor for most commercial strain gauges is 2. A gauge factor of 2 means that the resistance will change by 2% for every 1% change in length due to strain. For example, assume that we have a strain gauge with a normal resistance of 100 Ω and a gauge factor of 2. If the gauge is mounted on a bar and a force is applied to the bar so that its length increases by 0.5%, then the resistance of the gauge will increase by 1% ($2 \times 0.5\% = 1\%$). The resistance of the strain gauge will change from 100 Ω to 101 Ω. Since this change in resistance is so small, the strain gauge is normally incorporated into a Wheatstone bridge.

Wheatstone Bridge

Figure 10.78 shows a strain gauge connected as one resistor in a **Wheatstone bridge.** The resistance of resistor R1 equals the resis-

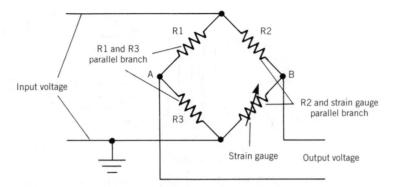

Figure 10.78   Wheatstone bridge with strain gauge

tance of resistor R2, and the resistance of resistor R3 equals the unstrained resistance of the strain gauge. The two parallel branches of the bridge (R1/R3 and R2/strain gauge) operate as voltage dividers. A bridge is balanced when the products of the opposite arms are equal—that is, when $R_1 R_{Strain} = R_3 R_2$. With the total resistances of the two parallel branches equal, the output voltage is 0 V. When the strain gauge resistance goes up due to strain, the voltage at point B becomes higher than the voltage at point A. This voltage differential can be read on a voltmeter, or the bridge can be connected to an amplifier and be used as an input to a robot.

Dummy Gauge

Strain gauges are sensitive to temperature changes. The resistance of the gauge changes with a change in temperature. The change of resistance affects the resistance balance of the bridge, and the bridge outputs a voltage even though the gauge is not being strained.

To eliminate the effects of temperature changes, a second gauge, called a **dummy gauge,** is mounted at right angles to the first gauge, which is called the **active gauge** (Figure 10.79). Since the dummy gauge is mounted at right angles to the force, its resistance is not significantly affected by the strain. The dummy gauge is connected to the opposite parallel leg of the Wheatstone bridge (Figure 10.80). A temperature change affects both gauges equally. When the resistance of the active gauge changes due to a change in temperature, the resistance of the dummy gauge also changes, the resistances of both legs remain equal, and the output voltage of the bridge is 0 V. When strain is applied, the resistance of the active gauge changes, the bridge becomes unbalanced, and a voltage appears at the output.

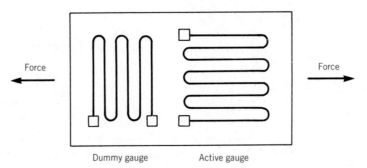

Figure 10.79 Dummy gauge mounted at right angles to active gauge to eliminate effects of changes in temperature

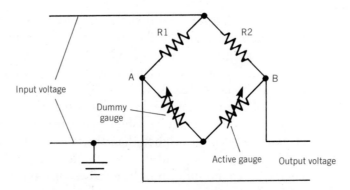

Figure 10.80 Wheatstone bridge with dummy gauge and active gauge

Semiconductor Strain Gauge

In addition to wire and foil strain gauges, there are also **semiconductor strain gauges.** The semiconductor strain gauge is made from silicon. As the semiconductor material is "strained," its resistance changes. This effect is often referred to as "piezoresistance." The semiconductor is normally connected into a Wheatstone bridge circuit in the same manner as a wire strain gauge.

The major advantage of the semiconductor strain gauge over the wire strain gauge is its increased gauge factor. The gauge factor for semiconductor strain gauges ranges from 50 to 200. This increased sensitivity to strain makes these gauges useful for measuring very low strains.

The major disadvantage of the semiconductor strain gauge is its lack of linearity. The output of a wire strain gauge is proportional to the strain over a wide range. The semiconductor strain gauge, on the other hand, is proportional to the strain over a very small range. Once outside this very small linear range, the resistance of the semiconductor strain gauge is no longer proportional to the applied force. Since the semiconductor strain gauge is not linear, the output voltage of the bridge circuit cannot be directly converted to strain. This problem can be overcome by using special circuits to convert the nonlinear output voltage of the bridge into a voltage that is proportional to the applied force.

Strain Gauge Applications

Strain gauges are used in industry to determine the magnitudes of strains exerted on machinery parts. This application is normally in the realm of the design engineer who wants to test a new design. In robotics, strain gauges are normally used to measure force. For example, suppose that a robot is being used to pour molten iron in a foundry. Since the robot cannot see, it does not know when its ladle is full. By mounting a strain gauge on the ladle, the amount of molten iron being poured into the ladle can be monitored. When the output of the strain gauge bridge circuit reaches a predetermined voltage, the robot shuts off the pour.

In another common application, the strain gauge is used as a "load cell." The load cell is made of a machined bar or plate. Normally, several strain gauges are mounted and wired into a bridge circuit. The output voltage of the bridge circuit is read as weight or force. Load cells are used as the measuring devices for many large industrial scales.

THERMOCOUPLES

The **thermocouple** is used for measuring temperature. The thermocouple was discovered by Thomas Seebeck in 1821. Seebeck, a German physicist, connected two wires made of dissimilar metals and found that heating of one end of the pair of wires caused a current to flow (Figure 10.81).

If the circuit is broken in half and a voltmeter is placed across the wires, a voltage will be read. This voltage is called the "Seebeck voltage" and is a function of the temperature at the junction and the compositions of the two metals.

A junction between any two dissimilar metals will output a voltage. The magnitude of this voltage relative to the temperature

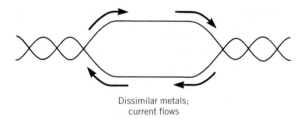

Dissimilar metals;
current flows

Figure 10.81 Seebeck effect

of the junction on the Kelvin scale (volts per kelvin) is called the "Seebeck coefficient." The Kelvin temperature scale assigns a value of 0 to the point at which no thermal energy remains. This point is referred to as "absolute zero" since it is not possible for any material to be colder. Absolute zero (0 K) is equal to -273.15 °C (degrees Celsius) or -459.6 °F (degrees Fahrenheit).

If an iron wire is connected to a copper-nickel wire, the output voltage of the junction between them will be 51 microvolts per kelvin (51 μV/K). The Seebeck coefficients for various junctions of dissimilar metals can be found by referring to thermocouple reference tables. Although wires of any two dissimilar metals connected together will output a voltage if the temperature of the junction is above 0 K, most thermocouple tables are based on the Celsius scale rather than on the Kelvin scale. Although use of the Celsius scale has become customary practice in industry, it is important to remember that all pairs of dissimilar metals output a voltage when they are at temperatures above absolute zero (0 K).

The formula for finding the change in output voltage, given the Seebeck coefficient and the temperature, is:

$$\Delta e = \alpha T$$

where Δe is the change in voltage from 0 °C, α is the Seebeck coefficient, and T is temperature in °C. For example, if the Seebeck coefficient is 51 μV and the temperature is 500 °C, then:

$$\Delta e = \alpha T$$
$$\Delta e = 51 \ \mu V \times 500$$
$$\Delta e = 25,500 \ \mu V = 0.0255 \ V$$

Now assume that the Seebeck coefficient for the selected thermocouple is 62 μV/°C and that the bridge voltage relative to 0 °C read on a voltmeter is 23,000 μV (0.023 V). The temperature can be found by rearranging the formula:

$$T = \frac{\Delta e}{\alpha}$$

$$T = \frac{23,000}{62}$$

$$T = 370.97 \text{ °C}$$

Although the thermocouple will generate the voltage calculated by use of the formula, the reading on the voltmeter will be higher than the calculated voltage. Recall that any two dissimilar metals that are joined together will generate a voltage when heated. When the voltmeter is hooked up to read the voltage, two additional junctions are formed (Figure 10.82). The thermocouple we have selected is a copper/iron thermocouple. The wires coming from the voltmeter are copper. The copper/copper junction does not generate a voltage. However, the copper/iron junction between the voltmeter lead and the iron thermocouple lead does generate a voltage. The voltages generated by the primary thermocouple and by the junction between the thermocouple and the voltmeter lead are added together, and the reading on the meter is higher than the predicted voltage.

One method of accurately determining the temperature of the primary junction is to place the secondary junction in an ice bath, which lowers its temperature to 0 °C. The secondary junction is still outputting a voltage since it is not at absolute 0, but the temperature of the secondary junction is known. The temperature can be read

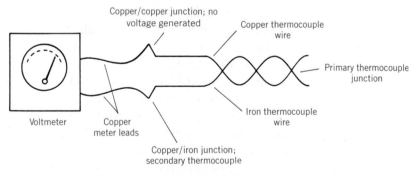

Figure 10.82 Illustration showing why a voltmeter connected to a thermocouple will indicate a voltage higher than the calculated voltage

directly from a voltage-to-temperature conversion table which takes into account the output voltage of the secondary junction at 0 °C.

It is obviously inconvenient to bring ice into a manufacturing plant to obtain an accurate temperature measurement, and so an electronic circuit has been developed that generates a small voltage to offset the voltage from the secondary junction. This circuit is called an **artificial ice point.** When an artificial ice point is employed, the standard voltage-to-temperature conversion table which assumes the secondary junction to be at 0 °C can be used.

Another possible solution is to establish the reference temperature for the secondary junction at 24 °C (75 °F), which is the normal temperature in a plant, and to use a table that takes this reference temperature into account. The temperature can be read directly from the table.

Thermocouple Identification

There is a virtually infinite number of different wires that could be used to form thermocouples. To simplify matters, a few standards have been established and are identified by letters of the alphabet. For example, the J-type thermocouple is made of iron and a copper-nickel alloy, and the T-type thermocouple is made of copper and copper-nickel. Seebeck coefficients, voltage-to-temperature conversion tables for secondary junctions at 0 °C and 24 °C, and recommended operating ranges are available for all standard thermocouples from the National Bureau of Standards.

Thermocouple Applications

Thermocouples have found many uses in industry for monitoring temperature. In a previous example, a strain gauge was used to determine if a ladle was full of molten iron. A thermocouple can be used to determine if the molten iron is at the correct temperature.

Another possible application of the thermocouple is monitoring the temperature of steel rods that are being heated in a furnace prior to forging. If a rod is not heated sufficiently before the robot places it in the forging die, the forging will not be formed properly or, even worse, the die may be ruined.

Any application that requires accurate and continual monitoring of temperature is a potential application for a thermocouple.

THERMISTORS

The **thermistor** is also a temperature-sensitive device. The resistance of a thermistor changes in response to a change in temperature. Although some thermistors increase in resistance in response to an increase in temperature, most thermistors exhibit an inverse relationship between temperature and resistance (an increase in temperature causes a decrease in the resistance of the thermistor).

Since a thermistor does not generate a voltage but rather a change in resistance in response to a change in temperature, it is normally used as one resistor in a Wheatstone bridge (Figure 10.83). As the resistance of the thermistor changes in response to a change in temperature, the bridge becomes unbalanced and a voltage appears at the output.

The major advantage of the thermistor is its dramatic change in resistance in response to a small change in temperature. As previously indicated, thermocouples generate very small output voltages and exhibit very small changes in output voltage in response to changes in temperature. When a more sensitive device (one that can sense very small changes in temperature) is needed, a thermistor is an excellent choice.

One major disadvantage of the thermistor is the nonlinearity of its response to changes in temperature (Figure 10.84). The resistance does not vary proportionately with a change in temperature. For this reason the thermistor is normally used as a monitor for maximum or minimum temperature. For example, a thermistor can be used to monitor the temperature of an electric motor. The

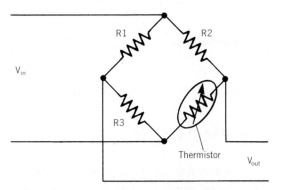

Figure 10.83   Wheatstone bridge with thermistor

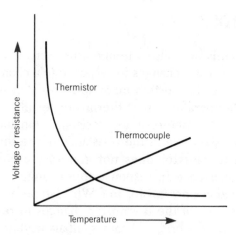

Figure 10.84   Graph showing the nonlinearity of the response of a thermistor to changes in temperature

thermistor is glued to the housing of the motor and is connected to a bridge circuit. The output voltage of the bridge circuit is compared with a reference voltage which represents the maximum safe operating temperature of the motor. If the output voltage of the bridge circuit exceeds the reference voltage, an alarm is sounded to indicate that the motor has overheated, or the motor is automatically turned off.

Recently, commercial compensation networks have become available. Compensation networks which linearize thermistors are taking over some of the applications that were previously reserved for thermocouples. It is also possible to connect the output voltage of a thermistor bridge to an A/D converter and to use the digital output of the A/D converter as an input to a computer. A program can be written which linearizes the thermistor and outputs the correct temperature reading on the CRT or printer.

The thermistor is a semiconductor device and is physically more sensitive than the thermocouple. Caution must be used when installing a thermistor to ensure that it is not crushed or damaged.

PROGRAMMABLE CONTROLLERS

A **programmable controller** is a specialized computer designed to control an automated system. Programmable controllers have replaced many of the relay control panels formerly used in industry. Figure 10.85 shows a programmable controller.

Figure 10.85   Programmable controller

Programmable Controller Components

Programmable controllers have three major components: a central processor, input/output modules, and a programmer.

The central processor of a programmable controller is a small computer. It has both RAM and ROM memories, a microprocessor, and ports which are connected to the input/output modules.

The input/output modules constitute the interface between the sensors that supply data to the central processor and the output of the central processor, which controls the automated system.

Various input modules and output modules can be plugged into the input/output rack depending on the requirements of the system (Figure 10.86). The most common input modules are designed for either a 120-V ac input signal or a low-voltage dc signal. Figure 10.87 shows a push-button switch connected to a 120-V ac input module. When the push button is closed, a 120-V ac signal is applied to the input module. It is important to remember that the input module is sensitive to a voltage input and not simply to the closing of a switch.

Figure 10.86   Input/output (I/O) rack

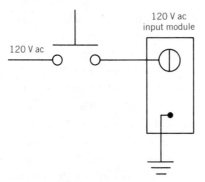

Figure 10.87   Input module with push-button switch

Output modules are available with either a 120-V ac output or a low-voltage dc output. When the output is switched on by the central processor, the output module outputs a voltage. A solenoid or small motor can be connected directly to the output module. The output current available from the output module is approximately

3 A. Since the output current is low, a large motor cannot be started and run from the output module alone. If an output module is to control a large motor, the output of the output module is connected to a **contactor** (a relay with large contacts for handling high current loads). The contacts on the contactor are connected between the main power lines and the motor. When the output module outputs a voltage, the contactor coil pulls the contacts closed, and power is supplied to the motor (Figure 10.88).

The programmer is used to enter the control program into the central processor. Programmable controllers are available with a CRT and keyboard for entering the program or with a teach pendant. Many programmable controllers also provide the option of storing the program on cassette tape. The programmer enters the program into the central processor through the keyboard or teach pendant and then records the program on tape. The cassette tape is a "backup" for the program. Should the program be destroyed due to a power failure or a malfunction of the central processor, the program can be read from the tape into the central processor without the need for complete reprogramming.

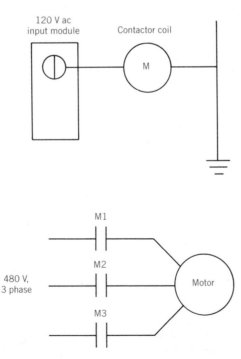

Figure 10.88 Output module with contactor used for control of a large motor

Programming the Programmable Controller

Programming a programmable controller is normally done by entering a ladder diagram through the controller's keyboard or teach pendant rather than by using a programming language such as **BASIC**.

Figure 10.89 shows a simple ladder diagram for controlling a solenoid. When both limit switches are closed at the same time, power is supplied to the solenoid. If either of the limit switches opens, the solenoid is turned off.

Figure 10.90 shows the same ladder diagram labeled in terms of a program to be entered into a programmable controller. To program the programmable controller, this diagram is drawn on the CRT of the controller by pushing keys on the controller's keyboard. The normally open (NO) contacts labeled 1IN and 2IN on the diagram are the inputs of the programmable controller. The limit switches LS1 and LS2 are connected to input terminals 1IN and 2IN on the input module. When limit switch LSI closes, an input voltage is applied to input terminal 1IN; and when limit switch LS2 closes, an input voltage is applied to input terminal 2IN. The programmable controller recognizes that both LS1 and LS2 are closed and turns on output 1OUT. Notice that the output is drawn as a circle, which is

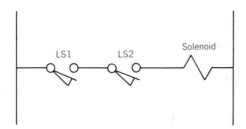

Figure 10.89 Ladder diagram of a solenoid control circuit

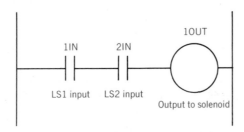

Figure 10.90 Same diagram as in Figure 10.89, but labeled in terms of a program for a programmable controller

the same symbol used for a relay coil in standard ladder diagrams. The number above the circle identifies the terminal on the output module that will be turned on. The solenoid is connected to output terminal 1OUT on the controller. When output 1OUT is turned on, the output supplies voltage to the solenoid.

Figure 10.91 shows another ladder diagram. This circuit includes a start button, a stop button, a control relay, an indicator lamp, and a solenoid. When the start button is pushed, the control relay (CR) is activated. Contact CR-1 closes, interlocking the start button. When the start button is released, the circuit continues to operate. Contact CR-2 closes, turning on a green indicator light. Contact CR-3 closes, turning on solenoid A. When the stop button is pushed, the current path to the CR relay is broken, all CR contacts open, and the solenoid is deactivated.

Figure 10.92 shows the same ladder diagram drawn as a program for a programmable controller. Normally closed (NC) contact 1IN is the input for the stop button, and normally open (NO) contact 2IN is the input for the start button. Output A is the controller's equivalent of the control relay in Figure 10.90. Output A is an "internal out-

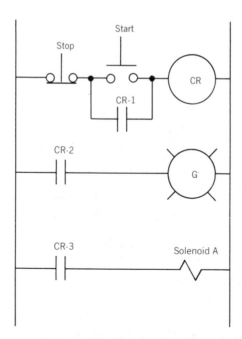

Figure 10.91 Ladder diagram of a solenoid control circuit that includes start and stop buttons, a control relay, and an indicator lamp

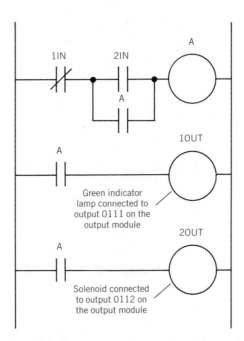

Figure 10.92   Same diagram as in Figure 10.91, but labeled in terms of a controller program

put," meaning that it is not connected to a terminal on the output module. Rather, it is an internal control which closes "internal switches" in the controller. When the start button is pushed, internal output A is activated, thus closing all internal switches labeled A. One of the A switches is connected as an interlock. When the start button is released, the circuit continues to operate. Another A switch is connected to output 1OUT. Output 1OUT is an "external output," meaning that it is connected to a terminal on the output module. The indicator lamp is connected to this terminal on the output module. A third A switch is connected to output 2OUT. When the start button is pushed, output 2OUT is also activated. The solenoid is connected to this output terminal. The operation of this circuit is identical to that of the ladder diagram circuit shown in Figure 10.90.

The simple circuits illustrated here are presented as samples of programming methods. When programmable controllers are used in industry, the program is normally more complex and the number of inputs and outputs is substantially greater. In fact, programmable controllers can have thousands of inputs and outputs and be in control of a totally automated manufacturing plant. Most programmable controller manufacturers also include programming options for timers (on-delay and off-delay) as well as for counters to count the number of times a cycle is operated.

In addition to simple ac or dc inputs and outputs, timers, and counters, many programmable controllers also include A/D input modules and D/A output modules. Some manufacturers also offer full PID (proportional plus integral plus derivative) control as an option on their controllers.

There are many programmable controller manufacturers. Each manufacturer uses his own numbering system for inputs, outputs, and internal inputs and outputs. It is not important to memorize the numbering system used in these examples. You should simply be familiar with the concepts of programming a programmable controller.

ROBOT INPUTS AND OUTPUTS

When a robot is installed, it assumes the central control of the work cell. All other machines in the work cell send information to the robot, and the robot outputs control signals to the other machines.

The inputs and outputs of a robot are similar to those of a programmable controller. The input signals to a robot are normally 120-V ac or low-voltage dc signals, and the outputs are also 120-V ac or low-voltage dc signals.

The robot can output a voltage to start conveyors, to cycle presses, to start turning centers, and to activate alarms when malfunctions occur in the cell. The robot receives signals from the work cell which tell the robot that a cycle is complete or that someone has invaded the work cell area.

The inputs and outputs are activated by the robot under the control of the robot program.

SAMPLE ROBOT INTERFACE

Figure 10.93 shows a plan view of a work cell. The work cell includes a robot, an infeed conveyor, a punch press, an outfeed conveyor, and a safety fence surrounding the entire work cell area. The robot picks up a piece of steel at the infeed conveyor, moves to the punch press, inserts the steel into the punch press, withdraws from the punch press, cycles the punch press, removes the finished part from the press, and delivers it to the outfeed conveyor.

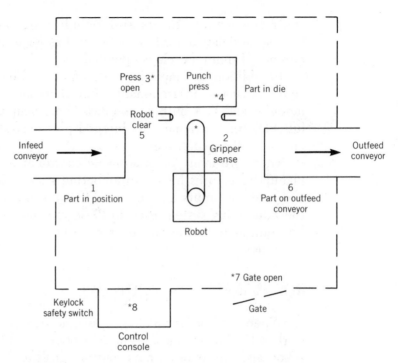

Figure 10.93   Plan view of a work cell

Robot Inputs

The sensors in the work cell that supply the robot with information are listed below. The ladder diagram for this work cell is shown in Figure 10.94.

1. *Part in Position.* A proximity switch is mounted on the infeed conveyor to tell the robot when a part is in the proper position for pickup by the robot. The proximity switch is connected to robot input 1IN.

2. *Gripper Sense Switch.* A vacuum switch is mounted on the gripper to tell the robot when it has picked up a part. The vacuum switch is connected to robot input 2IN.

3. *Press Open.* A limit switch is mounted on the top of the punch press. When the press is open (ram up), the limit switch is closed, telling the robot that it is safe to place the part in the press or remove it from the press. The limit switch is connected to robot input 3IN.

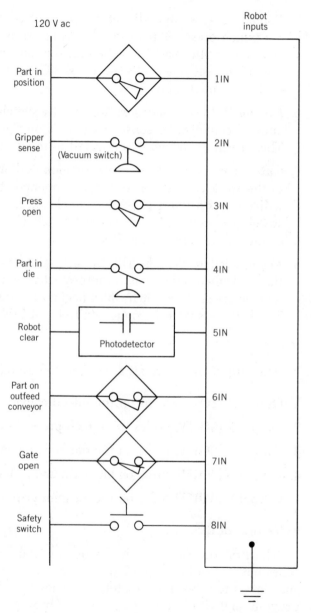

Figure 10.94 Ladder diagram of robot inputs for work cell shown in Figure 10.93

4. *Part in Die*. A pressure switch is mounted in the die to tell the robot that the part is properly positioned in the die. The pressure switch is connected to robot input 4IN.

5. *Robot Clear*. A photodetector (opposed) is mounted in front of the punch press. When the robot is inserting the part in the die or removing the part from the die, the light beam is broken and the robot will not cycle the press. The photodetector is connected to robot input 5IN.

6. *Part on Outfeed Conveyor*. A proximity switch is mounted on the outfeed conveyor to sense if a part is on the outfeed conveyor. This proximity switch is connected to robot input 6IN.

7. *Gate Open*. A proximity switch is mounted on the entrance gate to the work cell. When the gate is opened, the robot stops and will not restart until the gate is closed and the keylock start switch (see item 8, below) is reactivated. This proximity switch is connected to robot input 7IN.

8. *Keylock Safety Switch*. The keylock safety switch is mounted in the control console outside the protective fence. This switch is a momentary contact switch which begins or restarts the robot cycle. The keylock switch is connected to input 7IN.

Robot Outputs

The outputs of the robot are connected as follows (Figure 10.95):

1. Output 1 (1OUT) turns on the infeed conveyor.
2. Output 2 (2OUT) cycles the punch press.
3. Output 3 (3OUT) turns on the outfeed conveyor.
4. Output 4 (4OUT) turns on an alarm and a flashing light.
5. Output 5 (5OUT) turns on the vacuum gripper.

Program Sequence

The first statement in the robot's program directs the robot to continually examine input 7 (gate open sensor). If this input ever goes low, the robot immediately branches to a different part of the program called a "subroutine." The subroutine immediately halts the robot and turns on output 4 (alarm). The robot holds its current position until input 8 (keylock switch) is activated. When input 8 goes high momentarily, output 4 (alarm) is turned off and the robot returns to the main program and continues.

It is important when designing a robot work cell that the work cell be protected from inadvertent intrusion by workers into the

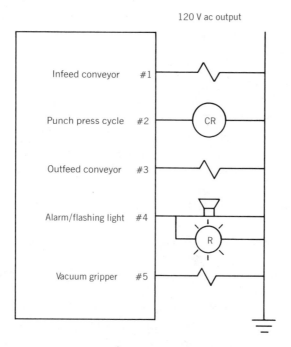

Figure 10.95   Ladder diagram of robot outputs for work cell shown in Figure 10.93

work envelope of the robot by a barrier such as a rail or chain link fence. A robot appears to be turned off when it is awaiting a signal from another machine in the work cell. When the robot receives the appropriate input, it proceeds with the next step in the program. Without warning, the robot will move to its next assigned position. A worker in the work cell may be pinned between the robot's arm and another machine in the work cell or may be struck by the robot's arm, which moves at high speed.

The second statement in the robot's program directs the robot to a point above the infeed conveyor. The robot then examines input 1 to determine whether or not the part is in position on the infeed conveyor. If the part is not in position, the robot turns on output 4, which activates the alarm and a flashing light. If the part is in position, the robot proceeds to the next step in the program.

The robot moves downward to the part on the conveyor and turns on output 5, which activates the solenoid valve of the vacuum gripper. The vacuum draws the part into the gripper. The robot then examines input 2. If the vacuum switch is closed, which indicates that a vacuum is being drawn and that the part is firmly gripped by the gripper, the robot proceeds. If, on the other hand, the vacuum switch is not closed, indicating that the part is not gripped firmly, the robot halts and turns on output 4, sounding the alarm.

If the vacuum switch is closed, the robot lifts the part off the conveyor and moves to a point in front of the punch press. The robot then examines input 3. The limit switch senses if the press is open and ready to receive the part. If the limit switch is closed, the robot proceeds. If the limit switch is open, then the ram of the punch press is down, and if the robot should proceed it would crash into the press. (In industry this is referred to as a "wreck.") With the limit switch open, the robot halts and closes output 4, which sounds the alarm.

If the limit switch is closed, the robot proceeds into the press. The robot then moves downward into the die and turns off output 5, which turns off the vacuum to the gripper, and the part is released in the die.

The robot then examines input 4. Input 4 is the pressure port in the die and the pressure switch. If the pressure switch is closed, indicating that the part is in position in the die, the robot proceeds. If the pressure switch is not closed, indicating that the part is not properly seated in the die, the robot halts and turns on output 4, which sounds an alarm.

If the part is properly seated in the die, the robot moves up and out of the press. The robot then examines the photodetector (input 5). If the photodetector is outputting a voltage, which indicates that the robot's arm is clear of the press, the robot proceeds. If the output of the photodetector is off, then the robot's arm is still in the press, and the robot closes output 4, which sounds the alarm.

If the photodetector is on, the robot proceeds with the program. The robot turns on output 2, which cycles the punch press. The robot immediately examines input 3. When the limit switch closes, indicating that the punch press cycle has been completed, the robot can safely enter the press to remove the finished part. The robot reaches into the press above the part and then moves down onto the part.

Output 5 turns on, opening the gripper's solenoid valve. The suction gripper turns on and the robot examines input 2, which senses whether a vacuum has been drawn. If a vacuum has been drawn, the vacuum switch closes, indicating that the part is firmly in the grip of the robot. If, however, the vacuum switch does not close, which indicates that the robot does not have a firm grip on the part, the robot turns on output 4, activating the alarm.

With the part firmly in the grip of the robot, the robot moves up and out of the press. The robot then examines input 5, which senses if a part is in the die. If the robot has successfully removed the entire part from the die, there will be very little pressure on the pressure switch and the switch will be open. The robot proceeds to the next program step. If, however, the part has not been successfully removed from the die, the pressure will cause the pressure switch to close. With the pressure switch closed, the robot turns on output 4, which sounds the alarm.

If the part has been successfully removed from the die, the robot moves to a point above the outfeed conveyor. The robot then examines input 6. If the proximity switch is on, indicating that a part is in the outfeed position, the robot holds its position and turns on output 4, which activates the alarm. If the proximity switch is off, indicating that the outfeed conveyor is clear, the robot moves the finished part to the conveyor.

The robot turns off output 5, which allows the vacuum solenoid to return to the off position, and the part is deposited on the outfeed conveyor. The robot then turns on input 6. If input 6 is low, indicating that the part is not properly positioned on the conveyor, the robot turns on output 4, which sounds the alarm. If input 6 is high, indicating that the part is properly positioned on the conveyor, the robot moves up and turns on output 3, which turns on the outfeed conveyor, and the part is carried out of the work cell. The robot also turns on output 1, which starts the infeed conveyor. The robot moves to a point above the infeed conveyor, and the entire cycle is repeated.

Recall that during the entire cycle the robot has been sampling input 7, which is the "gate open" sensor. If anyone opens the gate during the cycle, the robot immediately halts and turns on output 4, which sounds the alarm.

QUESTIONS FOR CHAPTER 10

1. Draw the ladder diagram symbols for NO and NC push-button switches.

2. What is the most common application of a mushroom head switch?

3. Draw the ladder diagram symbol for, and explain the purpose of, a "press to test" indicator lamp.

4. What are the horizontal lines on a ladder diagram called?

5. Draw the symbols for NO on-delay and NO off-delay relays, and explain their operation.

6. Draw the symbols for NO operated, NO unoperated, NC operated, and NC unoperated limit switches.

7. Explain the terms "pretravel," "overtravel," and "differential travel." Use diagrams in your explanation.

8. What is the advantage of a proximity switch over a mechanical limit switch?

9. Explain the difference between a photoconductive cell and a photovoltaic cell.

10. Are photodiodes and phototransistors interchangeable? If not — why not?

11. What is the advantage of opposed scanning?

12. Explain the operation of retroreflective scanning, convergent scanning, and specular scanning. Use diagrams in your explanation.

13. What is the primary advantage of modulated light over non-modulated light?

14. What are the four types of pressure switches?

15. Define the terms "tensile stress" and "compression stress."

16. Why does the resistance of a strain gauge change when a tensile or compression stress is applied?

17. Define the term "gauge factor."

18. What is the purpose of a dummy gauge?

19. Define "Seebeck coefficient."

20. What are the advantage and the disadvantage of a thermistor in comparison with a thermocouple?

21. Draw a work cell which includes a CNC turning center, an infeed conveyor, an outfeed conveyor, sensors, and a safety fence. Also include the robot program listing and a ladder diagram.

CHAPTER ELEVEN

Automated Manufacturing—The Second Industrial Revolution

The introduction of computer controlled automation into the manufacturing process has been called the "second industrial revolution." In this chapter we will examine the ways in which computer controlled automation is changing both the manufacture of products and the products themselves. Before examining the second industrial revolution it is important that we gain a perspective of manufacturing by reviewing the first industrial revolution.

THE FIRST INDUSTRIAL REVOLUTION

The beginning of mass production can be traced to Eli Whitney. In 1798, Whitney won a contract to produce 10,000 muskets for the U.S. Army. If these muskets had been produced by the traditional manufacturing method of the time, they would have been built by individual gunsmiths each working independently. Each craftsman would have produced weapons of similar design, but each part would have fit only the specific weapon for which it was built. Whitney's contract specified not only that he deliver 10,000 muskets but also that the parts be interchangeable among all the weapons.

Whitney fulfilled his contract by building jigs. A jig was built for each type of part, and each part was mounted in the appropriate jig and hand filed until it matched the jig. This painstaking process was the first step in the conversion from manufacture of products by individual craftsmen to mass production.

Invented during the same period were machine tools which substituted machine power for muscle power and provided an accuracy in manufacturing that was previously impossible. In 1775 in England, John Wilkinson built the first horizontal boring machine. The horizontal boring machine was used to machine the bores of cylinders used in piston pumps. Wilkinson's horizontal boring machine is reported to have produced 57-in. cylinders with an accuracy such that the clearance between the piston and the cylinder was less than a "shilling" (approximately $\frac{1}{16}$ in.).

The next development in machine tools occurred in 1797 when Henry Maudsley built the first lathe. In 1800, the lathe was improved through the addition of a lead screw and interchangeable gears which made thread cutting on a lathe possible. In 1818, Eli Whitney developed the first milling machine.

The development of machine tools was changing the manufacturing process, but a major problem still existed. Precision measurement of the parts produced on the machine tools was far from an exact science, and interchangeability of parts was limited. Starting in about 1830, Joseph Whitmore began the development of precision measuring instruments and standards for threads.

In 1840, John Nasnyth added an automatic feed to the drill press, and automation had begun.

By the end of the 19th century, most of the major machine tools had been developed. During the next 50 years, machine tools were improved. Machine tool builders added cams, gears, and lead screws to make the operation of the machine tool more automatic, thus increasing productivity and the consistency of the parts. Although these developments were significant in improving productivity, setup and operation of the machine tools still required a high degree of skill. The machines also were limited in the type of parts they could produce.

In 1947, John Parsons of Parsons Corporation in Traverse City, Michigan began work on a machine that would come to be known as "numerical control." Parson's objective was to build a machine that was versatile—one that could be "programmed" to produce a large variety of complex parts with an unskilled operator. Parsons was successful in connecting a computer to a jig boring machine.

By 1949 the U.S. military needed a large volume of complex parts for its aircraft and missiles. Parsons was given a contract to continue the development of his machine. In 1951, Massachusetts Institute of Technology (MIT) took over the work that Parsons had begun, and in 1952 demonstrated a working system. MIT had connected a computer to a Cincinnati Hydrotel milling machine.

In 1955, numerical control (NC) machines were on display at the National Machine Tool Show. A few years later, NC machines were being used to produce parts in industry. The NC machines of the mid-1950's were crude by today's standards. The computers were built with vacuum tubes. The life of the tubes was short, and the machines seemed to be down for repair more than they were operating. The programming of these early machines was so difficult that it required specially trained engineers and mathematicians to produce the simplest part.

At the same time that industry was struggling with the new NC machines, the electronics industry was putting transistors to work. The developments in electronics quickly entered the machine tool industry, and the NC machines became faster, easier to program, and more reliable.

At the same time, George Devol was working on connecting a computer to a manipulator. In 1956, Devol met Joseph Engelberger and explained to him a patent application he had filed called a "programmed article transfer." Devol and Engelberger combined their talents and continued their work, and in 1961 the first industrial robot was installed.

Development of NC machines and robots offered potentials for productivity never before possible, but industry was slow to adopt this combined technology. It wasn't until the mid-1970's that industry began to take the robot seriously. Gradually, the robot started finding its way into the shop, and now, with perfect 20-20 hindsight, we can look back on the work of Parsons, Devol, and Engelberger, and declare them as the founders of the second industrial revolution.

MANUFACTURING METHODS

Along with the development of machine tools for manufacturing, three separate manufacturing systems were developed: the **job shop system,** the **flow shop system,** and the **project shop system.** Each of these systems is still used today.

Job Shop System

The **job shop system** is the most prevalent manufacturing method. The job shop brings together a variety of general-purpose machines. When a product is to be manufactured, the first step is the establishment of the manufacturing process. When a job shop receives a work order to produce a shaft, the manufacturing engineer determines the manufacturing sequence and orders the steel for the job. When the steel is received it is routed to the lathes for turning of the shafts. The turned shafts are then routed to drilling, grinding, heat treating, regrinding, and shipping. The routing of the job is shown in Figure 11.1.

A second job enters the shop at the same time as the first job. The second job is the production of valve bodies for hydraulic valves. The routing of this job is shown in Figure 11.2. Figure 11.3 combines the two previous figures, which showed the routing of each individual job. Figure 11.3 clearly shows conflicts in the production of the two jobs. For example, both jobs arrive at the drills at the same time. Either the first or the second job will be delayed by the production of the other. Also notice that at times machines are idle while waiting for a job to arrive.

From this simple example one can conclude that job shop manufacturing is extremely inefficient. Conflicts in the use of machines are inevitable, causing delays in the manufacture of parts and thus

Figure 11.1   Job shop routing chart for production of shafts

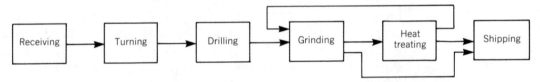

Figure 11.2   Job shop routing chart for production of valve bodies

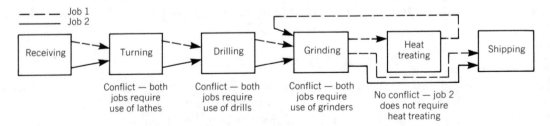

Figure 11.3   Combined job shop routing chart for production of shafts and valve bodies

increasing their cost. However, the job shop has an advantage over other forms of manufacturing, and this advantage is versatility. The collection of machines in the job shop allows the shop to accept a large variety of projects.

Flow Shop System

The **flow shop system** is an alternative production process. Figure 11.4 shows the flow shop manufacturing system for making the same parts as those described for the job shop.

The flow shop is designed to produce limited numbers of different products. A specific manufacturing line is designed and constructed to process each particular type of product.

A limited amount of flexibility is designed into the flow shop. The line which is designed for shaft production (line 1 in Figure 11.4) can produce a large variety of shafts of various lengths, diameters, and complexities, but this line can produce only shafts, and no other product.

Another line (line 2) is specifically designed for producing valve bodies, with some of the machines being specifically designed to optimize the production process. Even though this line can be used only for production of valve bodies, it can produce bodies in a large variety of sizes.

In the flow shop, there is no conflict for the use of machinery, since each line contains all of the machines needed to produce the product.

Most automotive production is done in flow shops. For example, a production line is designed for the manufacture of engines. Although much of the equipment on this line is custom designed for

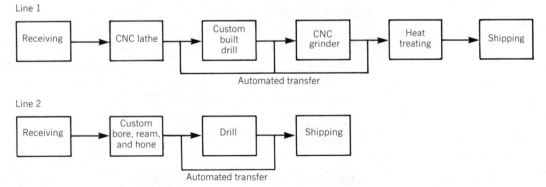

Figure 11.4 Routing chart for flow shop using automated machines and automated transfer

the manufacture of engines, a variety of engines can be produced using the same line. Through changes in the fixtures that hold the engine blocks during manufacture, changes in the tooling in the various machines, and adjustments in the positions of the machines, various engines can be produced.

The flow shop lacks much of the flexibility of the job shop. Since the job shop is a collection of general-purpose machines and material is moved with forklift trucks in bins or on pallets from one machine to another, the job shop can produce the largest variety of parts. Inherent in the job shop production method is its lack of efficiency. The bottlenecking of jobs and the continual changing of machine setups cause delays and cost money. The advantage of the flow shop is its efficiency. After tooling of the machines in a flow shop, a high volume of parts can be produced with speed and efficiency.

In reality, most flow shops incorporate some aspects of the job shop. The customized tooling for the various machines on the line in the flow shop are built in the job shop. Also, since the line in a flow shop is used to produce a variety of similar parts, at times a special feature of a particular job cannot be produced on the line. When this situation occurs, the parts are removed from the line after the standard processes have been completed and the special features are completed in a job shop.

Project Shop System

In the **project shop system,** a group of machines is brought together at a particular location to complete a large project. The project shop system is used extensively in the shipbuilding industry. The required machines are brought to a site where they are set up to produce one project. Each piece for the ship is made and custom fit to the ship. When the project has been completed, the machinery is moved to another site and another project is begun.

The project shop is unique to the production of large projects and does not offer much opportunity for automation. It is a process that has existed since man began to build products and will continue to be used in spite of its high cost and low efficiency.

FLEXIBLE MANUFACTURING SYSTEMS

Recently, a new manufacturing system has been emerging. This system, called the **Flexible Manufacturing System,** or **FMS,** attempts to combine much of the versatility of the job shop with the efficiency and cost savings of the flow shop.

A flexible manufacturing system is a grouping of computer controlled machines and a material handling system for delivering the in-process parts to the machines. Since computer numeric controlled (CNC) machines are used, the operation can be changed quickly, thus offering the flexibility of the job shop. Since the CNC operation is changed by simply changing its program, we have the efficiency of the flow shop.

In theory, FMS appears to be the answer for which industry has been searching. Unfortunately, the simplicity of the FMS concept masks its difficulty of execution.

Group Technology

Implementation of FMS requires precise identification of the manufacturing process needed to complete the part. **Group technology** identifies the similarities among parts and utilizes these similarities to maximize production. For example, all shafts will fit into a group since they all require turning. This oversimplification is similar to saying that all Americans are the same since they all live in America. Americans can be divided into groups on the bases of race, religion, occupation, sex, age, etc. When examining various shafts and attempting to group them into families, we must ask questions such as: "What is the diameter of the shaft?" (a shaft for a wristwatch and a drive shaft for a truck would be unlikely to be manufactured by the same method); "Does the shaft require cutting of a spline?"; "Does the shaft require heat treating?"; "Does the shaft require boring, reaming, or grinding?".

Identifying groups of parts is a difficult and time-consuming process. Many shops produce thousands of different parts in a year. One method of identifying groups of parts is through use of a matrix. Figure 11.5 shows a list of jobs and the machining processes required to complete each type of part. At first glance, the manufacture of these parts does not seem to lend itself to identification by groups. In fact, rearrangement of the same list (Figure 11.6) shows that three manufacturing groups exist within the list, along with a few exceptions. If the quantity of parts required within each group is large, machines can be grouped and automated material handling equipment can be installed.

Figure 11.7 shows an arrangement of machines that can be used to produce the parts in group 1 of Figure 11.6. Note that part 17 re-

Manufacturing codes

Job No.	A	B	C	D	E	F	G	H	I	J
1	X	X	X							
2					X	X	X			
3								X	X	
4								X		X
5					X	X				
6								X		
7								X	X	X
8		X	X							
9					X					
10								X	X	
11					X		X			
12			X					X	X	
13					X	X	X			
14	X	X	X							
15								X	X	X
16								X	X	
17		X	X	X				X		
18								X		
19								X	X	
20	X	X	X	X						

Figure 11.5   Chart of manufacturing processes for 20 different jobs

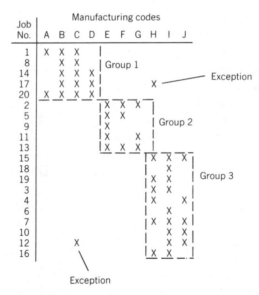

Figure 11.6   Rearrangement of chart shown in Figure 11.5 to form manufacturing groups

quires broaching and that a broach has not been included in the machine grouping. When part 17 is run, it is moved to another part of the plant for the broaching operation.

When machines are grouped as shown in Figure 11.7, the grouping is normally referred to as a "cell" or a "work cell." In the sim-

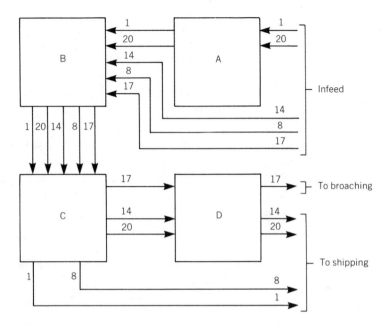

Figure 11.7 Work cell for producing parts from group 1 in Figure 11.6

plest cell, a machine operator loads the machines and calls the program stored in the machine's computer that will perform the prescribed operation. Although the cell still requires an operator, it is far more efficient than production of the parts by the job shop method. A single operator loads and operates all the machines in the cell, inspects the parts, and makes any adjustments necessary to produce all parts within the specified tolerances.

Transfer Mechanisms

The next step in automation is the addition of **transfer mechanisms** for moving the parts between the various machines. The transfer mechanisms are custom designed conveyors which move the parts from machine to machine. In addition to being moved, the parts must also be loaded into the machines. Designing a transfer mechanism that can move a variety of parts between machines and also perform the loading operations is difficult. The machine tools normally used in industry were originally designed to be loaded by human operators rather than by transfer devices. Often the parts must be lifted and moved into the machine around obstructions. For this reason, transfer mechanisms have been limited to handling only a few different parts and thus are not truly "flexible."

Figure 11.8 shows a work cell in which a robot is used as the transfer device. Raw material is delivered to the work cell on a pallet. An operator enters the program number into the robot's control console and the CNC machine tools. The robot's program directs the robot to pick raw material from the pallet and load it into the first machine in the process. The robot activates the clamps that lock the part in position and then moves clear of the machine and begins the machine cycle. When the first machine has completed its cycle, the robot unloads the partially completed part and loads it into the next machine for processing. The robot begins the cycle of the second machine and immediately returns to the pallet to pick up raw material and load it into the first machine. The robot is operating the cell as a human would.

The operation of reaching up and into the machine around obstructions and loading the part into the machine, which is a problem for a transfer mechanism, is not a problem for a robot. Robots have been designed for exactly this capability (Figure 11.9).

From Work Cells to FMS

The work cell described above can effectively manufacture many different parts within a group and thus is flexible by definition.

Figure 11.8   Work cell in which robot is used for transfer of parts

Figure 11.9   Robot loading a part into a machine

However, because it requires a forklift operator to deliver the raw material to the cell and another operator to enter the program number into the robot and the CNC machine tools, this work cell is not truly an FMS.

The system illustrated in Figure 11.10 is truly a flexible manufacturing system. The raw material is stored in an automated warehouse and is moved from the warehouse to the work cell under the control of the computer in the central control room. The material is carried from the warehouse to the designated work cell on an Automated Guided Vehicle (AGV). An AGV is a motorized cart which follows the program direction of the central computer. In moving between the warehouse and the work cell, some AGV's follow tracks in the plant floor, others follow a wire embedded in the plant floor, while still others are laser controlled. Regardless of how the AGV is guided, it delivers the raw material to the work cell. The master control computer then outputs data to the robot and to the CNC machine tools in the work cell to establish the correct program for them to follow. The parts are machined and returned to a pallet by the robot. An AGV picks up the parts and brings them to the warehouse for storage or on to another work cell for further machining.

For the system to be truly flexible, the work cell must be re-tooled either to replace worn and broken tools or to change the tooling so that the cell can produce a variety of parts. The tooling is also carried from the tool storage area to the work cell on AGV's.

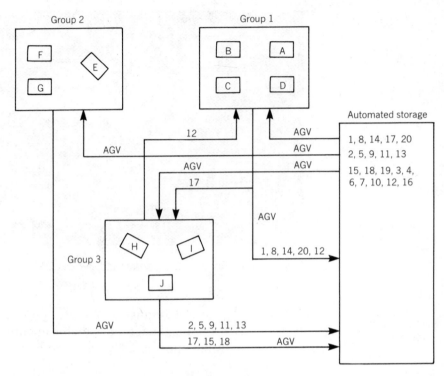

Figure 11.10   Flexible manufacturing system

The truly flexible manufacturing system operates unattended except for the system programmers who monitor and modify programs and maintenance personnel who repair the system when a failure occurs.

ROBOTS IN MANUFACTURING

With a basic understanding of FMS, we will direct our attention to the robot's function in the work cell. The robot can perform two basic functions within the work cell: it can perform **value added work** and **non-value added work.**

Value added work is any work which increases the value of the raw material. For example, a robot can have a grinder as its end effector. When the robot grinds a part, the value of the part is increased. Another example of value added work is spray painting. If the robot is fitted with a spray painting gun and paints a part, the value of the part is increased.

Non-value added work is also performed by robots. One non-value added job for robots is removal of the finished product from a conveyor and stacking of the product on a pallet. Although stacking of the finished parts must be done before the product can be shipped, the value of the product is not increased by the stacking operation.

In this section we will briefly examine a variety of primary jobs that robots can perform in industry. Some of these jobs are primarily value added jobs whereas others are primarily non-value added jobs. Some jobs incorporate both value added and non-value added aspects: these jobs are classified by their primary function.

Non-value Added Jobs

The non-value added jobs for robots discussed here include applications in material handling, machine loading, press loading, die casting, plastic injection molding, and forging.

Material Handling. Material handling is the process of moving materials about a manufacturing plant. "Material" may be the raw material that goes into the production of the part or product, the partially completed part or product, or the finished part or product ready for shipment. Material handling has been a major problem in industry because it is a truly non-value added function — moving of material does not add any value to the product. For this reason, industry has worked hard to minimize material handling costs. The effective use of conveyors is certainly one important means of minimizing these costs. Connecting programmable controllers to a conveyor system allows flexibility in moving material in and out of storage and between the various machines that will perform the value added work.

Material often arrives at a plant on pallets (or skids), which are wooden or steel platforms on which the material is stacked. The pallets are moved from the delivery truck to the storage area by forklift truck. At some point, the material must be brought out of storage and taken off the pallet. The job of removing the material from the pallet often requires human labor. The material is removed from the pallet and placed on a conveyor. This process is referred to as "depalletizing." Depalletizing is a boring and potentially dangerous job. The constant lifting of boxes from the pallet can cause back or other muscle injuries. Depalletizing is thus a potential application for robots.

Figure 11.11   Robot unloading a pallet ("depalletizing")

Figure 11.11 shows a robot lifting material from a pallet and placing it on a conveyor. Since a large variety of materials needs to be depalletized, the robot must be "flexible." The robot must be able to store a program for every different type of pallet load. There are several possible ways of telling the robot which program to execute. The simplest method is to have an operator key in the program number on the robot's control console. Although this is a simple method, it requires that a decision be made by an operator and thus is subject to human error.

Another possible solution is the use of bar code scanning. You have probably seen an example of bar code scanning in your local grocery store. Most products have a code printed on the package. When the product is passed over a laser scanner at the checkout counter, the code is read and the price is automatically recorded in the cash register. Figure 11.12 shows a bar code scanner and a bar code label.

Figure 11.12   Bar code scanner and label

If a bar code label is mounted on the pallet, the code can be read and the robot can be automatically switched to the correct program for depalletizing.

Another possible solution to identification of pallets is an automatic storage and retrieval system. When the material is unloaded from the truck it is placed in the automated storage system. The product is identified when it is stored, and the number that identifies it is entered into the master computer. When the product is needed, the computer calls the product from the storage bin. The product is brought to the robot on a conveyor or on an automatic guided vehicle (AGV), and the master computer sends the program number to the robot.

Depalletizing enters the product into the manufacturing process. After the product has been completed it is normally "palletized" for shipment. This operation can be performed by an automatic palletizing system provided that all units of the product coming down the line are identical. In single-purpose manufacturing, an automated palletizing system is fast and efficient. However, most manufacturing plants in the U.S. produce large varieties of products, and automated palletizers do not have the versatility required for such plants.

The robot, on the other hand, can work very well in this environment. If the products coming down the conveyor have bar code labels, the robot can identify them, pick them up, and put them on the correct pallet. The robot also keeps track of the number of pieces on the pallet. When the pallet is full, it is removed from the cell by a forklift truck or a conveyor, and an empty pallet is put in its place. Using the robot for palletizing sorts the product and eliminates the need for human labor to lift the product and place it on the pallets.

Machine Loading.   With the introduction of NC/CNC machines into industry, it is no longer necessary to have a highly skilled machinist as the machine operator. After the NC/CNC machine is set up, the operator has little to do but load parts into the machine, press the start button, and stand back. The NC machine goes through its complex operations without the aid of the operator. When the part is complete, the machine comes to a stop. The operator removes the finished part and inserts the next part to be machined.

The machine operator is often responsible for checking the finished parts and making any required adjustments in the machining process to ensure that all parts are produced within specifications.

Machine loading is a natural application for robots. The robot can pick up a part from a pallet or an infeed conveyor, load the part into the machine tool, send a signal to the machine to clamp the part, pull clear of the machine, and send a signal to begin the machine cycle. When the machine cycle is complete, the robot reaches into the machine, removes the finished part, and reloads the machine with another part to be machined (Figure 11.13).

The robot can also perform inspection of parts. Included in the work cell is a laser gauge which accurately measures the finished part (Figure 11.14). If the dimensions of the part are within specifi-

Figure 11.13 Machine loading utilizing a robot

Figure 11.14 Automatic gauging of finished part (laser gauge)

cations but not optimum, signals are sent to the machine for the necessary adjustments. In this way, all parts produced in the work cell will be within the designer's specified tolerances.

The NC machines in the work cell often have cycle times as long as several minutes. Having a robot idle for such periods of time while parts are being machined is obviously poor utilization of the robot. For this reason, a robot is normally positioned among several NC machines and is used for loading and unloading all the machines, thus making better use of the robot.

Press Loading.   An application of a robot in press loading and unloading was described in Chapter 10. When robots were first introduced in industry it was thought that press loading would be a significant robot application. This, however, is not the case.

Press loading is a dangerous job because it requires a worker to reach into the press to place a part in the die. Ever present is the risk that the press may accidentally cycle, crushing the worker's hands. Removing workers from this risk is laudable, but application of robots to press loading is often not cost justifiable. A less costly and equally effective method of protecting the worker is the installation of safety interlock controls that will not allow the press to cycle until the worker is fully clear of the press.

Recall that currently robots are not very fast. At best the robot operates at or near the speed of a human operator. In many press loading applications, the robot actually produces from 10 to 25% fewer parts per hour than would a human operator. However, a portion of the loss in production due to the low speed of the robot is compensated for by the consistency of the robot. If the parts being loaded into the press are heavy, the worker becomes tired toward the end of the shift, whereas the robot continues to produce parts at the same rate hour after hour.

If a press loading application is chosen for a robot, the application designer should consider nonservo robots. The nonservo robot is less expensive and often faster than the servo robot. The nonservo robot also has excellent repeatability since it "bangs" into end stops. The combination of speed and repeatability is necessary to make the robot competitive with a human operator.

Another problem that must be overcome in all applications but that often presents very real problems in press loading is part orientation. The parts delivered to presses are often loosely stacked on a pallet. Exact location of parts on a pallet is not a problem for a human press operator. The press operator simply picks up the part and places it correctly in the die. If the part is upside down on the pallet, the operator flips the part over and places it in the die. Since the robot is blind, it cannot load parts correctly if the parts are not precisely located on the pallet. A sensor in the die can sense that the part is not properly located and interrupt the cycle before the material or the die is damaged, but this constant interruption of the cycle slows the process even further.

Figure 11.15 shows a successful press loading application that provides a solution to the problem described above. The robot has been fitted with a double gripper. The robot picks up a part from the infeed conveyor in one of the grippers, moves to the punch press, picks a finished part out of the press, and immediately places the unfinished part in the die. The robot moves to the outfeed conveyor, deposits the finished part on the outfeed conveyor, and moves to the infeed conveyor where it picks up another unfinished part. Use of a double gripper allows the speed of the robot to match more closely the speed of the punch press.

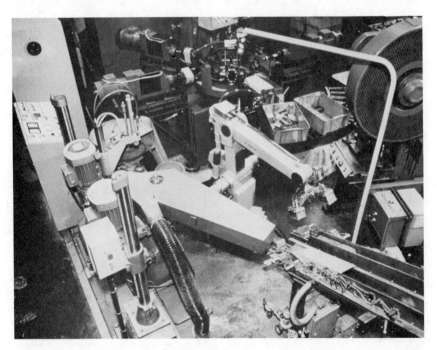

Figure 11.15 Robot press loading. Robot uses a double gripper for increased speed

Die Casting. The die casting industry provided one of the earliest applications of robots. Die casting is the process of injecting molten metal into a die to form a part. The molten metal injected into the die is a nonferrous metal such as copper, zinc, or aluminum. Recently the process has been extended to some ferrous metals.

In a basic die casting machine, the dies used to form the parts are in two halves. The dies are forced together under hydraulic pressure. When the dies are together, molten metal is injected into the die. After the metal cools, the dies are opened and the parts are removed from the dies.

Die casting machines are rated by the amount of clamping force that holds the dies together while the molten metal is being injected. Die casting machines are available with clamping forces from as low as 5000 lb to as high as 5000 tons. Die casting machines with clamping forces from 100 to 500 tons are common in industry.

There are two types of die casting machines: **hot chamber machines** and **cold chamber machines.** Figure 11.16 shows a hot chamber die casting machine. The hot chamber machine is also known as a "gooseneck" machine. The gooseneck machine melts the metal in a furnace which is part of the machine. The machine is cycled by opening a valve and forcing molten metal through the gooseneck into the dies by means of a plunger or high-pressure air. The gooseneck machine is the faster of the two types. However, it can be used only with metals that have relatively low melting points. If the melting point of the metal is too high, the hot metal will pick up iron

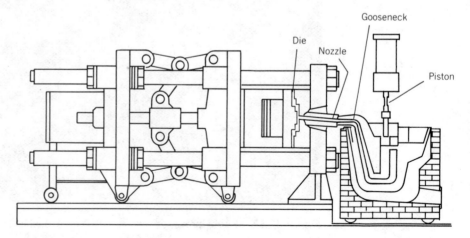

Figure 11.16   Hot chamber die casting machine

from the gooseneck and plunger, contaminating the metal and damaging the machine. Hot chamber machines are normally used with zinc- and tin-base alloys.

Figure 11.17 shows a cold chamber die casting machine. The metal is melted in a separate furnace and brought to the machine in a ladle. The molten metal is poured into the cold chamber and is forced into the dies by a piston. The cold chamber machine can be used with higher-melting-point metals such as bronze and brass.

Robot Applications in Die Casting.   There are many possible applications for robots in the die casting industry. All of these applications have the objective of removing a human worker from the hazardous environment of die casting as well as increasing productivity.

Some of the operations that may be performed by human die cast operators are die lubrication, removal of cast parts from the dies, inspection of parts, inspection of dies to ensure that the entire casting has been removed from the dies, quenching of castings, loading of castings into trimming dies, and loading of inserts into the dies. Although all of these operations can be automated using hard automation, the high cost and lack of flexibility when dies are changed to produce a new part make hard automation an uneconomical solution.

Die Lubrication.   Die lubrication is important because it allows the molten metal to flow evenly in the cavity of the die. Lubrication of dies also prevents sticking of the casting in the die when it cools. A

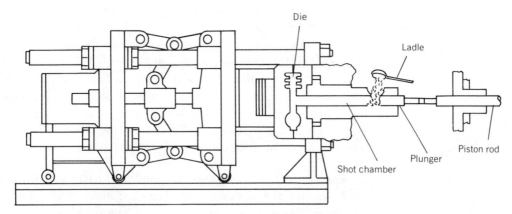

Figure 11.17   Cold chamber die casting machine

robot can be programmed to enter the opening between the dies and to spray lubricant into the die cavities.

Removal of Castings.   To facilitate removal of castings from the dies, **stripper pins** are built into the dies. When the dies open, the stripper pins are pushed out, forcing the part out of the die. The part drops out of the die into a tank where it is cooled (quenched). It is not uncommon for the die cast part to hang up in the die. An operator must then reach between the dies and remove the part. Reaching between the dies exposes the operator to the danger of the press accidentally closing. Also, while the operator is removing the part from the dies, the dies cool. This constant heating and cooling of the dies shortens die life.

A robot can be used to unload a die casting machine. Use of the robot ensures that the part is removed from the die at a constant rate, which keeps the dies at a constant temperature. The quality of the parts produced is improved, and die life is extended. Use of the robot also eliminates the need for an operator to reach between the dies to remove parts. Figure 11.18 shows a robot reaching into a die casting machine to remove a part.

Inspection.   After a part has been cast, it is important that the entire casting be removed from the die. When robots are used for unloading die casting machines, some form of inspection system must be installed to ensure that the entire casting has been removed before the machine is again cycled.

Inspection can be as simple as passing the removed part over a limit switch. Contact between the part and the limit switch indicates that the part has been removed. The limit switch can only sense contact between itself and one part of the casting, and it is possible that contact can be made between the casting and the limit switch while part of the casting remains in the die. An alternate method of inspecting for complete removal of the casting is the use of infrared sensors. The sensors are mounted in an array. If all the sensors have high outputs, the robot continues the program. Failure of any sensor to go high indicates that part of the casting has not been removed from the die. The robot halts and sounds an alarm.

Quenching.   Quenching (quick cooling) of castings, which changes the hardness of the metal, is often desirable. After removing the part from the dies and inspecting to ensure that the entire part

Figure 11.18 Robot removing a part from a die casting machine

has been removed, the robot can dip the part into a quench tank (Figure 11.19).

Trim Press Loading.   Often two or more parts are cast at one time in a die. This multiple casting requires that the cast parts be separated. Often the robot will load the die casting into a trimming press for separation of the individual parts.

Insert Loading.   Die castings are often produced with an insert cast into the part. For example, a pulley is to be cast from zinc and the center of the pulley hub requires a bronze bushing (Figure 11.20). The robot picks up the bronze bushing from an infeed system and inserts it onto steel pins in the die. When the dies are

Figure 11.19   Robot dipping a cast part into a quench tank

closed and the zinc is injected, the pulley is formed around the bronze bushing.

By using a custom designed gripper, the robot can remove a finished part from the die, and then roll its wrist and load an insert into the die before pulling its arm clear. While the next part is being injected and allowed to cool, the robot can quench the finished pulley, place it in a trim press, and pick up another insert for the next cycle.

Plastic Injection Molding.   Plastic injection molding is similar to the die casting process described in the previous section. The plastic used in injection molding is called a "thermoplastic." A thermoplastic becomes soft when first heated. If the thermoplastic is heated further it becomes fluid and can be injected into dies.

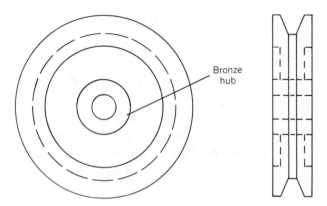

Figure 11.20   Pulley with bronze hub insert

The injection molding machine is similar to the die casting machine. Granules of plastic are loaded into a hopper and conveyed to a heated chamber where they are melted. The fluid plastic is injected into the closed dies under very high pressure. When the part has cooled in the dies, the dies open and stripper pins push the finished part out of the dies. Just as in die casting, the part can hang up in the dies, and an operator must reach between the dies to remove the part. A robot can be installed to remove the finished part from the molding machine, improving productivity and keeping operators from being at risk while reaching between the dies.

Having a robot remove the finished parts from the injection molding machine also has another advantage. When the robot removes the part, the orientation of the part is maintained. If the part does not need any secondary operations, the robot can package the finished part. If the part does require secondary operations, it can be placed on a conveyor which maintains the orientation of the part. The part can be carried to another automated operation without the necessity of reorienting the part, which can be difficult and costly.

Often a metal insert is placed in the dies and the part is cast around the insert. A robot can be used to place the insert in the dies. Using a robot for this application has two advantages. First, the robot, with its consistent speed, operates the injection molding machine at the highest possible rate. (If a human operator is responsible for placing the inserts in the dies, the operator may slow down toward the end of the shift, resulting in a loss in production.) The second advantage is that no human operator needs to be at risk while placing inserts into the dies. Figure 11.21 shows a robot unloading an injection molding machine.

Forging.   The forging process is similar to the press operations described earlier. The major difference between press work and forging is that forging is done while the metal is hot whereas press work is done with the metal at room temperature.

In the method of forging referred to as "press forging," bars of steel are heated in a furnace, and when the metal is red hot it is placed between two dies and formed into shape under very high pressure.

In another forging process, called "drop forging," one of the two dies is lifted above the other die. A heated bar of steel is placed in the stationary die and the movable die is "dropped," forming the part. To create a high force, the movable die is normally accelerated with steam or hydraulic pressure.

Attempts have been made to use a robot to move the heated bars of steel from the furnace to the forge. Most of these applications have failed, because the heat causes failure of seals in the robot and causes the robot's wrist and gripper to change shape due to expansion, thus reducing the robot's repeatability.

Figure 11.21   Robot unloading an injection molding machine

Since good solutions to these problems have not been found, the robot is normally used for secondary operations such as picking up the forged part and loading it into a trimming press where the excess steel, called "flash," is trimmed from the forged part.

Work is continuing on overcoming the problems of using robots in forging applications since skilled forge operators are retiring and young people are not choosing this occupation as a career.

Value Added Applications

The applications examined up to this point have primarily consisted of moving material or parts so that value added operations could be performed. The following examples are truly value added applications since the robot has a tool mounted as its end effector and is performing a job that increases the value of the part.

Arc Welding.   Arc welding is one of the fastest growing applications for robots. The robot is fitted with a welding torch, moves the welding torch into position, and welds the workpieces together.

Welding is a highly skilled craft. The skilled welder chooses the setting of the welding machine based on the type and thickness of the material to be welded. The welder also controls the angle of the welding torch, the amount of metal that is added, and the speed at which the torch is moved, in order to make a weld that is as strong as or stronger than the material being welded.

Often the workpieces to be welded do not fit together very tightly. When the workpieces do not fit properly and a gap is formed between them, the welder weaves the welding torch back and forth, filling in the gap. This requires both judgment and skill.

Because of the high degree of skill that it requires, arc welding is a difficult application for robots. However, the motivation to succeed in using robots for arc welding is high. An experienced arc welder welds for only two to three hours per shift. The welder spends the remainder of the shift adjusting parts to be welded, waiting for welds to cool before continuing, or simply taking a break to get away from the fumes and heat of the welding process. A robot, on the other hand, can weld for more than six hours per shift. This dramatic increase in productivity makes arc welding an excellent application for robots.

Successful robot arc welding requires excellent fit between workpieces, which in turn often requires redesign of the product and improved quality control standards in the shop. If excellent fit is maintained, the robot can be programmed to weld the work-pieces together.

Even with redesign of the product and improved quality control standards, it is still difficult to ensure sufficiently close fits. To overcome this problem, many different systems have been developed that allow the robot to sense the location of the seam to be welded and to modify its program so as to make a good weld.

One method is called "through the arc sensing." Through the arc sensing monitors the electric current being drawn while the robot is welding. The robot is programmed to weave (Figure 11.22). As the robot moves the welding torch over the gap in the seam, the current goes down. By monitoring the current, the robot can sense the width of the seam and adjust its forward speed to allow the gap to fill.

In another system, a camera is mounted on the robot's arm. The camera "looks" at the seam and adjusts the robot's program accordingly. A third system uses a laser beam to sense the weld seam.

All of these systems add to the cost of the arc welding operation, but the resulting increase in productivity can justify the added cost.

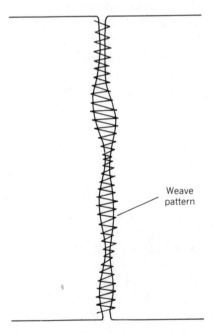

Figure 11.22 Typical weave pattern used to fill gaps in robot arc welding ("through the arc sensing" method)

Even with these systems the skilled welder is still needed. The skilled welder is responsible for programming the robot, adjusting the welding equipment, and inspecting the welds that the robot is making.

Figures 11.23 and 11.24 show robots arc welding.

Spot Welding.   Spot welding, also called resistance welding, differs from arc welding in that no metal is added during the welding process. Spot welding is used to weld together thin sheets of metal such as those used in automobile bodies.

Figure 11.25 shows a spot welding gun. The tips of the gun are called electrodes. The spot welding gun is positioned so that the sheet metal is between the two welding electrodes. When the gun is actuated, the electrodes force the sheets of metal against each other. With the sheet metal clamped between the electrodes, electric current is passed between the electrodes, heating the sheet metal to the melting point. The molten metal bonds the sheets of steel. The electric current is turned off, and the joint is allowed to cool so that the metal solidifies. When the joint is cool, the electrodes are opened and the gun is moved to the next spot to be welded.

Spot welding is the primary method of assembling auto bodies. Spot welding requires little skill. However, spot welding guns are

Figure 11.23   Robot arc welding

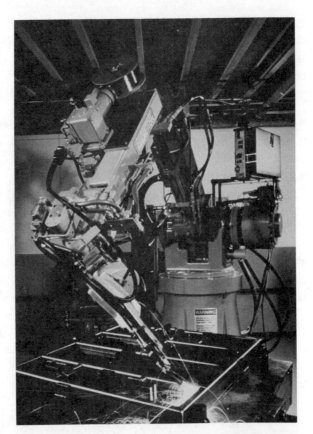

Figure 11.24   Another arc welding application

heavy, and the operator soon tires from wrestling the gun from point to point. The automobile industry recognized the advantages of programming a robot to make the hundreds of spot welds on a car, and today virtually all spot welds on American-made cars are made by robots. Since so many welds must be made to hold a car body together, spot welding lines in automotive plants have as many as 40 robots. However, not all of the robots on the line are needed to make the welds. If one or two robots should fail, the remaining robots are reprogrammed by a master computer and all of the welds are completed without the need to slow the assembly line.

Figure 11.26 shows robots making spot welds on an automobile assembly line.

Spray Painting.   Another value added job in which robots have proved themselves useful is spray painting. Spray painting has al-

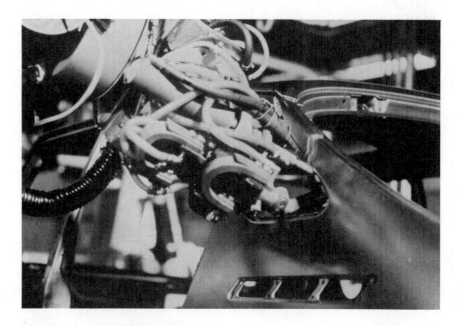

Figure 11.25 Spot welding gun

Figure 11.26 Robots making spot welds on an automobile assembly line

ways been a necessary but undesirable job in manufacturing. Often the spray booth is considered an entry level job in industry. The worker who is assigned to the spray booth attempts to be promoted into a more desirable job as quickly as possible. This constant changeover of personnel in the spray booth causes quality control problems.

In addition to the personnel problems, the spray booth is a dangerous place to work. The airborne paint particles, if inhaled, cause serious health problems. Some of the newer paints being used in industry are cancer-causing. For protection from this hazard, the painters dress in suits akin to the space suits worn by astronauts. Fresh air is pumped to the painter through a hose connected to the suit.

With all of the dangers of the paint booth, robots were quickly adopted for spray painting applications, but this is not to say that spray painting applications are easy. Spray painting presented difficult problems that had to be overcome. Two of these problems were maintaining the viscosity of the paint being sprayed and accurately locating the parts in front of the robot. Since the robot is blind, the part must always be in the same place at the same time. Parts to be sprayed are often carried on hooks suspended from an overhead conveyor. As the parts are moved into the spray booth, they swing on the hooks, and if they are to be consistently sprayed by the robot they must first be stabilized.

All of these problems have been overcome in many applications, and spray painting is now a significant application for robots in industry. Figure 11.27 shows a robot in a successful spray painting application on a truck assembly line.

Buffing, Grinding, and Drilling.   By mounting of the appropriate tools as robot end effectors, robots are now being used for grinding, buffing, and drilling operations, as well as for many other similar applications.

Figure 11.28 shows a small grinder mounted as the end effector of a robot. The robot removes burrs from castings.

Figure 11.29 shows a robot that has picked a stainless steel sink off an infeed conveyor and is passing the sink over a buffer to remove scratches before shipping.

Figure 11.30 shows a robot grinding parts. In this installation, the robot senses the pressure between the part and the grinding

Figure 11.27   Robot spray painting a truck on an assembly line

Figure 11.28   Robot used for grinding (burr removal)

Figure 11.29   Robot passing a stainless steel sink over a buffer

Figure 11.30   Robot grinding parts

wheel and maintains this pressure at an optimum level that is adequate for proper grinding but not high enough to damage the part or crack the grinding wheel.

Figure 11.31 shows a robot fitted with a drill. In this installation the robot drills hundreds of holes. Before this robot was installed, all the holes were drilled by hand.

Assembly.   Assembly of products in industry often requires a disproportionate amount of labor. In many industries, assembly accounts for 80% of all direct labor employed in producing the product. With labor costs increasing rapidly, assembly has become a major area of development for robot applications.

Assembly of products by robots presents significant problems, none the least of which is the problem of accurate fit between the components being assembled. Even the simplest assembly operation requires subtle judgments on the part of the assembler. In the process of snapping parts together, or of aligning parts and inserting screws to complete the assembly, an assembler must continually

Figure 11.31   Robot used for drilling

make precise adjustments in alignment and fit in order to produce acceptable assemblies. These adjustments require the use of sight, touch, and reason.

When the robot was introduced in Chapter 1, you were told that the robot is "blind." With such a handicap, successful use of robots in assembly applications seems impossible. However, with the development of visual systems to give the robot sight, as well as tactile sensing systems to give the robot touch, and artificial intelligence systems to give the robot reasoning ability, assembly by robots becomes possible. The addition of vision, touch, and reasoning ability increases the cost of the robot system and makes it questionable whether or not the resulting savings in labor costs outweigh the additional capital outlay.

An alternative approach to robot assembly is also possible. Remember that the products being assembled were designed for ease of assembly by humans and not by robots. If the product is completely redesigned for robot assembly, many of the problems of using a deaf, blind, and dumb machine can be eliminated. For example, many of today's plastic products are assembled by snapping one part onto another; these parts are held together with plastic hooks called "toy tabs." The alternative approach is to use no tabs at all but rather to place one part on top of the other and then join them by sonic welding. This approach has been used to a limited extent in industry and has been successful. As more and more products are redesigned for robot assembly, robots will become a significant factor in assembly applications.

One type of assembly operation to which robots have been successfully applied is electronic circuit board assembly. When thousands of circuit boards are to be assembled, a custom machine is set up to do the assembly work. Often, however, electronic circuit boards are assembled in batches of several hundred. In such instances, setup of a large custom assembly machine is not cost justifiable, and "stuffing" of the electronic components into the boards by hand is too expensive. The alternative is to use robots. The electronic components are of a consistent size, and the locations of the holes in the circuit boards are also consistent. A robot assembly line can be quickly programmed to stuff the components into the boards, eliminating the labor cost and the potential for error that existed when the boards were assembled by hand. Figure 11.32 shows an IBM robot assembling circuit boards.

Figure 11.32   Robot assembling a circuit board

THE FUTURE OF ROBOTS

In this section we have examined a few of the existing and potential applications of robots in industry. Robots offer reduced cost, more versatility in production, and improved worker safety. Since they offer effective answers to the difficult problems facing industry in today's competitive economy, robots will most certainly be employed to an increasing extent in manufacturing plants around the world, creating a need for well-trained application designers and service personnel to maintain these complex installations.

QUESTIONS FOR CHAPTER 11

1. What was John Parsons' contribution to the development of FMS?

2. Who were the founders of the second industrial revolution?

3. Describe the manufacturing methods used in the job shop, the flow shop, and the project shop.

4. Why is group technology important to FMS?

5. What is the primary advantage of robots over transfer mechanisms in a work cell.

6. Describe the difference between value added and non-value added operations.

7. What is the major problem associated with the use of robots in press loading?

8. What is the difference between cold chamber and hot chamber die casting?

9. List and explain the possible applications of robots in the die casting industry.

10. Describe the problems involved in using robots for arc welding.

11. What are the advantages of using robots for spray painting?

12. List additional value added operations that can be considered as robot applications.

GLOSSARY

Accuracy. 1. Quality, state, or degree of conformance to a recognized standard or specification. 2. Degree to which actual position corresponds to desired or commanded position.

Active accommodation. Integration of sensors, control, and robot motion to achieve alteration of a robot's preprogrammed motions in response to sensed forces. Used to stop a robot when forces reach set levels, or to perform force feedback tasks like insertions, door opening and edge tracing.

Actuator. A motor or transducer which converts electrical, hydraulic or pneumatic energy to effect motion of the robot.

Adaptable. Capable of making self-directed corrections. In a robot this is often accomplished with the aid of visual, force or tactile sensors.

Adaptive. (See *Adaptable*.)

Adaptive control. (See *Control, adaptive*.)

Address. A label or number specifying where a unit of information is stored in the computer's memory.

Advanced Data Communications Control Procedure (ADCCP). The proposed American National Standard for a bit oriented synchronous data communications protocol. Supports full or half duplex, point-to-point or multipoint communication. Employs a cyclic redundancy error check algorithm.

Air motor. A device that converts pneumatic pressure and flow into continuous rotary or reciprocating motion.

Algorithm. A prescribed set of well-defined rules or processes or mathematical equations for the solution of a problem in a finite number of steps.

Alphanumeric. Alphabetic, numeric and usually other characters such as punctuation marks.

Ambient temperature. A temperature within a given volume; e.g., room or building.

This glossary is a reprint of the terms and definitions included in the *RIA Robotics Glossary* published by Robotic Industries Association (RIA). Founded in 1974, RIA is the only trade association in North America organized specifically to serve the field of robotics. RIA plays an active role in establishing industry standards, reporting statistics, publishing and distributing robotics literature, and sponsoring trade-related meetings, workshops and expositions. For more information, contact Robotic Industries Association, P.O. Box 1366, Dearborn, Michigan, 48121, Telephone 313/271-7800.

Analog. An expression of values which can vary continuously; e.g., translation, rotation, voltage, or resistance. (Contrasted with *Digital*.)

Analog communications. Transmission of information in the form of a continuously varying physical quantity.

Analog control. Control involving analog signal processing devices (electronic, hydraulic, pneumatic, etc.).

Android. A robot which resembles a human in physical appearance.

ANSI. Abbreviation for American National Standards Institute.

Anthropomorphic. An adjective with the literal meaning "of human shape". An anthropomorphic robot is one which performs tasks with motions similar to a human.

Anthropomorphic robot. A robot with all rotary joints and motions similar to a human's arm. (Also called *jointed-arm robot*.)

Architecture. Physical and logical structure of a computer or manufacturing process.

Arm. An interconnected set of links and powered joints comprising a manipulator which supports or moves a wrist and hand or end-effector.

Artificial intelligence. The capability of a machine to perform human-like intelligence functions such as learning, adapting, reasoning and self correction. (See *Control, adaptive*.)

ASCII. Abbreviation for American Standard Code for Information Interchange. It is an eight-bit (7 bits plus a parity bit) code for representing alphanumerics, punctuation marks, and certain special characters for control purposes.

Asynchronous system. A system whose processes occur as needed by input data instead of a system timing control.

Automatic operation. That time when the robot is performing its programmed tasks through continuous program execution. (Supersedes *Normal operation*.)

Automation. Automatically controlled operation of an apparatus, process or system by mechanical or electronic devices that take the place of human observation, effort and decision.

Axis. A traveled path in space, usually referred to as a linear direction of travel in any three dimensions. Labels of "X" and "Y" and "Z" are commonly used to depict directions relative to the Earth. "X" refers to a directional plane or line parallel to the Earth. "Y" refers to a directional plane or line that is parallel to Earth and perpendicular to "X"; "Z" refers to a directional plane or line that is vertical to and perpendicular to the Earth surface.

Azimuth. Direction of a straight line to a point in a horizontal plane, expressed as the angular distance from a reference line, such as the observer's line of view.

Backlash. Free play in a power transmission system, such as a gear train, resulting in a characteristic form of hysteresis.

Bang-bang control. A binary control system which rapidly changes from one mode or state to the other (in motion systems, this applies to direction only).

Bang-bang robot. (See *Pick-and-place robot*.)

Barriers. A physical means of separating persons from the robot restricted work envelope.

Base. The platform or structure to which the shoulder of a robot arm is attached; the end of a kinematic chain of arm links and joints opposite to that which grasps or processes external objects.

Batch manufacturing. The production of parts or material in discrete runs, or batches, interspersed with other production operations or runs of other parts or materials.

Baud. A unit of signalling speed equal to the number of discrete conditions of signal events per second.

Bend. (See *Pitch.*)

Bilateral manipulator. A master-slave manipulator with symmetric force reflection where both master and slave arms have sensors and actuators such that in any degree of freedom a positional error between the master and slave results in equal and opposing forces applied to the master and the slave arms.

Binary. The basis for calculations in all digital computers. This two-digit numbering system consists of the digits 0 and 1, in contrast to the ten-digit decimal system.

Binary picture. A video picture with only two shades of brightness (white and black).

Bit rate. The rate at which binary digits, or pulses representing them, pass a given point in a communication line.

Block diagram. A simplified schematic drawing.

Branching. Changing of the normal sequential execution of statements in a program.

Breakaway force. Same as *Static friction,* though this term implies more strongly that the resistive force is not constant as the relative velocity increases.

Buffer. Memory area in a computer or peripheral used for temporary storage of information that has just been received. The information is held in the buffer until the computer or device is ready to process it.

Byte. A sequence of binary digits usually operated upon as a unit. (The exact number depends on the system.)

Cabinet. An enclosure for mounting equipment.

Cable. 1. A stranded conductor (single-conductor cable) or a combination of conductors insulated from one another (multipleconductor cable). 2. A group of twisted wires, usually steel, used to support or move a load.

Cable drive. Transmission of power from an actuator to a remote mechanism by means of a flexible cable and pulleys.

CAD/CAM. An acronym for Computer Aided Design and Computer Aided Manufacturing.

Calibration. Determination of the deviation from a standard so as to ascertain the proper corrections.

Cam. 1. A device with one or more lobes (projections) which, as it moves, operates levers or switches that cause mechanical or electrical functions. 2. An acronym for Computer Aided Manufacturing (all capitals).

Cartesian coordinate. All robot motions travel in right angle lines to each other. There are no radial motions. The profile of its envelope represents a rectangular shape.

Cartesian coordinate robot. A robot whose manipulator arm degrees of freedom are defined primarily by cartesian coordinates.

Cartesian coordinate system. A coordinate system whose axes or dimensions are three intersecting perpendicular straight lines and whose origin is the intersection. (Also described as *rectilinear.*)

Cassette recorder. A peripheral device for transferring information between the computer or robot memory and magnetic tape. In the record mode it is used to make a permanent record of a program existing in the CPU memory. In the playback mode it is used to enter a previously recorded program into the CPU memory.

Cassette tape. A magnetic recording tape permanently enclosed in a protective housing.

CCD camera. A solid state television camera which uses charge coupled device (CCD) technology.

Cell. A manufacturing unit consisting of two or more work stations or machines and the materials transport mechanisms and storage buffers which interconnect them.

Center. A manufacturing unit consisting of two or more cells and the materials transport and storage buffers which interconnect them.

Center of gravity. That point in a rigid body at which the entire mass of the body could be concentrated and produce the same gravity resultant as that for the body itself.

Central processing unit (CPU). 1. Another term for *Processor*. It includes the circuits controlling the interpretation and execution of the user-inserted program instructions stored in the computer or robot memory. 2. The hardware part (CPU) of a computer which directs the sequence of operations, interprets the coded instructions, performs arithmetic and logical operations, and initiates the proper commands to the computer circuits for execution. The arithmetic and logic unit and the control unit of a digital computer. Controls the computer operation as directed by the program it is executing.

Chain drive. Transmission of power from an actuator to a remote mechanism by means of a flexible chain and mating toothed sprocket wheels.

Character. One symbol of a set of elementary symbols, such as a letter of the alphabet or a decimal numeral.

Chip. A small piece of silicon impregnated with impurities in a pattern to form

transistors, diodes, and resistors. Electrical paths are formed on it by depositing thin layers of aluminum or gold.

CID camera. A solid state television camera which uses charge injection device (CID) technology.

Circular interpolation. A function automatically performed in the control of defining the continuum of points in a radius based on a minimum of three taught coordinate positions.

Closed loop. A method of control in which feedback is used to link a controlled process back to the original command signal.

Closed loop control. Control achieved by feedback, i.e., by measuring the degree to which actual system response conforms to desired system response and utilizing the difference to drive the system into conformance.

Compensation. Logical operations employed in a control scheme to counteract dynamic lags or otherwise to modify the transformation between measured signals and controller output to produce prompt stable response.

Compliance. 1. The quality or state of bending or deforming to stresses within the elastic limit. 2. The amount of displacement per unit of applied force.

Computed path control. A control scheme wherein the path of the manipulator endpoint is computed to achieve a desired result in conformance to a given criterion, such as an acceleration limit, a minimum time, etc.

Computer aided design (CAD). The use of a computer to assist in the creation or modification of a design.

Computer aided manufacture (CAM). The use of computer technology in the management, control, and operation of manufacturing.

Computer control. Control involving one or more electronic digital computers.

Computer numerical control (CNC). The use of a dedicated mini or microcomputer to implement the numerical control function. Uses local data input from devices such as paper tape, magnetic tape cassette or floppy disk.

Conductor. A wire, or strands of wire not insulated from one another, suitable for carrying an electric current.

Conduit. Solid or flexible metal or other tubing through which insulated electric wires are run.

Contact sensor. A device capable of sensing mechanical contact.

Continuous path control. A control scheme whereby the inputs or commands specify every point along a desired path of motion.

Control. The process of making a variable or system of variables conform to what is desired. **1.** A device to achieve such conformance automatically. **2.** A device by which a person may communicate commands to a machine.

Control, adaptive. A control algorithm or technique where the controller changes its control parameters and performance characteristics in response to its environment and experience.

Control enclosure. (See *Enclosure.*)

Control hierarchy. A relationship of sensory processing elements whereby the results of lower level elements are utilized as inputs by higher level elements.

Control system. Sensors, manual input and mode selection elements, interlocking and decision-making circuitry, and output elements to the operating mechanism.

Controller. 1. An information processing device whose inputs are both desired and measured position, velocity or other pertinent variables in a process and whose outputs are drive signals to a controlling motor or actuator. **2.** A communication device through which a person introduces commands to a control system. **3.** A person who does the same.

Coordinated axis control. 1. Control wherein the axes of the robot arrive at their respective endpoints simultaneously, giving a smooth appearance to the motion. **2.** Control wherein the motions of the axes are such that the endpoint moves along a prespecified type of path (line, circle, etc.). Also called *End point control*.

CP. Acronym for Continuous Path.

CPU. (See *Central processing unit.*)

CRT terminal. A terminal containing a cathode ray tube to display program data.

Cumulative lost time. (See *Downtime.*)

Cursor. A means for indicating on a CRT screen the point at which data entry or editing will occur. The intensified element may be at constant high intensity or flashing (alternate high intensity and normal intensity).

Cycle. A sequence of operations that is repeated regularly.

Cycle time. The period of time from starting one machine operation to starting another (in a pattern of continuous repetition).

Cylindrical coordinate robot. A robot whose manipulator arm degrees of freedom are defined primarily by cylindrical coordinates.

Cylindrical coordinate system. A coordinate system which defines the position of any point in terms of an angular dimension, a radial dimension, and a height from a reference plane. These three dimensions specify a point on a cylinder.

Damping. The absorption of energy from a moving body for the purpose of controlling oscillatory vibrations. This term has an electrical analog.

Danger. Situation in which there is a reasonable foreseeable risk of injury from mechanical or electrical hazards.

Data. A general term for any type of information.

Database. A large collection of records stored on a computer system from which specialized data may be extracted, organized, and manipulated by a program. Any organized and structured collection of data in memory.

Dead zone. A range within which a non-zero input causes no output.

Debug. (See *Debugging*.)

Debugging. Process of detecting, locating, and correcting mistakes in hardware or software (program). Synonymous with trouble-shooting.

Degree of freedom. One of a limited number of ways in which a point or a body may move or in which a dynamic system may change, each way being expressed by an independent variable and all required to be specified if the physical state of the body or system is to be completely defined.

Derivative control. Control scheme whereby the actuator drive signal is proportional to the time derivative of the difference between the input (desired output) and the measured actual output.

Diagnostic program. A user-inserted test program to help isolate hardware malfunctions in the computer or robot and the application equipment.

Digital control. Control involving digital logic devices which may or may not be complete digital computers.

Disk. A secondary memory device in which information is stored on one or both sides of a magnetically sensitive, rotating disk. The disk is rotated by a disk drive and information is retrieved/stored by means of one or more read/write heads mounted on movable or fixed arms. Disks are rigid (hard) or flexible (floppy). Disk memory has a much faster access time than magnetic tape but is slower than semiconductor memory.

Disk, dual floppy. Two floppy disk drives packaged together.

Disk, fixed head. A disk memory unit on which a separate read/write head is provided for each track of each disk surface. A typical disk has about 800 tracks per side. Access time on a fixed head disk is typically about 10 milliseconds and much faster than access time for a moving head disk which is typically 25 to 50 milliseconds.

Disk, floppy. A circular, flexible material with a magnetic film on both sides. Digital information is stored on one or both sides of the floppy disks in concentric circles or tracks. The term "floppy" is used because the disk is soft and bends easily. The disk is usually contained in a rigid, protective envelope.

Disk, hard. Disk memory that uses rigid disks rather than flexible disks as the storage medium. Hard-disk devices can generally store more information and access it faster. Cost considerations, however, currently restrict their usage to medium and large-scale applications.

Disk memory unit. A secondary memory device often used in real-time control systems. A disk memory unit consists of one or more disks stacked on top of one another. The disks are rotated continuously at a uniform speed. Information is stored on the magnetic disks in concentric tracks. A typical disk surface can record between 1MM and 100MM bytes of information.

Disk storage. A memory system which uses a revolving magnetic disk to store information. (See *Disk memory unit*.)

Disk, Winchester. A hard, fixed head disk memory unit which is hermetically sealed to improve reliability and increase its useful life. A separate device is required for loading and dumping software from the disk because it is not removable.

Diskette. A floppy disk. (See *Disk, floppy.*)

Distal. Away from the base, toward the end-effector of the arm.

Distributed control. A control technique whereby portions of a single control process are located in two or more places.

Documentation. An orderly collection of recorded hardware and software data such as tables, listing, diagrams, etc., to provide reference information for any application operation and maintenance.

Downtime. The time when a system is not available for production.

Drift. The tendency of a system's response to gradually move away from the desired response.

Drive power. The source or means of supplying energy to the robot actuators to produce motion.

Droop. Same as *Static deflection*.

Duty cycle. The fraction of time during which a device or system will be active, or at full power.

Dynamic accuracy. 1. Degree of conformance to the true value when relevant variables are changing with time. 2. Degree to which actual motion corresponds to desired or commanded motion.

Dynamic range. The extent of any variable of a system.

EAROM. Electrically Alterable ROM. A type of memory that combines the characteristics of RAM and ROM. It is nonvolatile (like ROM) but can be written into by the processor (like RAM). The EAROM, however, has a substantially longer writing time (currently about 2 microseconds vs. 400 nano-

seconds) as well as a limited number of writes (about 1,000,000) before the chip can no longer be reprogrammed.

EBCDIC. Extended Binary Coded Decimal Interchange Code. A specific code using eight bits to represent a character. (Contrast with *ASCII*.)

Echo check. A method of checking the accuracy of transmission of data in which the received data is returned to the sending end for comparison with the original data.

Edit. To modify the form or format of data, e.g., to insert or delete characters.

EIA. Electronics Industries Association is an organization which publishes various recommended standards.

Elbow. The joint which connects the upperarm and forearm.

Electrical-optical isolator. A device which couples input to output using a light source and detector in the same package. It is used to provide electrical isolation between input circuitry and output circuitry.

Elevation. Direction of a straight line to a point in a vertical plane, expressed as the angular distance from a reference line, such as the observer's line of view.

Emergency stop. A method using hardware-based components that overrides all other robot controls and removes drive power from the robot actuators and brings all moving parts to a stop.

Empirical. Based on experience rather than deductive logic.

Enclosure. A surrounding case designed to provide a degree of protection for equipment against a specified environment and to protect personnel against accidental contact with the enclosed equipment.

Encode. To put into code.

Encoder. 1. A rotary feedback device which transmits a specific code for each position. **2.** A device which transmits a fixed amount of pulses for each revolution.

Encoder accuracy. The maximum positional difference between the input to an encoder and the position indicated by its output; includes both deviation from theoretical code transition positions and quantizing uncertainty caused by converting from a scale having an infinite number of points to a digital representation containing a finite number of points. (See *Resolution.*)

Encoder ambiguity. Inherent error caused by multiple bit changes at code transition positions, which is eliminated by various scanning techniques. Used in relation to encoders.

End-effector. An actuator, gripper, or mechanical device attached to the wrist of a manipulator by which objects can be grasped or otherwise acted upon.

End of axis control. Controlling the delivery of tooling through a path or to a point by driving each axis of a robot in sequence. The joints arrive at their pre-programmed positions in a given axis before the next joint sequence is actuated.

End point control. Any control scheme in which only the motion of the manipulator end point may be controlled and the computer can control the actuators at the various degrees of freedom to achieve the desired result.

End point rigidity. The resistance of the hand, tool or end point of a manipulator arm to motion under applied force.

Envelope. (See *Work envelope.*)

EPROM. Electrically Programmable ROM. A nonvolatile semiconductor memory that can be erased and reprogrammed. Programming is done off-line with an EPROM or PROM burner. Erasure is accomplished by exposure to ultraviolet light.

Error control procedure. A method of detecting and recovering from errors occurring in transmitted data. Typically employs methods such as parity, Cyclic Redundancy checks and Frame Sequence Numbering to detect errors, plus a Request for Retransmission to recover.

Error signal. The difference between desired response and actual response.

Even parity. The condition that occurs when the sum of the number of "1's" in a binary word is always even.

Execute. To perform a computer instruction or run a program.

Execution. The performance of a specific operation such as would be accomplished through processing one instruction, a series of instructions, or a complete program.

Execution time. The total time required for the execution of one specific operation. (See *Execution.*)

Exoskeleton. An articulated mechanism whose joints correspond to those of a human arm and, when attached to the arm of a human operator, will move in correspondence to his/her arm. Exoskeletal devices are sometimes instrumented and used for master-slave control of manipulators.

Extension. A linear motion in the direction of travel of the sliding motion mechanism, or an equivalent linear motion produced by two or more angular displacements of a linkage mechanism.

External sensor. A feedback device that is outside the inherent makeup of a robot system or a device used to effect the actions of a robot system that are used to source a signal independent of the robot's internal design.

Factory. A manufacturing unit consisting of two or more centers and the materials transport, storage buffers, and communications which interconnect them.

Fail safe. Failure of a device without danger to personnel or damage to product or plant facilities.

Fail soft. Failure in performance of some component part of a system without immediate major interruption or failure of performance of the system as a whole and/or sacrifice in quality of the product.

Fault. Any malfunction which interferes with normal application operation.

Feedback. The signal or data sent to the control system from a controlled machine or process to denote its response to the command signal.

Feeding. The process of placing or removing material within or from the point of operation.

Fettling. The removal (automatic or manual) of flash, sprues, parting lines, etc. from castings by grinding, milling, nibbling, etc.

Fiber optics. A communication technique where information is transmitted in the form of light over a transparent material (fiber) such as a strand of glass. Advantages are noise free communication not susceptible to electromagnetic interference.

Fixed coordinate system. A coordinate system fixed in time and space.

Fixed stop robot. A robot with stop point control but no trajectory control. That is, each of its axes has a fixed limit at each end of its stroke and cannot stop except at one or the other of these limits. Such a robot with N degrees of freedom can therefore stop at no more than 2N locations (where location includes position and orientation). Some controllers do offer the capability of program selection of one of several mechanical stops to be used. Often very good repeatability can be obtained with a fixed stop robot. Also called *nonservo robot*.

Flexible. 1. Pliable or capable of bending. In robot mechanisms this may be due to

joints, links, or transmission elements. Flexibility allows the endpoint of the robot to sag or deflect under load and to vibrate as a result of acceleration or deceleration. 2. Multipurpose; adaptable; capable of being redirected, trained or used for new purposes. Refers to the reprogrammability or multi-task capability of robots.

Flexible manufacturing. Production with machines capable of making a different product without retooling or any similar changeover. Flexible manufacturing is usually carried out with numerically controlled machine tools, robots, and conveyors under the control of a central computer.

Flip-flop. A circuit that changes its logical state when signaled to do so by another device.

Floor-to-floor time. The total time elapsed for picking up a part, loading it into a machine, carrying out operations and unloading it (back to the floor, or bin or pallet, etc.). Generally applies to batch production.

Forearm. That portion of a jointed arm which is connected to the wrist.

Frequency. The number of times a given event occurs within a specified period. It most commonly refers to the number of pulses per second occurring in various electronics devices. The standard unit of measure is Hz (Hertz), for cycles per second.

Frequency response. 1. The output of a system with a periodic input. Frequency response may be defined in terms of the fourier coefficients or the gain and phase at each multiple of the period. 2. The characterization of system output to a continuous spectral input according to a continuous plot of gain and phase as a function of frequency.

Friction. The resistive forces resulting from two bodies sliding relative to one another.

Gantry. A bridge-like frame along which a suspended robot moves. A gantry creates a much larger work envelope than the robot would have if it were pedestal mounted. (See *Work envelope*.)

GPM. Acronym for gallons per minute.

Graceful degradation. Decline in performance of some component part of a system without immediate and significant decline in performance of the system as a whole and/or decline in the quality of the product.

Graphics. The use of diagrams or other graphic means to present operating data, curve plots, answers to inquiries and other computer output.

Graphics terminal. A CRT terminal capable of displaying user-programmed graphics.

Grey scale picture. A video picture with many shades of brightness.

Gripper. (See *End-effector*.)

Ground. A conducting connection, intentional or accidental, between an electric circuit or equipment chassis and the earth ground.

Ground potential. Zero voltage potential with respect to earth ground.

Group technology. 1. A system for coding parts based on similarities in geometrical shape or other characteristics of the parts. 2. The grouping of parts into families based on similarities in their production so that the parts of a particular family could then be processed together.

Guard. A physical means of separating persons from danger.

Guard-distance. The guard may take the form of a fixed barrier or fence designed to a height so as to prevent normal access to a danger area, or a fixed tunnel which prevents access to a danger point by reason of the relationship of the opening dimensions of the guard to the length of tunnel.

Guard-distance/perimeter. A fixed barrier or fence designed to prevent normal access to the danger or prohibited area.

Guard-fixed. A barrier not readily removable to prevent entry of personnel into potentially dangerous areas.

Hand. (See *End-effector*.)

Hard copy. Any form of printed document such as ladder diagram program listing, paper tape, or punched cards.

Hardware. The mechanical, electrical, electronic, pneumatic, and hydraulic devices which compose a computer, controller, robot or panels.

Hardwired. Electrical devices interconnected through physical wiring.

Hazard. A condition or changing set of circumstances that presents a potential for injury, illness, or property damage. The potential or inherent characteristics of an activity, condition, or circumstance which can produce adverse or harmful consequences.

Hertz. (Abbreviated *Hz*.) Cycles per second.

Hierarchical control. A distributed control technique in which the controlling processes are arranged in a hierarchy. (See *Hierarchy*.)

Hierarchy. A relationship of elements in a structure divided into levels, with those at higher levels having priority or precedence over those at lower levels. (See *Control hierarchy* and *Sensory hierarchy*.)

High-level language. Programming language that generates machine codes from problem or function oriented statements. ALGOL, FORTRAN, PASCAL and BASIC are four commonly used high-level languages. A single functional statement may translate into a series of instructions or subroutines in machine language, in contrast to a low-level (assembly) language in which statements translate on a one-for-one basis.

Hold. A stopping of all movement of the robot during its sequence in which some power is maintained on the robot, e.g., on hydraulically-driven robots power is shut off to the servo valves but is present on the main electrical and hydraulic systems.

Host computer. A computer which monitors and controls other computers.

Housing. (See *Enclosure.*)

Hydraulic motor. An actuator consisting of interconnected valves and pistons or vanes which converts high pressure hydraulic or pneumatic fluid into mechanical shaft translation or rotation.

Hysteresis. The lag in an instrument's or process response when a force acting on it is abruptly changed.

ICAM. Acronym for Integrated Computer Aided Manufacturing.

IEEE. Acronym for Institute of Electrical and Electronics Engineers.

Impact resistance. Resistance to fracture under shock force.

Induction motor. An alternating current motor in which torque is produced by the reaction between a varying or rotating magnetic field that is generated in stationary field magnets and the current that is induced in the coils or circuits of the rotor.

Industrial robot. A reprogrammable, multifunctional manipulator designed to move material, parts, tools, or specialized devices through variable programmed motions for the performance of a variety of tasks. (RIA)

Inertia. 1. The tendency of a mass at rest to remain at rest, and of a mass in motion to remain in motion. 2. The Newtonian property of a physical mass that a force is required to change the velocity, proportional to the mass and the time rate of change of velocity.

Input devices. Devices such as limit switches, pressure switches, push buttons, etc., that supply data to a robot controller. These discrete inputs are of two types: those with common return, and those with individual returns (referred to as *isolated inputs*). Other inputs include analog devices and digital encoders.

Instruction set. The list of machine language instructions which a computer can perform.

Integral control. Control scheme whereby the signal which drives the actuator equals the time integral of the error signal.

Integrated circuit. A complex electronic circuit fabricated on a single chip of semiconductor material such as silicon.

Intelligent robot. A robot which can be programmed to make performance choices contingent on sensory inputs.

Interface. A shared boundary. An interface might be a mechanical or electrical connection between two devices; it might be a portion of computer storage accessed by two or more programs; or it might be a device for communication to or from a human operator.

Interference. Any undesired electrical signal induced into a conductor by electrostatic or electromagnetic means.

Interlock. To arrange the control of machines or devices so that their operation is interdependent in order to assure their proper coordination.

Interrupt. To stop the normal processing of a computer or robot controller for the purpose of input/output of data, or for making an inquiry or receiving a reply, or carrying out interactive processes or procedures.

I/O. Abbreviation for input/output.

I/O electrical isolation. Separation of the field wiring circuits from the logic level circuits of the control system. This is typically achieved using electrical-optical isolators mounted in the I/O module.

Job shop. A discrete parts manufacturing facility characterized by a mix of products of relatively low volume production in batch lots.

Joint. A rotational or translational degree of freedom in a manipulator system.

Joint space. The space defined by a vector whose components are the angular or translational displacement of each joint of a multi-degree-of-freedom linkage relative to a reference displacement for each such joint.

Joystick. (Also called *syntaxer*). A movable handle which a human operator may grasp and rotate to a limited extent in one or more degrees of freedom, and whose variable position or applied force is measured and results in commands to a control system.

Jumper. A short length of conductor used to make a connection between terminals, around a break in a circuit, or around an instrument.

K. Abbreviation for 2 to the 10th power = 1024. This abbreviation is used to denote sizes of memory, e.g., 2K = 2048.

Lag. 1. The tendency of the dynamic response of a physical system to be delayed. 2. The phase difference between input and response sinusoids. 3. Any time parameter which characterizes the delay of a response relative to an input.

Language. A set of symbols and rules for representing and communicating information (data) among people, or between people and machines.

Latching relay. A relay constructed so that it maintains a given position by mechani-cal means until released mechanically or electrically.

Lead screw. A precision machine screw which, when turned, drives a sliding nut or mating part in translation.

Lead through. Programming or teaching by physically guiding the robot through the desired actions. The speed of the robot is increased when programming is complete.

Learning control. A control scheme whereby experience is automatically used to change control parameters or algorithms.

LED display. An illuminated visual read-out composed of light emitting diode (LED) alphanumeric character segments.

Level of automation. The degree to which a process has been made automatic. Relevant to the level of automation are questions of automatic failure recovery, the variety of situations which will be automatically handled, and the situation under which manual intervention or action by humans is required.

Limit detecting hardware. A device for stopping robot motion independently from control logic.

Limit switch. A switch which is actuated by some part or motion of a machine or equipment to alter the electrical circuit associated with it.

Limited-degree-of-freedom robot. A robot able to position and orient its end effector in fewer than six degrees of freedom.

Limiting devices. To qualify as a means for restricting the work envelope, these devices must stop all motion of the robot independent of control logic.

Linear. The quality of an input-output relationship in which there is direct proportionality.

Linear array camera. A solid state television camera which has only one row of photosensitive elements.

Linear interpolation. A function automatically performed in the control that defines the continuum of points in a straight line based on only two taught coordinate positions. All calculated points are automatically inserted between the taught coordinate positions upon playback.

Linearity. The degree to which input-output is a directly proportional relationship.

Liquid crystal display (LCD). A reflective visual readout of alphanumeric characters. Since its segments are displayed only by reflected light, it has extremely low power consumption — as contrasted with an *LED display,* which emits light.

Listing, program. A printed list of computer instructions, usually prepared by an assembler or compiler.

Load. 1. The power delivered to a machine or apparatus. **2.** The weight (force) applied to the end of the robot arm. **3.** A device intentionally placed in a circuit or connected to a machine or apparatus to absorb power and convert it into the desired useful form. **4.** To insert data into memory storage.

Load capacity. The maximum total weight that can be applied to the end of the robot arm without sacrifice of any of the applicable published specifications of the robot.

Load deflection. 1. The difference in position of some point in a body between a non-loaded and an externally loaded condition. **2.** The difference in position of a manipulator hand or tool, usually with the arm extended, between a non-loaded condition (other than gravity) and an externally loaded condition. Either or both static and dynamic (inertial) loads may be considered.

Logic. A means of solving complex problems through the repeated use of simple functions which define basic concepts. Three basic logic functions are AND, OR, and NOT.

Logic diagram. A drawing which represents the logic functions AND, OR, NOT, etc.

Logic level. The voltage magnitude associated with signal pulses representing ones and zeros ("1" and "0") in binary computation.

Long-term repeatability. (See *Repeatability.*)

Loop. A sequence of instructions that is executed repeatedly until some specified condition is met.

LSI. Large Scale Integration. The process of integrating 1000 to 100,000 (not well defined) circuits or equivalent components on a single chip of semiconductor material. Up to 2000 logic gates or up to 16,000 bits on a chip. Microprocessors employ LSI technology.

LSI chip. An integrated circuit with LSI circuit/component density.

Magnetic core memory. A configuration of magnetic beads strung on current carrying conductors which retain magnetic polarization for the purpose of storing and retrieving data.

Magnetic tape. Tape made of plastic and coated with magnetic material; used to store information.

Maintenance. The act of keeping the robot system in its proper operating condition.

Major axes (motions). These axes may be described as the number of independent directions the arm can move the attached wrist and end-effector relative to a point of origin of the manipulator such as the base. The number of robot arm axes required to reach world coordinate points is dependent on the design of robot arm configuration.

Malfunction. Any incorrect functioning within electronic, electrical, or mechanical hardware. (See *Fault.*)

Manipulation. 1. The process of controlling and monitoring data table bits or words by means of the user's program in order to vary application functions. **2.** The movement or reorientation of objects, such as parts or tools.

Manipulator. A mechanism, usually consisting of a series of segments, jointed or sliding relative to one another, for the purpose of grasping and moving objects usually in several degrees of freedom. It may be remotely controlled by a computer or PC, or by a human.

Manual manipulator. A manipulator worked (controlled) by a human operator.

Mass production. The large scale production of parts or material in a continuous process uninterrupted by the production of other parts or material.

Master control relay. A **mandatory** hardwired relay which can be de-energized by any hardwired series-connected emergency stop switch. Whenever the master control relay is de-energized, its contacts **must** open to de-energize **all** application I/O devices and power source.

Master-slave manipulator. A class of teleoperator having geometrically isomorphic "master and slave" arms. The master is held and positioned by a person; the slave duplicates the motions, sometimes with a change of scale in displacement or force.

Matrix. 1. A two-dimensional array of circuit elements, such as wires, diodes, etc., which can transform a digital code from one type to another. **2.** A group of numbers organized on a rectangular grid and treated as a unit. The numbers can be referenced by their position in the grid.

Matrix array camera. A solid state television camera which has a rectangular array of photosensitive elements.

Maximum operating temperature. The maximum temperature at which any point within an application system may be safely maintained during continuous use.

Maximum speed. 1. The greatest rate at which an operation can be accomplished according to some criterion of satisfaction.

2. The greatest velocity of movement of a tool or end-effector which can be achieved in producing a satisfactory result.

Mean-time-between-failures (MTBF). The average time that a device will operate before failure.

Mean-time-to-repair (MTTR). The average time needed to repair or service a device after failure.

Memory. A device into which data can be entered, in which it can be stored, and from which it can be retrieved at a later time.

Memory module. A processor module consisting of memory storage and capable of storing a finite number of words.

Message. A meaningful combination of alphanumeric characters which establishes the content and format of a report.

Metal-oxide semiconductor (MOS). A semiconductor manufacturing technology used to produce large scale integrated (LSI) circuit logic components.

Microcomputer. (See *Microprocessor.*)

Microprocessor. An electronic computer processor section implemented in relatively few IC chips (typically LSI) which contain arithmetic, logic, register, control, and memory functions. The microprocessor is characterized by having instructions which reference micro operations. Functional equivalence of minicomputer instructions accomplished by programming series of micro instructions.

Military-type connector. A round threaded receptacle containing female contacts which mates with a threaded cable plug containing male contacts. It protects the connections against physical damage and contamination by dust, oil, or water. When not in use, the receptacle and plug may be covered with threaded watertight dust caps.

Minicomputer. Larger in size than a microcomputer and having a typical word length of

16 or 32 bits. Addressable main memory is typically from 64K to over 1M. Compared with a microcomputer, the minicomputer is typically characterized by higher performance, a richer instruction set, higher price, a variety of high level languages, several operating systems and networking software.

Minor axes (motions). The axes may be described as the number of independent attitudes the wrist can orient the attached end-effector. Relative to the mounting point of the wrist assembly on the arm.

Mobile robot. A robot mounted on a movable platform.

Mode. A selected method of operation (e.g., RUN, TEST, or PROGRAM).

Modem. Acronym for MOdulator/DEModulator. A device used to transmit and receive data by frequency-shift-keying (FSK). It converts FSK tones into their digital equivalent and vice versa.

Modern control. A general term used to encompass both the description of systems in terms of state variables and canonical state equations, and the ideas of optimal control.

Modular. Made up of subunits which can be combined in various ways. **1.** In robots, a robot constructed from a number of interchangeable subunits each of which can be one of a range of sizes or have one of several possible motion styles (rectangular, cylindrical, etc.) and number of axes. **2.** "Modular design" permits assembly of products, or software or hardware, from standardized components.

Module. **1.** An interchangeable "plug-in" item containing electronic components which may be combined with other interchangeable items to form a complete unit. **2.** A mechanical component having a single degree of freedom which can be combined with other components to form a multi-axis manipulator or robot.

Monitor. **1.** CRT display package. **2.** To observe an operation.

Motion hold. A means of externally interrupting continuance of motion of the robot from any further sequence or action steps without dissipating stored energy. (See *Hold.*)

Motor controller. A device or group of devices which serves to govern, in a predetermined manner, the electrical power delivered to a motor.

Motor starter. (See *Motor controller.*)

Multimeter. A portable test instrument which can be used to measure voltage, current, and resistance.

Multiplex. To interleave or simultaneously transmit two or more messages on a single channel.

Multiplexing. The time-shared scanning of a number of data lines into a single channel. Only one data line is enabled at any instant.

Multiprocessing. The employment of multiple interconnected CPU's to execute two or more different programs simultaneously.

Multiprocessor control. A control scheme which employs more than one central processing unit in simultaneous parallel computation.

Nanosecond. One billionth of a second. Abbreviated *ns*.

National Electrical Code. A consensus standard for the construction and installation of electrical wiring and apparatus, established by the National Fire Protection Association.

Natural binary. A number system to the base (radix) 2, in which the ones and zeros have weighted value in accordance with their **relative position** in the binary word. Carries may affect many digits. (Contrasted with Gray code, which permits only one digit to change state.)

NBS. Acronym for National Bureau of Standards.

NEMA standards. Consensus standards for electrical equipment approved by the majority of the members of the National Electric Manufacturers Association (NEMA).

NEMA type 12. A category of industrial enclosures intended for indoor use primarily to provide a degree of protection against dust, falling dirt, and dripping non-corrosive liquids. They are designed to meet drip, dust, and rust-resistance tests. They are not intended to provide protection against conditions such as internal condensation.

Net load capacity. The additional weight or mass of a material that can be handled by a machine or process without failure over and above that required for a container, pallet or other device which necessarily accompanies the material.

Noise. Extraneous signals; any disturbance which causes interference with the desired signal or operation. (See *Interference.*)

Noise-free environment. A space theoretically containing neither electrostatic nor electromagnetic radiation. Since this is an industrial impossibility, users attempt to approximate it in direct relation to application requirements.

Noise immunity. The ability of the computer or robot controller to reject unwanted noise signals.

Noise spike. Voltage or current surge produced in the industrial operating environment.

Nominal. Identifying value of a measurable property. This standard value is halfway between maximum and minimum limits of the tolerance range.

Nonvolatile memory. A memory that is designed to retain its information while its power supply is turned off.

Normal operation. Superseded by *Automatic operation.*

Numerical control (NC). A technique that provides for the automatic control of a machine tool from information prerecorded in symbolic form representing every detail of the machining sequence.

Objective function. An equation defining a scalar quantity (to be minimized under given constraints by an optimal controller) in terms of such performance variables as error, energy and time. The objective function defines a trade-off relation between the variables.

Odd parity. Condition existing when the sum of the number of "1's" in a binary word is always odd.

Off-line. Describes equipment or devices which are not connected to the communications line.

Off-line programming. Defining the sequences and conditions of actions on a computer system that is independent of the robot's "on-board" control. The pre-packaged program is loaded into the robot's controller for subsequent automatic action of the manipulator.

On-line. Describes equipment or devices which are connected to the communications line.

Open loop. 1. A method of control in which there is no self-correcting action for the error of the desired operational condition. **2.** Without feedback. (See *Feedback*.)

Open loop control. Control achieved by driving control actuators with a sequence of preprogrammed signals without measuring actual system response and closing the feedback loop.

Open loop robot. A robot which incorporates no feedback, i.e., no means of comparing actual output to command input of position or rate. (See *Fixed stop robot*.)

Optic sensor. A device or system that converts light into an electrical signal.

Optimal control. A control scheme whereby the system response to a commanded input is optimal according to a specified objective function or criterion of performance, given the dynamics of the process to be controlled and the constraints on measuring.

Optional features. Additional capabilities available at a cost above a basic price.

Orientation. The movement or manipulation of an object consistently into a controlled position in space.

OSHA. Acronym for Occupational Safety and Health Act and Occupational Safety & Health Administration. The act is a federal law for the protection of employees in the workplace. The Administration is the agency which enforces the Act.

Output. Information transferred from the robot controller through output modules to control output devices.

Output devices. Devices such as solenoids, motor starters, etc., that receive data from the computer or robot controller.

Overload. A load greater than that which a device is designated to handle.

Overshoot. The degree to which a system response to a step change in reference input goes beyond the desired value.

Pan. 1. Orientation of a view, as with a video camera, in azimuth. **2.** Motion in the azimuth direction.

Parallel output. Simultaneous availability of two or more bits, channels, or digits.

Parity. A method of testing the accuracy of binary numbers used in recorded, transmitted, or received data.

Parity bit. An additional bit added to a binary word to make the sum of the number of "1's" in a word always even or odd.

Parity check. A check that tests whether the number of "1's" in an array of binary digits is odd or even.

Part classification. A coding scheme, typically involving four or more digits, which specifies a discrete product as belonging to a part family.

Passive accommodation. Compliant behavior of a robot's endpoint in response to forces exerted on it. No sensors, controls or actuators are involved. The remote center compliance (RCC) provides this in a coordinate system acting at the tip of a gripped part.

Pattern recognition. Description or classification of pictures or other data structures into a set of classes or categories; a subset of the subject artificial intelligence.

Pause/interrupt. A stopping of all movement of the robot during its sequence in which all power is maintained on the robot arm. (See *Hold*.)

Payload. The maximum weight or mass of a material that can be handled satisfactorily by a machine or process in normal and continuous operation.

PC. Abbreviation for *Programmable controller* (which see).

PC board. Printed circuit board. A circuit board whose electrical connections are made through conductive material that is contained on the board itself, rather than with individual wires.

Peak. The maximum or minimum instantaneous value of a changing quantity.

Performance. 1. The quality of behavior. **2.** The degree to which a specified result is achieved. **3.** A quantitative index of such behavior or achievement, such as speed, power, or accuracy.

Peripheral equipment. A device(s) used for storing data, entering it into or retrieving it from the computer or robot controller (e.g., TTY, CRT, recorder, etc.).

Photo-isolator. A solid state device which allows complete electrical isolation between the field wiring and the controller.

Pick-and-place robot. A simple robot, often with only two or three degrees of freedom, which transfers items from place to place by means of point-to-point moves. Little or no trajectory control is available. Often referred to as a "bang-bang" robot.

Pilot type device. A device used in a circuit for control apparatus that carries electrical signals for directing the performance, but does not carry the main power current.

Pin. 1. The conductive post, contact, or fitting for each wire within a connector. **2.** A connection point on the edge of a printed circuit board. **3.** A mechanical device for blocking motion.

Pinch point. Any point where it is possible for a part of the body to be injured between the moving or stationary parts of a robot and the moving or stationary parts of associated equipment or between the material and moving parts of the robot or associated equipment.

Pitch. The angular rotation of a moving body about an axis perpendicular to its direction of motion and in the same plane as its top side.

Pixel: Picture element or sensor element — also called *photoelement* or *photosite*.

Plated wire memory. A memory consisting of wires which are coated with a magnetic material. The magnetic material may be magnetized in either of two directions to represent ones and zeros.

Playback accuracy. 1. Difference between a position command recorded in an automatic control system and that actually produced at a later time when the recorded position is used to execute control. **2.** Difference between actual position response of an automatic control system during a programming or teaching run and that corresponding response in a subsequent run.

Point-to-point control. A control scheme whereby the inputs or commands specify only a limited number of points along a desired path of motion. The control system determines the intervening path segments.

Polar coordinate system. (See *Spherical coordinate system.*)

Position control. Control by a system in which the input command is the desired position of a body.

Positioning. (See *Orientation.*)

Potentiometer. An encoder based upon tapping the voltage at a point along a continuous electrical resistive element.

Power supply. In general a device which converts AC line voltage to one or more DC voltages. **1.** A robot power supply provides only the DC voltages required by the electronic circuits internal to the robot controller. **2.** Separate power supply, installed by the user, to provide any DC voltages required by the application input and output devices.

Precision. The standard deviation, or root-mean-square deviation, of values around their mean.

Presence sensing device. A device designed, constructed and installed to create a sensing field or area around a robot and which will detect an intrusion into such field or area by a person, robot, etc.

Priority. 1. Order of importance. **2.** An order of precedence established for competing events.

Procedure. The course of action taken for the solution of a problem.

Process. 1. Continuous and regular production executed in a definite uninterrupted manner. **2.** A computer application which primarily requires data comparison and manipulation. The CPU monitors the input parameters in order to vary the output values. (As generally contrasted with a machine, a process of a CPU does **not** cause mechanical motion.)

Process control. Control of the product and associated variables of processes (such as oil refining, chemical manufacture, water supply, and electrical power generation) which are continuous in time.

Processor. 1. A unit in the computer or robot controller which scans all the inputs and outputs in a predetermined order. The processor monitors the status of the inputs and outputs in response to the user programmed instructions in memory, and it energizes or de-energizes outputs as results of the logical comparisons made through these instructions. 2. A device capable of receiving data, manipulating it, and supplying results, usually of an internally stored program. A program that assembles, compiles, interprets or translates. Also used synonymously with *CPU*.

Program. 1. n A sequence of instructions to be executed by the computer or robot controller to control a machine or process. 2. v To furnish (a computer) with a code of instructions. 3. v To teach a robot system a specific set of movements and instructions to accomplish a task. 4. (See *Teach*.)

Program panel. A device for inserting, monitoring, and editing a program in a robot controller.

Programmable. Capable of being instructed to operate in a specified manner or of accepting setpoints or other commands from a remote source.

Programmable controller (PC). A solid state control system which has a user programmable memory for storage of instructions to implement specific functions such as: I/O control logic, timing, counting, arithmetic, and data manipulation. A PC consists of a central processor, input/output interface, memory, and programming device which typically uses relay-equivalent symbols. The PC is purposely designed as an industrial control system which can perform functions equivalent to a relay panel or a wired solid state logic control system.

Programmable manipulator. A device which is capable of manipulating objects by executing a stored program resident in its memory.

Programmer. 1. One who prepares programs for a computer. 2. One who "teaches" a robot a complete set of instructions to accomplish a specific task.

Programming. The act of providing the control instruction(s) required for the robot to perform its intended task.

Prohibited area. Any location within the total machine area to which access is prohibited during normal working cycles of the system.

PROM. Acronym for Programmable Read Only Memory. A type of ROM that requires an electrical operation to generate the desired bit or word pattern. In use, bits or words are accessed on demand, but not changed.

Pronation. Orientation or motion toward a position with the back, or protected side, facing up or exposed.

Proportional control. Control scheme whereby the signal which drives the actuator is monotonically related to the difference between the input command (desired output) and the measured actual output.

Proportional-integral-derivative control (PID). Control scheme whereby the signal which drives the actuator equals a weighted sum of the difference, time integral of the difference, and time derivative of the difference between the input (desired output) and the measured actual output.

Prosthetic device. A device which substitutes for lost manipulative or mobility functions of the human limbs.

Protected memory. Storage (memory) locations reserved for special purposes in which data cannot be entered directly by the user.

Protocol. A defined means of establishing criteria for receiving and transmitting data through communication channels.

Proximal. Close to the base, away from the end-effector of the arm.

Proximity sensor. A device which senses that an object is only a short distance (e.g., a few inches or feet) away, and/or measures how far away it is. Proximity sensors work on the principles of triangulation of reflected light, elapsed time for reflected sound, intensity induced eddy currents, magnetic fields, back pressure from air jets, and others.

Pulse. A brief voltage or current surge of measurable duration.

Punched card. A piece of lightweight cardboard on which information is represented by holes punched in specific positions.

Punched tape. A strip of paper on which characters are represented by combinations of holes.

RAM. Acronym for Random Access Memory.

Rate control. Control system in which the input is the desired velocity of the controlled object.

Rated load capacity. A specified weight or mass of a material that can be handled by a machine or process which allows for some margin of safety relative to the point of expected failure.

Rated voltage. That maximum voltage at which an electrical component can operate for extended periods without undue degradation.

Read only memory (ROM). A data storage device generally used for a control program, whose content is not alterable by normal operating procedures.

Read/write memory. A data storage device which can be both read from and written into (i.e., modified) during the normal execution of a program.

Real-time. Pertaining to computation performed in synchronization with the related physical process.

Real-time clock. A program-accessible clock which indicates the passage of actual time. The clock may be updated by hardware or software.

Record-playback robot. A manipulator for which the critical points along desired trajectories are stored in sequence by recording the actual values of the joint position encoders of the robot as it is moved under operator control. To perform the task, these points are played back to the robot servo system.

Rectangular coordinate system. Same as *Cartesian coordinate system* but applied to points in a plane (only two axes used).

Redundancy. Duplication of information or devices in order to improve reliability.

Register. A memory word or area used for temporary storage of data used within mathematical, logical, or transferral functions.

Relative coordinate system. A coordinate system whose origin moves relative to world or fixed coordinates.

Relative humidity. The ratio of the mol fraction of water vapor present in the air, to the mol fraction of water vapor present in saturated air at the same temperature and barometric pressure; approximately, it equals the ratio of the partial pressure or density of the water vapor in the air, to the saturation pressure or density, respectively, of water vapor at the same temperature.

Reliability. The probability that a device will function without failure over a specified time period or amount of usage.

Remote center compliance (RCC). A compliant device used to interface a robot or other mechanical workhead to its tool or working medium. The RCC allows a gripped part to rotate about its tip or to translate without rotating when pushed laterally at its tip. The RCC thus provides general lateral and rotational "float" and greatly eases robot or other mechanical assembly in the presence of errors in parts, jigs, pallets and robots. It is especially useful in performing very close clearance or interference insertions.

Repair. To restore the robot system to operating condition after damage or malfunction.

Repeatability. Closeness of agreement of repeated position movements, under the same conditions, to the same location.

Replica master. A control device which duplicates a manipulator arm in shape. Control is achieved by servoing each joint of the manipulator to the corresponding joint of the replica master.

Resolution. A measure of the smallest possible increment of change in the variable output of a device.

Resolved motion rate control. 1. A control scheme whereby the velocity vector of the endpoint of a manipulator arm is commanded and the computer determines the joint angular velocities to achieve the desired result. 2. Coordination of a robot's axes so that the velocity vector of the endpoint is under direct control. Motion in the coordinate system of the endpoint along specified directions or trajectories (line, circle, etc.) is possible. Used in manual control of manipulators and as a computational method for achieving programmed coordinate axis control in robots.

Resolver. A transducer which converts rotary or linear mechanical position into an analog electrical signal by means of the interaction of electromagnetic fields between the movable and the stationary parts of the transducer.

Restricted envelope. (See *Work envelope*.)

Revision. A change which does not greatly affect the main function of the basic system.

Robot. A mechanical device which can be programmed to perform some task of manipulation or locomotion under automatic control.

Robot systems. A "robot system" includes the robot(s) (hardware and software) consisting of the manipulator, power supply, and controller; the end-effector(s); any equipment, devices, and sensors the robot is directly interfacing with; any equipment, devices and sensors required for the robot to perform its task; and any communications interface that is operating and monitoring the robot, equipment, and sensors. (This definition excludes the rest of the operating system hardware and software.)

Robotic. Pertaining to robots.

Robotics. The science of designing, building, and applying robots.

Roll. The angular displacement of a moving body around the principle axis of its motion.

ROM. (See *Read only memory*.)

Routine. A sequence of a computer or robot controller instructions which monitors and controls a specific application function.

RPS. Acronym for Random Program Selection.

RS-232, RS-449, RS-422 and RS-423. Technical specifications published by the Electronic Industries Association establishing the interface requirements between modems and terminals or computers.

Run-length encoding. A method of denoting the length of a string of elements each of which has the same property. Usually used in denoting the length of strings of picture elements with the same brightness.

Safeguard. A guard, device or procedure designed to protect persons from danger.

Saturation. The extremes of operating range wherein the output is constant regardless of changes in input.

Schematic. A diagram of a circuit in which symbols illustrate circuit components.

Search function. A computer or robot controller programming equipment feature which allows the user to quickly display and/or edit any part of the software program.

Search routine. A robot function by which an axis or axes move in one direction until terminated by an external signal. Used in stacking and unstacking of parts or to locate workpieces.

Segmentation. The dividing up of a picture into regions according to some property of the picture within regions or along the region boundaries.

Semiconductors. (See *Solid state devices*.)

Sensor. A transducer whose input is a physical phenomenon and whose output is a quantitative measure of that physical phenomenon.

Sensory control. Control of a robot based on sensor readings. Several types can be employed. **1.** Sensors used in threshold tests to terminate robot activity or branch to other activity. **2.** Sensors used in a continuous way to guide or direct changes in robot motions. (See *Active accommodation*.) **3.** Sensors used to monitor robot progress and to check for task completion or unsafe conditions. **4.** Sensors used to retrospectively update robot motion plans prior to the next cycle.

Servo-controlled robot. A robot driven by servomechanisms, i.e., motors whose driving tion of information sensed from its environment. Robot can be servoed or non-servoed. (See *Intelligent robot*.)

Sensory hierarchy. A relationship of sensory processing elements whereby the results of lower level elements are utilized as inputs by higher level elements.

Sequencer. A controller which operates an application through a fixed sequence of events.

Service. To make fit for use; adjust; repair; maintain.

Servo-controlled robot. A robot driven by servomechanisms, i.e., motors whose driving signal is a function of the difference between commanded position and/or rate and measured actual position and/or rate. Such a robot is capable of stopping at or moving through a practically unlimited number of points in executing a programmed trajectory.

Servomechanism. An automatic control mechanism consisting of a motor driven by a signal which is a function of the difference between commanded position and/or rate and measured actual position and/or rate.

Servovalve. A transducer whose input is a low-energy signal and whose output is a higher energy fluid flow which is proportional to the low-energy signal.

Settling time. The time for a damped oscillatory response to decay to within some given limit.

Shaft encoder. An encoder used to measure shaft angle position.

Shake. Vibration of a robot's "arm" during or at the end of a movement.

Shield. Any barrier to the passage of interference-causing electrostatic or electromagnetic fields. An electrostatic shield is formed by a conductive layer (usually foil) surrounding a cable core. An electromagnetic shield is a ferrous metal cabinet or wireway.

Shielding. The practice of confining the electrical field around a conductor to the primary insulation of the cable by putting a conducting layer over and/or under the cable insulation. (External shielding is a conducting layer on the outside of the cable insulation. Strand or internal shielding is a conducting layer over the wire insulation.)

Short-term repeatability. (See *Repeatability*.)

Shoulder. The joint, or pair of joints, which connect the arm to the base.

Signal. The event, phenomenon, or electrical quantity, that conveys information from one point to another.

Single point control of motion. To be a safeguarding method for certain maintenance operations a single point control of motion (for the robot system) when used for entry into the restricted work envelope of the robot shall be such that it cannot be overridden at any location, in a manner which would adversely affect the safety of the person(s) performing the maintenance function. Before the robot system can be returned to its "automatic" operation, it shall require a deliberate separate action involving the single point control or by the person in possession of the single point control.

Slew rate. The maximum rate at which a system can follow a commanded motion.

Smart sensor. A sensing device whose output signal is contingent upon mathematical or logical operations which are based upon internal data or additional sensing devices.

Software. The program which controls the operation of a computer or robot controller.

Solenoid. An electromagnet with a movable core, or plunger, which, when it is energized, can move a small mechanical part a short distance.

Solid state camera. A television camera which has an array of photosensitive semiconductor elements to produce a TV output.

Solid state devices (semiconductors). Electronic components that control electron flow through solid materials, e.g., transistors, diodes, integrated circuits.

Sonic sensor. A device or system that converts sound into an electrical signal.

Space robot. A robot used for manipulation or inspection in an earth orbit or deep space environment.

Speed-payload tradeoff. The relationship between corresponding values of maximum speed and payload with which an operation can be accomplished to some criterion of satisfaction, and with all other factors remaining the same. (See *Maximum speed* and *Payload*.)

Speed-reliability tradeoff. The relationship between corresponding values of maximum speed and reliability with which an operation can be accomplished to some criterion of satisfaction, and with all other factors remaining the same. (See *Maximum speed* and *Reliability*.)

Spherical coordinate robot. A robot whose manipulator arm degrees of freedom are defined primarily by spherical coordinates.

Spherical coordinate system. A coordinate system, two of whose dimensions are angles, the third being a linear distance from the point of origin. These three coordinates specify a point on a sphere.

Spike. (See *Noise spike*.)

Springback. The deflection of a body when external load is removed. Usually refers to deflection of the end-effector of a manipulator arm.

Standard features. All capabilities included in the stated cost.

Startup. The time between equipment installation and the full specified operation of the system.

State. The logic "0" or "1" condition in computer or robot controller memory or at a circuit's input or output.

Static. Refers to a state in which a quantity does not change appreciably within an arbitrarily long time interval.

Static deflection. Load deflection considering only static loads, i.e., excluding inertial loads. Sometimes static deflection is meant to include the effects of gravity loads.

Static error. 1. Deviation from true value when relevant variables are not changing with time. **2.** Difference between actual position response and position desired or commanded for an automatic control system as determined in the steady state, i.e., when all transient responses have died out.

Static friction. The force required to initiate sliding or rolling motion between two contacting bodies, also called *stiction*.

Station. 1. (See *Work station*.) **2.** Any PC, computer, or data terminal connected to, and communicating by means of, a data highway.

Steadiness. Relative absence of high frequency vibration or jerk.

Steady state. 1. General term referring to a value which does not change with time. **2.** Response of a dynamic system in accordance with its characteristic behavior, i.e., after transient response has died out; the steady state response may be either a constant or periodic signal.

Stepping motor. A bi-directional, permanent magnet motor which turns through one angular increment for each pulse applied to it.

Stiffness. The amount of applied force per unit of displacement of a compliant body.

Stop. A mechanical constraint or limit on some motion which can be set to stop the motion at a desired point.

Storage media. Materials on which data may be recorded. (See *Cassette tape, Disk storage, Punched card*, and *Punched tape*.)

Strain gage. A sensor which when cemented to elastic materials measures very small amounts of stretch by the change in its electrical resistance. When used on materials with high modulus of elasticity strain gages become force sensors.

Strength. (See *Load capacity*.)

Structured light. Illumination which is projected in a particular geometrical pattern.

Subprogram. A part of a larger program that can be compiled independently. Often used synonymously with *Subroutine* and *Routine*.

Subroutine. A computer program portion which performs a secondary or repeated function such as printing or sorting. A subroutine is executed repeatedly as required by the main program.

Supervisory control. A control scheme whereby a person or computer monitors and intermittently reprograms, sets subgoals or adjusts control parameters of a lower level automatic controller, while the lower level controller·performs the control task continuously in real time.

Supination. Orientation or motion toward a position with the front, or unprotected side, facing up or exposed.

Suppression device. A unit that attenuates the high intensity electrical noise caused by inductive loads operated through hard contacts.

Surge. A transient variation in the current and/or potential at a point in the circuit.

Switching. The action of turning on and off a device.

Synchro. A shaft encoder based upon differential inductive coupling between an energized rotor coil and field coils positioned at different shaft angles.

Synchronous. Happening at the same time. Having the same period between movements or occurrences.

Synchronous data link control (SDLC). A bit oriented protocol for managing the flow of information on a data communications link. Supports full or half-duplex, point-to-point, multipoint, and loop communications using synchronous data transmission techniques. Employs a cyclic redundancy error check algorithm.

Synchronous shift register. Shift register which uses a clock for timing of a system operation and where one state change per clock pulse occurs.

System. A collection of units combined to work as a larger integrated unit having the capabilities of all the separate units.

Tachometer. A rotational velocity sensor.

Tactile. Perceived by the touch, or having the sense of touch.

Tactile sensor. A transducer which is sensitive to touch.

Tape punch. A peripheral device which generates (punches) holes in a paper tape to produce a hard copy of robot controller memory contents.

Tape reader. A peripheral device for converting information stored on punched paper tape to electrical signals for entry into computer or robot controller memory.

Teach. 1. To move a robot to or through a series of points to be stored for the robot to perform its intended task. 2. (See *Program*.)

Teach control. (See *Teach pendant*.)

Teach pendant. A hand held control unit usually connected by a cable to the control system with which a robot can be programmed or moved.

Teach restrict. A facility whereby the speed of movement of a robot during teaching, which during normal operation would be considered dangerous, is restricted to a safe speed.

Teleoperator. A device having sensors and actuators for mobility and/or manipulation, remotely controlled by a human operator. A teleoperator can extend the human's sensory-motor function to remote or hazardous environments.

Teletype (TTY). Teletype is a registered trademark of a type of input/output device originally manufactured by Teletype Corp.;

now often incorrectly used as a generic term to indicate any similar piece of equipment.

Template matching. The comparison of a picture, or other data set, against a stored pattern, or template.

Terminal. 1. Any fitting attached to a circuit or device for convenience in making electrical connections. 2. An interface device containing a cathode ray tube and a keyboard, for communication with a computer or robot controller.

Thresholding. The comparison of an element value, such as pixel brightness, against a set point value, or threshold. All elements whose values are above threshold are set to the binary value "1". All elements below threshold are set to the binary value "0".

Thumbwheel switch. A rotating numeric switch used to input numeric information to a controller.

Tilt. 1. Orientation of a view, as with a video camera, in elevation. 2. Motion in the elevation direction.

Time constant. 1. Any of a number of parameters of a dynamic function which have units of time. 2. Parameters which particularly characterize the temporal properties of a dynamic function, such as the period of a periodic function or the inverse of the initial slope of a first order exponential response to a step.

Timer. In relay-panel hardware, an electromechanical device which can be wired and preset to control the operating interval of other devices.

Toggle switch. A panel-mounted switch with an extended lever; normally used for on/off switching.

Tolerance. A specified allowance for error from a desired or measured quantity.

Tracking. Continuous position control response to a continuously changing input.

Transducer. A device used to convert physical parameters, such as temperature, pressure, and weight, into electrical signals.

Transfer line. A manufacturing system in which individual stations, indirectly connected together, are used for dedicated purposes. The indirect connecting device performs no function other than the movement of work pieces from one station to another. Generally a synchronous system.

Transformer coupling. One method of isolating I/O devices from the computer or robot controller.

Transient. 1. A value which changes, decays, or disappears in time. 2. Momentary response of a dynamic system to a rapid input variation such as a step or a pulse.

Transistor. A solid state component capable of switching direct current.

Translation. Movement of a body such that all axes remain parallel to what they were, i.e., without rotation.

Triac. A solid state component capable of switching alternating current.

TTL. Abbreviation for Transistor/Transistor Logic. A family of integrated circuit logic.

TTY. Abbreviation for *Teletype*.

Twist. (See *Roll*.)

Type 12. (See *NEMA type 12*.)

Undershoot. The degree to which a system response to a step change in reference input falls short of the desired value.

Underwriter's Laboratories (UL). An organization chartered as a nonprofit corporation to establish, maintain, and operate laboratories for the examination and testing of devices, systems, and materials.

Upperarm. That portion of a jointed arm which is connected to the shoulder.

Variable. A factor which can be altered, measured, or controlled.

Velocity. A measure of speed or rate of motion.

Vibration test. A test to determine the ability of a device to withstand physical oscillations of specified frequency, duration, and magnitude.

Videcon. An electron tube device used in a television camera to convert an optical image into an electrical signal through the scanning of an electron beam over a photo sensitive window.

Viscous friction. 1. The resistive force on a body moving through a fluid. 2. Ideally, a resistive force proportional to relative velocities of a body and a fluid.

Vision. A sensory capability involving the image of an object or scene.

Vision optical system. A device, such as a camera, which is designed, constructed and installed to detect intrusion by a person into the robot restricted work envelope and which could also serve to restrict a robot's work envelope.

Volatile memory. A memory that loses its information if the power is removed from it.

Voltage rating. The maximum voltage at which a given device may be safely maintained during continuous use in a normal manner. It is also called *working voltage*.

Warning device. An audible or visual device such as a bell, horn, flasher, etc., used to alert persons to expect robot movement.

Weighted value. The numerical value assigned to any single bit as a function of its position in the code word.

Windup. Colloquial term describing the twisting of a shaft under torsional load, so called because the twist usually unwinds, sometimes causing vibration or other undesirable effects.

Word. A grouping or a number of bits in a sequence that is treated as a unit.

Word length. The number of bits in a word, a computer or robot controller.

Work cell. A manufacturing unit consisting of one or more stations. (See *Work station*.)

Work envelope. (Also *working envelope*, *robot operating envelope*.) The set of points representing the maximum extent or reach of the robot hand or working tool in all directions. The work envelope can be reduced or restricted by limiting devices which establish limits that will not be exceeded in the event of any foreseeable failure of the robot or its controls. The maximum distance which the robot can travel after the limit device is actuated will be considered the basis for defining the restricted (or reduced) work envelope.

Work station. A manufacturing unit consisting of one robot and the machine tools, conveyors, and other equipment with which it interacts.

Working range. **1.** The volume of space which can be reached by maximum extensions of the robot's axes. **2.** The range of any variable within which the system normally operates. (See *Work envelope*.)

Working space or volume. The physical space bounded by the working envelope in physical space (See *Work envelope*.)

World coordinates. A coordinate system referenced to earth, or a shop floor.

Wrist. A set of rotary joints between the arm and end-effector which allow the end-effector to be oriented to the workpiece.

Write. The process of loading information into computer or robot controller memory.

Yaw. The angular displacement of a moving body about an axis which is perpendicular to the line of motion and to the top side of the body.

Zero point. The origin of a coordinate system.

INDEX

665